Research Foresight

Research Foresight
Priority-Setting in Science

Ben R. Martin and John Irvine

Pinter Publishers
London and New York

First published in Great Britain in 1989 by
Pinter Publishers Limited
25 Floral Street, London WC2E 9DS

British Library Cataloguing in Publication Data

A CIP catalogue record for this book is available from the British Library
ISBN 0 86187 510 9

Library of Congress Cataloguing-in-Publication Data

Research foresight : priority setting in science / edited by Ben R. Martin and John Irvine
p. cm.
ISBN 0-86187-510-9
1. Research. 2. Science and state. I. Martin, Ben R. II. Irvine, John.
Q180.A1R443 1989 89-22783
338.9'26--dc20 CIP

Typeset by Saxon Printing Ltd, Derby
Printed and bound in Great Britain by SRP Limited, Exeter

Contents

List of figures and tables

Preface and acknowledgements

This book presents the findings of an international review of experience with research foresight commissioned from the Science Policy Research Unit, University of Sussex, by the Directorate-General for Science Policy of the Netherlands Ministry of Education and Science.* The authors are grateful to the Ministry for inviting them to carry out the study, and especially for advice from the Sounding Board which was convened to oversee progress and to identify lessons for possible future Dutch foresight activities.

The unforeseen complexity of the issues addressed unfortunately led to lengthy delays in completing the review, and we hope that the final results go some way to repaying the patience of both the Sounding Board and officials in the Ministry, particularly Dr John Marks and Dr Joop Schriever, who undertook the unenviable task of coordinating the study. It should be stressed that no official endorsement is implied of the data and findings contained in this book. While we are jointly responsible for the overall conclusions drawn, Ben Martin prepared the chapters dealing with foresight in France, West Germany and the United States. John Irvine prepared the remaining chapters on Japan, Australia, Canada, Sweden and Norway, and also managed the study.

The authors are extremely grateful to several hundred officials in government departments, funding agencies and companies who so kindly gave up their time to be interviewed. Many in addition subsequently made valuable comments on early drafts of individual country chapters. We should also like to express our thanks to Judy Baker, Sue Burnham and Vanessa Smith for secretarial assistance, and to our families, friends and colleagues for their forbearance and support without which this book would never have been completed. Those deserving special mention include Valerie, Paul, Sarah and David Martin, Dot Griffiths (and Polly and Minnie), Aant Elzinga, Ian Miles, Phoebe Isard, Keith Pavitt, Hilary Walker, Chris Freeman and Geoff Oldham. Finally, we must acknowledge the two Economic and Social Research Council programmes on 'Research Evaluation' and 'Public Sector Science and Commercial Enterprise' which financed part of the more theoretical work involved in preparing Chapters 2 and 11.

*The findings of the study have also been published as a government report for circulation within the Netherlands — see Irvine, J. and Martin, B.R. (1989), *Research Foresight: Creating the Future*, Zoetermeer, Ministry of Education and Science.

Foreword

The way in which innovation in research does or does not come about is of central interest to all those involved in science and technology irrespective of whether they are laboratory managers, researchers, policy-makers, or concerned with applying new knowledge in industry and elsewhere. Research is by its nature innovative since it time and again leads to the identification of novel scientific questions, sometimes surprising, sometimes predictable. Increasingly, too, the socio-economic, environmental and other problems facing all segments of society are presenting fundamental challenges for new research, and indeed constitute an important factor shaping the choice of topics and priorities in science. Moreover, the nature of these demands on science is such that they can no longer be met primarily by research groups performing shorter-term applied work. It is now necessary to give greater emphasis to research with a more distant time-horizon and of an often strongly fundamental nature, but at the same time addressing concrete problems. This is widely recognized nowadays and most countries are therefore funding future-oriented programmes of basic mission-oriented or strategic research, with universities and academic laboratories generally playing a key role.

The growing importance of strategic research has focussed attention on the urgent need for improved mechanisms to ensure that the correct science-policy choices are made. While this is the case even for large nations able to support major research efforts across a broad range of fields, it holds even more for a relatively small country like the Netherlands which has to make the most of its rather modest funds and talents. In this respect, the awareness is growing that the implementation of priorities is strongly foreshadowed in the process of making choices.

In their 1984 book, *Foresight in Science: Picking the Winners*, John Irvine and Ben Martin reported the findings of a study for the British government concerning the use of foresight to set research priorities in the United States, France, the Federal Republic of Germany and Japan. They came to the conclusion that in Japan a particularly successful approach had been developed. Arguing that the cultural context of Japanese society was an important but not decisive factor in that success, they suggested a similar approach might profitably be employed in other countries. These findings led the Directorate-General for Science Policy of the Netherlands Ministry of Education and Science to commission a follow-up study with a rather broader remit. At our request, Irvine and Martin extended their coverage to include not only four smaller countries — Australia, Canada, Norway and Sweden — but also foresight activities in the social sciences and humanities. In addition, we asked that they explore in greater depth the relationship of foresight to the policy-making process, and investigate the factors determining success and failure in prospective activities. It was intended that the findings of the study form an input to the 'Prospectives Project' which was launched by the Ministry to develop an improved understanding of the foresight process, especially in relation to longer-term research. As a result, our hope was that new impetus would be given to stimulating the use of foresight in priority-setting and budget allocation by government bodies, research councils and scientific organizations in the Netherlands.

This volume contains the findings of the new study by Irvine and Martin. It is a fascinating book, containing a wealth of information on strategic policy-making and the setting of priorities in research, rounded up with valuable analyses of the main issues in foresight. The authors have done a remarkable job in synthesizing the most important elements in the foresight activities of the eight countries covered. Three of their conclusions are of particular importance and are of general relevance to any nation, funding agency or research organization which has decided to use foresight techniques to aid making the difficult choices necessary on priorities for science and technology. These are:

— analysis and discussion of long-term needs and opportunities for research should be part of a well-designed process which includes implementation of the choices made;
— the increased pace of scientific progress means that it is essential, especially for smaller countries, to carry out systematic global monitoring of developments in research;
— a policy of *nurturing the scientific winners* offers a rather better chance of success than one merely emphasizing *picking* the winners.

The third of these conclusions points in my opinion to the crucial contribution that foresight can make in ensuring research policy articulates responsiveness to longer-term societal priorities with an understanding of the changing internal dynamics of science and technology.

I am convinced that this book will make a valuable contribution to the further development of science and technology policy for the 1990s.

Dr E. van Spiegel
Director-General for Science Policy
Ministry of Education and Science
Zoetermeer
The Netherlands

1 Introduction: research foresight and priority-setting in science

1. THE NEED FOR RESEARCH FORESIGHT

There is now remarkable international consensus that emerging technologies within the fields of electronics, information and communications, advanced materials and biotechnology will have a revolutionary impact on both economic activity and society more generally over coming years. These generic technologies are, in the view of many (e.g. Freeman and Perez, 1988; Hirooka, 1986), driving a new Kondratiev 'long-wave' of economic development. What is less accepted, although views have been changing rapidly, is the extent to which progress in such technologies will be science-dependent. It has been industry (in Europe, the United States and particularly Japan) rather than government which has been first to emphasize the fundamental and pervasive impact that basic science is likely to make over coming decades — especially research in interdisciplinary life-sciences, chemistry and physics areas:

> Today, we are witnessing *waves of important basic discoveries.* They are so significant that some people even compare them to those of the First Industrial Revolution. The discoveries and innovations derived therefrom are likely to gain speed as we enter into the twenty-first century, giving far-reaching impact on the industrial structures of many countries and the world economy as a whole. More importantly, some of these technological innovations are key to improving the quality of life of the peoples of the world. The discoveries and innovations are most noticeable in areas such as electronics and information, advanced materials, biotechnology and new energy sources. But many of them *are based* on molecular- and atomic-level analysis of substances, an area that inevitably requires a chemical approach. Future innovations *will depend largely on the combination of chemistry with various other sciences,* such as advanced materials sciences ... We are entering into 'the Age of New Chemistry' in which chemistry will play a major role *in creating new technologies, in fusion with other scientific disciplines.* [ASPRONC, 1987, pp.1–2, emphasis added

Consequently, basic science[1] and intellectual capital seem certain to play a more fundamental role in relation to socio-economic development than hitherto. Recognizing this, high-technology companies have been channelling increased resources to basic research, especially in areas where fundamental scientific breakthroughs are needed if emerging generic technologies are to realize their full commercial potential. At the same time, government research priorities have begun to shift in response to new (and qualitatively different) demands being placed on the public R&D infrastructure, with the academic system gaining in importance relative to departmental laboratories undertaking more applied programmes. Not only has there been a sharp growth in demand for specialized graduate scientists, but the interface between academic and industrial research has also assumed greater significance as companies have begun to exploit the results of work in specialties previously funded only for their intrinsic intellectual interest.

Much of the current interest in foresight stems from a recognition within government that longer-term research planning is now unavoidable in an era of increasing global economic competition. The locus of economic power has moved from the United States and Europe towards Japan and the Pacific-rim nations, and a large part of the responsibility for this can be attributed to R&D and technological innovation. In Japan, for example, it has recently been concluded that 'technological progress has contributed more than 50% to the annual growth in the gross national product (GNP) ... since the mid-1970s' (STA, 1987, p.15). If the technologies underpinning an emerging Kondratiev long-wave are indeed science-dependent, this gives Western countries with their strong traditions in basic research a chance to wrest back the advantage from the Far East. That threat is well understood in Japan which has been investing heavily in upgrading its basic-science system and restructuring public-sector research institutions to inject new vitality and creativity. Some European nations now recognize that they must respond by also treating basic research as a strategic resource. For this, they need an early warning of the most promising developments in science and technology on which to target funds if economic prosperity and social progress are to be secured over coming decades (Elzinga, 1986). However, awareness of potential research opportunities is not sufficient. Information is also required on three other sets of factors: (i) likely trends in socio-economic needs and demands for research; (ii) internal strengths and weaknesses in R&D, and relative international standing across strategic scientific and technological fields; and (iii) the domestic capacity to exploit, commercially or otherwise, the results of promising research. As we shall see in Chapter 2, these represent the basic inputs to foresight (although their precise form and the balance between them will vary according to such factors as the nature of the research under review and the types of organization involved).

While the growing strategic importance of basic science is the principal factor underlying the current international interest in research foresight, two other developments have also contributed. One relates to the changed internal and external 'boundary conditions' under which the research system now operates, with Ziman (1987) characterizing science as having entered a 'steady-state' phase in which greater selectivity and planning is required. The 'internal' conditions relate to the nature, structure and dynamics of research, with increasing technical sophistication and reliance upon advanced instrumentation in many disciplines resulting in science becoming more capital-intensive. Resource constraints mean that only limited numbers of groups can be provided with the funds to acquire 'state-of-the-art' equipment, which introduces the need for clearer priorities within and between fields. Yet, at the same time, the pace of progress in science seems to be quickening; not only are new disciplines proliferating (this being one of the factors underlying the phenomenon of 'literature inflation'), but there is also growing technical and conceptual complexity in many fields (Papon, 1988, p.406). This makes it ever more difficult to identify individual scientists with the breadth and depth of knowledge necessary to make the choices required, a problem accentuated by the interdisciplinary nature of many developing research areas. Existing science-policy mechanisms based on peer-review tend to be structured along disciplinary lines. Not surprisingly, therefore, an impasse is often reached when discussing priorities within and between such hybrid areas. This is reflected especially in the now common institutional disputes among funding agencies over lead responsibility for fields like biotechnology, strengthening the pressure for more systematic approaches to planning within the research system.

The changed 'external' boundary conditions, in contrast, relate to the budgetary and political constraints being imposed by government upon its support for research. The rapid increases in funding witnessed during the 1950s and 1960s have generally been replaced by reduced growth rates, while in some countries level budgets or even cuts have become the norm. Coupled with the changed 'internal' conditions, this means that increased support for one area must generally be matched by reductions in others (the recent growth in Japanese basic science, for example, has involved shifting resources away from applied R&D). In addition, many governments are now laying greater stress on public accountability in relation to research expenditure (as with all major state-funded activities). This has made it essential for the scientific community to demonstrate 'value-for-money', and has often led to preferential financial treatment for fields able to combine intellectual merit with the promise of socio-economic benefits. In turn, the trend towards the use in priority-setting of external criteria (such as likely future technological impact) has further strained the traditional mechanisms of scientific decision-making based on peer-review and disciplinary reputational systems.[2] The effects of 'steady-state' conditions have now become a global phenomenon so that by 1985 some were beginning to argue that 'priority-setting was ... inescapable ... even in the USA' (Shaw, 1985, p.1). Research foresight is one response — and in our view the only plausible response — to resolving conflicts over priority-setting caused by escalating experimental costs, limited resources, complexity in scientific decision-making and pressures to achieve 'value-for-money' and socio-economic relevance.

Finally, a third set of wider societal developments can be seen as reinforcing the demand for increased research foresight. A common structural tendency within most social and economic sectors has been a move towards greater complexity and diversity coupled with accelerating rates of change. At the same time, progress in information technology in particular, together with a radical restructuring of production, administrative and even political systems, has led to increased interconnectedness both nationally and globally (Coates, 1985, pp.34–5). One facet of this is the growing integration of science and technology (Narin and Noma, 1986), with the easier and more rapid availability of information on basic research advances responsible in part for a shortening of the innovation cycle (Ishizaka, 1986, p.4). An inevitable result of the interdependencies within and among nations is that individual decisions have more far-reaching consequences (Elzinga, 1983, p.58). Furthermore, the costs of errors in decision-making stemming from inadequate information or analysis are rising (Solem, 1986, p.5). Foresight provides, at least in principle, a systematic mechanism for coping with that complexity and interdependence as it affects long-term decisions on research, in particular facilitating policy-making where integration of activities across several fields is vital.

2. WHAT ROLE FOR FORESIGHT?

(a) From forecasting to foresight

There are certainly many who, while accepting the need for research foresight, nevertheless doubt whether such an endeavour is possible in practice. 'You cannot forecast science — it is too unpredictable' is likely to be their reaction. The response to this objection by advocates of foresight is well summed up in Bright's rhetorical question: 'Does it make sense to commit a nation's research resources on the basis of total inexplicability and lack of an *anticipatory rationale*? ... Are we content to be "scientific" about everything but science itself?' (Bright, 1986, pp.1 and 12, emphasis added).

The use of the term 'anticipatory rationale' is significant since it represents a conceptual shift on the part of technological forecasters away from their previous emphasis on 'prediction'.[3] Forecasting has been defined by Wills (1972, p.263) as the task of making 'a probabalistic statement, on a relatively high confidence level, about the future,' and accuracy is therefore of paramount importance. Hence, the failure to predict the 1973 'oil-shock' led to considerable scepticism concerning the validity and utility of forecasting, and the boom in futurology which had begun in the mid-1960s consequently ended equally rapidly during the latter part of 1970s when many firms disbanded their longer-term corporate planning groups. The scientific community has thus not been alone in its resistance to the 'rational' management tools offered by forecasters being imposed by policy-makers to assist in setting research priorities, and the new emphasis on 'anticipation' in part represents an attempt to escape from the empirically impractical and conceptually dubious notion of 'prediction'. Anticipation or foresight involves an explicit recognition that the choices made today can shape or create the future, and that there is little point in making deterministic predictions in spheres (including science and technology) where social and political processes exercise a major influence. There has consequently been a move away from forecasting and prediction towards activities variously labelled as 'outlook', 'foresight', 'issues management' and '*la prospective*'. Of these terms, 'outlook' and 'foresight' tend to be used in government circles, while 'issues management'[4] and the related task of 'strategic thinking' are more commonly employed in industry (Elzinga, 1985a; 1985b; 1987).

In short, it should be clear that research foresight cannot to be equated with forecasting, nor even with long-term research planning; rather it denotes a specific conception of the role of prospective analysis in policy-making. It is important to be unambiguous from the outset about the terminology used here, and we shall follow Coates (1985, pp.30 and 33) in defining 'foresight' as:

> A process by which one comes to a fuller understanding of the forces shaping the long-term future which should be taken into account in policy formulation, planning and decision-making... Foresight includes qualitative and quantitative means for monitoring clues and indicators of evolving trends and developments and is best and most useful when directly linked to the analysis of policy implications. Foresight prepares us to meet the needs and opportunities of the future. Foresight in government cannot define policy, but it can help condition policies to be more appropriate, more flexible, and more robust in their implementation, as times and circumstances change.

Foresight is, therefore, closely tied to planning. It is not planning — merely a step in planning.

Several points should be noted concerning this definition. The first is the emphasis placed on foresight as a *process* rather than as a set of techniques. Although making use of techniques like modelling and scenario-analysis, foresight also encompasses the consultative procedures required to ensure feedback to and from relevant actors, including policy-makers, the scientific community, and research users in industry and elsewhere. As a result, foresight is generally more closely linked than forecasting to policy-making and planning, although, as Coates emphasizes, it constitutes only one of several inputs to decision-making.

Second, whereas forecasting techniques can be — and indeed often are — treated as a 'black box' for translating input assumptions into outputs taking the form of predictions about the future, foresight is much more concerned with creating an improved *understanding* of possible developments and the forces likely to shape them. As Wissema (1981, p.30) puts it, 'the aim ... is not so much to forecast as to show the possibilities for influencing, by laying bare the causal relationships of mutual influence patterns'. Central in most foresight, therefore, is the deployment of monitoring mechanisms to provide an early-warning of emerging trends and opportunities.

Third, the notions of forecasting and foresight involve very different ontological assumptions about the future. In conventional forecasting,[5] the aim is to arrive at predictions which can be justified 'scientifically' — i.e. demonstrating how, when starting from certain premises and using given methods and data inputs, the prediction was made (preferably with an assessment of its probability of being realized). The underlying assumption is that only one probable future exists, and that this can be linked in a unilinear and deterministic way to the present and past. In contrast, the goal in foresight is to survey as systematically as possible 'what chances for developments and what options for action are open at present, and then follow up analytically to determine to what alternative future outcomes the developments would lead' (Solem, 1986, p.4). Although there are differences in emphasis, the essentially Anglo-Saxon notion of foresight thus has a similar philosophical starting-point to *la prospective*, the approach developed by Michel Godet and others in France, namely that there exists not one but many possible futures (*les futuribles*): 'The future is not written, it remains to be carried out. The prospective process admits that, at any given moment, the future is multiple and that from the confrontation between the various actors [we] will derive one future rather than another' (Godet, 1986, p.136). Whereas predictive forecasting implies a rather passive attitude towards the future, foresight and *la prospective* involve a much more active stance – reflecting a belief that the future is there to be *created* through the actions we choose to take today.[6] The opportunities for active, participatory foresight have been stressed particularly in Sweden where 'choosing futures' has been a recurrent theme in government reports from the 1970s to the present (e.g. SFS, 1973).

The concepts of foresight and *la prospective* share several other features:

(1) In both, the aim is not so much to predict the details and timing of specific developments as to outline the range of possible futures which

stem from alternative sets of assumptions about emerging trends and opportunities.

(2) Besides enabling different possible futures to be explored, they provide a means of clarifying the scope for current action and its implications for potential developments (often identified in alternative scenarios). In a world characterized by increasing uncertainty, we should not wait for crises to occur, but should take preventive action to forestall their emergence (Godet, 1986).

(3) Since the phenomena and activities subject to foresight/prospective analysis (in our case scientific and technological research) are invariably complex and interdependent, a holistic or systemic approach is essential, integrating and assessing the cross-impacts among all the constituent processes, conflicts and challenges.

(4) Different actors have their own, sometimes mutually irreconcilable needs, and these will be reflected in their alternative visions of desirable futures. Socio-cultural, political and motivational factors are therefore often crucial parameters in foresight, and failure to take them sufficiently into account can lead to the types of problems encountered in conventional forecasting where the emphasis tends to be on mechanistic projections. Related to this is the explicit recognition in the *prospective* approach that 'information about the future, as with the present, is rarely neutral; by omission as much as anything it usually serves very specific interests. Information is corrupted because it is power' (ibid., pp.137–8). From this, it follows that the foresight or prospective process should be transparent to enable the underlying assumptions, analytical framework and data inputs to be subject to external scrutiny. Such openness also helps prevent nonconformist views from being obscured by conventional wisdom.

(5) Finally, both approaches recognize that the information available to make judgements about future prospects is invariably incomplete, frequently unquantifiable and intrinsically uncertain. Consequently, a range of complementary approaches ideally needs to be employed bringing contrasting perspectives to bear on the foresight process. Furthermore, in selecting which techniques to use, one should take into account the fact that:

> Analytical forecasting and decision methods are, first and foremost, tools to promote rigour, coherence, imagination and dialogue. A method can be considered a good and useful one if it can improve coherence and stimulate imagination; the [actual] method matters little provided that the required stimulation can be achieved. [ibid., p.138]

We shall see that this shift from predictive forecasting to foresight is not just confined to countries like France or Sweden, nor to the more socially-oriented parts of the public sector. It is also evident in the private sector, especially among Japanese companies, as is clear from the following statement on 'anticipating changes':

> The rapid advancement of modern technology makes *it difficult to predict* the future of our societies. Yet it behoves us in the chemical industry to *anticipate* changes in the

social structure, market needs and public sentiment towards the industry. We should offer our answers to future challenges before we are asked to do so. Technological advancement is not an objective per se: it should be supported by *a clear vision of the future and a strong will to contribute positively* in enhancing the well-being of the people in the world. [ASPRONC, 1987, p.6, emphasis added]

(b) Is foresight possible?

Returning to Bright's rhetorical question, let us consider the extent to which foresight can make more effective the decision-making process for distributing national or organizational resources among competing research initiatives. Could it not be argued that the deployment of foresight in science policy-making is ultimately subject to the same criticisms levelled against traditional forecasting techniques, in particular concerning the intrinsic unpredictability of research? (Van Vught, 1988).[7]

Certainly, the unpredictability of much scientific progress poses difficulties for foresight. However, the extent of the problem varies with the type of research, in particular its 'orientation'. Here, the traditional three-fold distinction between 'basic research', 'applied research' and 'experimental development' is now recognized as inadequate. The 'basic' category is especially problematic in that it covers a disparate variety of activities ranging from curiosity-oriented, proposal-driven research through long-term targeted programmes supported by sectoral government agencies, to speculative work in industry where no specific application is yet in mind. It is therefore useful to subdivide 'basic research' into 'curiosity-oriented research' and 'strategic research':[8]

> The former corresponds to the traditional notion of academic research carried out with the aim of producing new knowledge primarily for its own sake ... 'Strategic research' differs most importantly from pure or curiosity-oriented research in the rationale behind its support, there being at least some expectation that it will contribute background knowledge required in the development of new technologies. [Irvine and Martin, 1984, p.3]

Furthermore, the latter is carried out not just in universities but also in science-based firms and mission-oriented government laboratories.

The degree of unpredictability is lowest in applied research which we define as 'original investigation undertaken in order to acquire new knowledge, and directed primarily towards *specific* practical aims or objectives' (ibid., p.4, emphasis added). The time-horizon between execution and implementation of research is generally relatively short (up to five years), and a significant proportion of likely advances can be anticipated reasonably reliably. Well-established methods exist, particularly in industry, for planning and managing applied research (see e.g. Irvine, 1988), but these fall largely outside the scope of this volume.

Unpredictability is greatest in curiosity-oriented research, and it would be foolish in the extreme to pretend that it is possible in any reliable way to foresee the fundamental breakthroughs which periodically revolutionize our understanding of the world, as well as opening up entirely new technological possibilities over the longer term. However, some aspects of the future can be anticipated, in particular the likely impact of one field on others (for example, the potential influence of particle physics on astronomy and cosmology in the event that the long-heralded 'grand unified theory' can be established). Although a major focus of interest, our coverage of foresight in curiosity-oriented research is necessarily limited by the intrinsic difficulties of prediction, with most activities studied relating to the selection and planning of centralized experimental facilities (like accelerators) where a longer-term perspective is essential. Our principal concern, therefore, is with strategic research where the time-scale for implementation of results in new products, processes or services is typically a decade (although it can be as little as five years in some areas and up to twenty or thirty in others). We shall show in the chapters which follow that sufficient possibilities for useful foresight exist to merit investment of the resources required, especially in respect of: (i) identifying areas of curiosity-oriented science currently in the process of being transformed into strategic research; (ii) assessing the long-term technological and socio-economic prospects for different research areas competing for support; and (iii) determining the R&D infrastructure government needs to set in place to ensure that industry and other 'users' will be in a position to exploit future results from strategic research.

Aside from intrinsic unpredictability, another fundamental criticism frequently levelled at forecasting is that different studies on the same or similar subjects often yield inconsistent projections of the future. This commonly arises as a result of technical problems, and those engaged in foresight should certainly endeavour to avoid creating inconsistencies through ill-considered or arbitrary selection of variables and data-bases. However, it can also stem from a variety of other causes, not least the incorporation of implicit ideological (or even explicit self-interested) assumptions in forecasts. Indeed, it would be naïve to expect a high level of consistency in all cases unless one adheres to the belief that there is a single probable future (rather than a variety of possible futures). As Coates (1985, p.48) points out: 'Consistency in projections implies consistency in assumptions, which is not to be expected always and is even at times undesirable. Different models [of the future] are put together for different purposes, from different points of view.'

Coates goes on to observe that the very requirement of consistency may be ambiguous. On the one hand, the assumptions within any single model or forecast clearly need to be self-consistent. On the other, a major use of policy-related foresight is to explore a range of alternative futures, and this will generally involve making inconsistent assumptions. Consequently, 'if consistency means either uniformity or a single set of assumptions about the future, consistency is then at odds with a good policy formulation study' (ibid., p.49).

Objections against foresight on the grounds that projections are often 'inaccurate' can be dealt with in a similar manner. According primary importance to 'accuracy' implicitly assumes that foresight is successful if the forecasts made are realized within pre-specified levels of error over details like timing. As we have seen, however, 'predictive forecasting' failed precisely because of the emphasis it placed on 'accuracy' in relation to a single probable future, while ignoring the role forecasts themselves often play in making history (Miles and Irvine, 1979). As Martino (1977, p.114) has stressed, 'accuracy is important only in those cases in which the actor has no power to influence the outcome. Almost by definition, this excludes all policymakers'.

Indeed, the most successful forecasts can be those which succeed in mobilizing action to forestall a predicted negative future, and in such cases 'accuracy' is usually irrelevant.

While the concerns relating to 'unpredictability', 'inconsistency' and 'inaccuracy' are less fundamental than they at first sight appear, previous experience with technological forecasting has revealed other practical problems which may affect research foresight at least to some extent. Bright (1986) and Quinn (1987) distinguish several types of difficulty:

(1) The intrinsic conservatism introduced through the general inability or reluctance of experts to consider radical scientific or technological breakthroughs, especially when they run counter to existing research paradigms and conventional wisdom. While undoubtedly true, this is primarily a limitation with curiosity-oriented research or novel technologies unlikely to be applied for twenty or more years. In the case of strategic research, the typical time-horizon of a decade or so means that much of the science and technology likely to have a major impact over this period is already in place, and fundamental discontinuities are less probable.

(2) Parallel problems with conservatism stemming from difficulties in conceiving novel applications for emerging sciences and technologies. For example, while the development of radio was successfully predicted, it was seen as a replacement for telegraphy (removing the need for posts and wires) rather than as a means of mass communication (Bright, 1986, p.4). This highlights the importance of creative and counter-intuitive thought, and points to the dangers in longer-term foresight of relying only on scientists, engineers and research users. A valuable role can often be played by those with broader perspectives such as futurists, journalists and science-fiction writers (as in the Japanese Science and Technology Agency thirty-year forecasts considered later). Efforts also need be made to ensure that 'wild-card scenarios' are taken into account.

(3) The over-optimism of scientists (especially academics) concerning the prospects for exploitation of their results, in particular the time-horizon for commercialization. This may be because they fail to anticipate institutional constraints (such as resistance to investing in specialized equipment or infrastructure), or through their tendency to misjudge the social response to technological developments (for example, failing to foresee ethical problems or consumer resistance). Since scientists are rarely in a position to give due consideration to social and economic factors, other relevant actors must necessarily be involved in the foresight process. In the case of research likely to yield new technology suitable for commercialization, for example, participants should normally include not just scientists and industrial researchers, but those involved in the design, development, production, marketing and sales of potential products, as well as possible 'users'.

(4) The tendency of researchers to underestimate the rate of progress in rapidly developing areas of science and technology. As we shall see, this is a problem even in well-designed foresight exercises, but can be overcome in part by involving experts from related or neighbouring areas who are able to contribute differing perspectives and experience, as

well as being in a better position to identify potential cross-field interactions.

It should be evident, therefore, that foresight is neither simple nor unproblematic, and many of the case-studies discussed in later chapters reveal that success often only comes after a lengthy learning process involving costly trial and error. This is to some extent an inevitable consequence of the difficulty in achieving an appropriate balance in relation to certain inherent tensions or contradictory requirements in foresight. Coates (1985, pp.31–2) identifies nine such contradictions which need to be borne in mind:

(1) the insatiable demand of policy-makers for information versus its costs of acquisition — it takes time and resources to obtain, analyse, understand and present new data;
(2) the pressure for action versus the uncertainty of outcomes — although policy decisions can to some extent be deferred, eventually choices have to be made;
(3) rapid change and increasing complexity versus the limited capacity of individuals to think about large numbers of factors simultaneously — the latter results in a tendency to oversimplify not only when constructing scenarios and other models of the future, but also when transmitting the results of foresight from person to person;
(4) the pressure for a prompt response versus the need for more information — this is one reason foresight should ideally be conducted on a continuous basis;
(5) the preference of policy-makers for quantitative data (because of its presumed reliability, validity and public acceptability) versus the utility of complementary qualitative information and subjective evaluation;
(6) the need for efficiency versus the advantages of flexibility — while policy research must be capable of responding rapidly to changing circumstances, which is best guaranteed by the simultaneous exploration of alternative approaches by different individuals or groups, such pluralism is not conducive to the efficient implementation of results;
(7) attempts to render policy analyses objective and rational versus ideological and political factors — 'the risk of centralized analyses in a politically charged environment is that ideology and politics may too often swamp knowledge and objectivity' (ibid., p.32);
(8) short-term factors (the annual budget-setting process or electoral pressures, for example) versus often more important longer-term concerns;
(9) the pressure for certainty versus the contingent and hypothetical — 'good policy analysis ... must continually face the question of "what if?" and must continually traffic in "let us suppose" information and data. These needs are particularly at odds with rigid organizations, bureaucracy, narrow specialization and ideological preconception' (ibid., p.32).

The evident problems in resolving such tensions perhaps go some way to explaining why greater use is not made of research foresight despite the clear need which exists. To some extent, the limited use may reflect scepticism about forecasting, or even previous unsatisfactory experience. However, probably more important is widespread fear about two possible adverse consequences. The first is that institutionalized foresight can too easily become doctrinaire foresight (ibid., p.34). This may threaten or even destroy the creativity so essential in basic research, as well as stifling technological innovation and the flexible exploration of novel alternatives — aspects which are widely perceived as being strengths of Western market economies. The second is a fear among some that the institutionalization of foresight will promote centralized government planning: 'foresight by widening the span of considerations entering into decisions draws things together conceptually. This may lead to centralization of control ...' (ibid., p.34). Whether such fears are justified is an issue we shall address in our analysis of international experience with research foresight.

3. ORIGINS, AIMS AND STRUCTURE OF BOOK

It was concern with these and other issues which led the Netherlands Ministry of Education and Science to establish a review committee in late 1987 to consider the merits of employing systematic foresight techniques in the planning of longer-term basic research. Two studies were commissioned to evaluate recent experience with research foresight in the Netherlands (TNO, 1989), and in eight industrialized nations ranging in size from the United States to Norway. The Science Policy Research Unit (SPRU) was contracted to carry out the overseas survey reported here, primarily to take advantage of information gained during a previous project by the two authors for the British Advisory Council for Applied Research and Development (ACARD).[9] The main task in that earlier 1983 study was to establish what lessons could be drawn for the United Kingdom from research foresight undertaken in France, the Federal Republic of Germany, the United States and Japan over the preceding two decades, focussing in particular on the success achieved in identifying emerging areas of strategic research which subsequently had major industrial impact. The findings were published as *Foresight in Science: Picking the Winners* (Irvine and Martin, 1984), and provided an input to ACARD's own report on *Exploitable Areas of Science* (ACARD, 1986).

The remit for the new study is broader than that set by ACARD in several respects:

(1) disciplinary coverage encompasses the social sciences and humanities as well as science and engineering;
(2) it is concerned with foresight oriented not just to assessing the techno-economic prospects for research, but also its potential intellectual significance and likelihood of yielding wider social and cultural benefits;
(3) coverage is not restricted to larger countries (whose size may limit the general lessons which can be drawn) and includes four nations with overall levels of research activity more comparable to that of the Netherlands — Australia, Canada, Sweden and, to a lesser extent, Norway;[10]
(4) greater attention is devoted to issues concerning the organization and management of research foresight, including the resource, skill and information requirements;

(5) since a key lesson from the previous study was that the task of identifying promising research opportunities is generally simpler than translating foresight results into policy and research action, particular stress is placed on analysing the implementation process;
(6) more emphasis is given to theoretical questions, especially to developing a classification scheme for different types of research foresight, outlining a conceptual framework identifying and linking together the main elements and stages in the foresight process, and assessing the factors structuring success and failure.

In short, the review of research foresight reported here is considerably more ambitious in scope and potentially of wider policy utility than that carried out for ACARD, as even a cursory comparison of this volume with *Foresight in Science* will reveal.

As regards the methodological approach employed, information was drawn from two main sources. The first was a survey of the literature on research foresight, technological forecasting, science-technology links and policy-making in science. The strategic significance of foresight means that little relevant material is available in published books and journal articles, and it was therefore necessary to rely primarily on 'grey' literature obtained through personal contacts with overseas agencies over several years and during fieldwork for the study.

The second and main information source was in-depth interviews with several hundred officials in government departments, advisory councils, funding agencies and industrial companies in the eight countries studied. In total, some 110 organizations (listed in the Appendix) were visited which covered:

— high-level advisory bodies concerned with the overall co-ordination of national science policy (e.g. the Australian Science and Technology Council);
— university grants committees and other government agencies responsible for infrastructural and core support of academic research (e.g. the Japanese Ministry of Education, Science and Culture);
— research councils financing and/or executing academic and other basic research (e.g. the Natural Sciences and Engineering Research Council of Canada);
— ministries, departments and other mission-oriented agencies which support strategic research and basic technology (e.g. the Swedish National Board for Technical Development);
— government agencies and organizations whose specialist science-policy activities include research foresight (e.g. the French Centre de Prospective et d'Évaluation);
— academic and other science-policy groups with an interest in prospective analysis (e.g. the Institute for Systems and Innovation Research located in Karlsruhe, West Germany);
— professional scientific and learned societies (e.g. the American Association for the Advancement of Science);
— large science-based companies in industries experiencing major technological change (e.g. the Norwegian multinational, Norsk Hydro);

— commercial and industrial associations (e.g. the West German Chemical Industry Association);
— private and public-sector consultancy organizations and think-tanks (e.g. Arthur D. Little and the Japanese Institute for Future Technology).

Interviews typically lasted up to two hours, and most were conducted using a semi-structured questionnaire. Among the questions asked were the following:

(1) What research foresight activities have been undertaken or are under way? What new initiatives are planned?
(2) What are the main aims (both formal and informal) of foresight? Are they: (a) to identify areas of potential longer-term intellectual, commercial or social interest; (b) to establish consensus on future research priorities/policies among major actors or stakeholders (the scientific community, funding agencies, and likely research users in industry and elsewhere); or (c) oriented to other tasks such as identifying areas for international cooperation?
(3) In what research areas have foresight activities been undertaken, and are they: (a) existing fields or subfields; (b) newly emerging (and often interdisciplinary) fields; or (c) potential new areas in which science and technology can be applied? Are any fields considered unsuitable for foresight, and, if so, why?
(4) How are decisions taken to initiate foresight exercises and to identify areas for study: (a) what mechanisms are employed (for example, expert committees or political initiatives); (b) who participates (researchers, policy-makers, industrial companies, and so on); (c) what information inputs are used (for instance, forecasts of emerging social and economic needs, or views of scientists on 'hot' areas); and (d) what criteria are utilized?
(5) How is foresight organized and managed institutionally: (a) who is responsible for planning and initiating new activities, and for their subsequent execution; (b) what is the role of the research community and its professional organizations, and how is their participation organized; and (c) what is the relationship between the forecasters, the commissioning organization, other public or private-sector agencies, and potential users of the foresight results?
(6) What approaches to foresight have been adopted, and what factors govern the choice of methods and techniques employed (including trend-analysis, Delphi methods, expert panels, scenario techniques and cross-impact analysis): (a) how are these methods and techniques chosen — is selection just a technical matter, or is it governed by the objectives of the exercise or the resources available; (b) what is the relationship of forecasting approaches to the type of research under study, and to its 'orientation' — i.e. whether it is curiosity-oriented, strategic or applied research; (c) what are the links between research foresight and other longer-term social and economic forecasting; and (d) given that most forecasting techniques concentrate on evolutionary trends, how are possible discontinuities taken into account?

(7) What determines the time-horizon adopted in foresight: (a) does it vary with the nature of the research under study and associated technological, economic or social factors; and (b) how does it affect the selection of approach and techniques, and the mechanisms subsequently employed in implementing the results?

(8) What are the resource requirements for research foresight: (a) how much does a typical exercise cost; (b) what are the personnel requirements for managing and executing such an exercise; and (c) what skills and expertise are required, and are these available in-house or provided by consultants?

(9) What types of information inputs are utilized in foresight: (a) how are social and economic needs assessed (using demographic projections, for instance); (b) how are 'science and technology push' factors taken into account (for example, through peer-review or data-bases on research activities and outputs); (c) what inputs are made by industry and other potential research users, and how are 'signals' picked up concerning their views on emerging trends and priorities; (d) are international comparisons undertaken; and (e) are any other inputs employed (such as technology assessments or historical case-studies)?

(10) What form do the outputs from foresight activities take, and how are the results diffused (for instance, through conferences or popular journal articles)? How do the form and method of presentation of results affect their impact?

(11) How are foresight results used and implemented: (a) what organizations and individuals are involved, and what is their relationship to the foresight process; (b) how integrated is foresight with other policy-making and resource-allocation mechanisms (in particular, the peer-review process normally employed to determine priorities in basic research); (c) what use is made of the foresight results by industry and other prospective research users; (d) has much scepticism been encountered concerning the utility of foresight and, if so, how has this been overcome; and (e) how are the results from holistic and more micro-level forecasts integrated?

(12) Are foresight results evaluated either formally or informally? What success has been achieved in terms of: (a) the accuracy of forecasts; (b) identification of promising research areas; (c) establishment of initiatives to exploit the opportunities provided by those areas; and (d) wider impact on policies concerning restructuring of governmental R&D budgets, educational training programmes, stimulation of industry-university links and international research initiatives?

(13) What factors determine the impact of foresight on research policy, and to what extent is this a function of: (a) the technical quality of forecasts; (b) the methods and techniques employed; (c) the process by which foresight is carried out and the results implemented; and (d) the characteristics of the institutions and individuals involved (such as their expertise, reputation, status and authority)? Are these factors organizationally or country-specific, or more general in nature?

(14) Can any general lessons be learned about the factors affecting success and failure in research foresight? Are there any approaches which can be recommended or which should be particularly avoided?

(15) Finally, in the light of international experience, what conclusions can be drawn about how foresight might best be organized in a country like the Netherlands? To what extent do the size of the country, its specialization in particular industrial and technology sectors, and its social and cultural outlook limit the likely utility of foresight approaches imported from other nations?

The responses to these questions, along with the written material collected, were then subject to detailed analysis and the findings synthesized and written up in the form of eight national case-studies.[11] Before presenting the results, however, it is instructive to explore in greater depth the nature and organization of research foresight. This is the task of Chapter 2 which outlines a conceptual model of the foresight process, analysing the main elements and stages typically involved, and setting out a range of criteria for classifying different types of prospective activity. This provides a framework for Chapters 3 to 10 which consider in turn experience with foresight in the four larger scientific powers (France, the Federal Republic of Germany, the United States and Japan), followed by the four with medium-sized research efforts (Australia, Canada, Sweden and Norway). While each chapter attempts to draw lessons specific to the organizations and foresight initiatives reviewed, an overall assessment is left to Chapter 11. After summarizing the more general findings from the study, this employs the conceptual scheme set out in Chapter 2 to draw conclusions about how research foresight should be organized if it is to have the greatest chance of success. As we shall see, however, the high degree of intrinsic uncertainty involved means that there are no short-cuts to success, nor any simple routes to avoiding failure. Research foresight is a complex social and technical process, and the emphases which need to be placed on different elements and tasks in any exercise vary considerably according to the goals, field of study and numerous other factors. This said, it is still possible to identify a set of common criteria and guidelines to be taken into account when designing and executing research foresight. While these provide no guarantee of success, they can at least limit the likelihood of failure from the various problems which have plagued many of the foresight exercises reviewed in the eight country chapters.

NOTES

1. We are using 'science' in the broad sense denoted by the German term '*Wissenschaft*' (systematic knowledge) for which there is no direct English equivalent. 'Science' is therefore taken to include not just the 'hard' physical and biological fields, but also research in health, agriculture, social science and the humanities.
2. Elzinga (1987, p.30) has described the process involved in this transition as one of 'epistemic drift' — i.e. a move from reliance on criteria of scientific excellence and quality to a situation where external factors, in particular socio-economic relevance and political expediency, have become predominant. The early stages in the evolution of science policy in this direction are documented in Salomon (1973).
3. Prediction can be defined as 'a non-probabilistic statement, on an absolute level, about the future. As such, it is the objective of [a] forecasting exercise' (Wills, 1972, p.264).
4. Coates (1985, p.44) defines 'issues management' as 'the activity of anticipating public policy concerns which will emerge and affect the organization over the next

three to seven years'. See also the description given in Foresight Task Force (1983, pp.21–2).

5. I.e. 'predictive' as opposed to 'analytical' or 'normative' forecasting, both of which are closer to foresight.
6. In this respect, the subtitle of our earlier book — *Foresight in Science: Picking the Winners* (Irvine and Martin, 1984) — may have been somewhat misleading. Far better would have been 'nurturing the winners' (or, using another horse-racing metaphor, 'breeding the winners'), or simply 'creating the future'.
7. For another response to this question, see Papon (1988) who argues that prospective thinking in science is both possible and necessary.
8. There is some dispute over whether strategic research is best treated as a component of basic or applied research. The British government, for example, prefers to distinguish between 'specific' and 'strategic' applied research (see Cabinet Office, 1988). However, this difference in classification is not particularly significant — what *is* important is that 'curiosity-oriented' and 'strategic' research are clearly differentiated. Another term closely related to strategic research which we use extensively in subsequent chapters is 'basic technology'. This can be defined as 'technology which requires important scientific breakthroughs because the basic knowledge required for its development is usually not yet available' (Papon, 1988, p.407).
9. ACARD has since been replaced by the Advisory Council on Science and Technology (ACOST), which has a broader mandate covering basic science as well as applied research and development.
10. It was originally intended to include the United Kingdom in the study. Limitations of time and potential difficulties faced by the authors in commenting critically on foresight exercises in which they have been involved ultimately made this impossible.
11. Organizations consulted were sent a draft of the relevant country chapter or section for comment. Although this led to considerable revision, it was not possible to incorporate all proposed changes since views on success and failure, for example, often varied markedly among those commissioning, executing and using individual foresight studies.

REFERENCES

ACARD (1986), *Exploitable Areas of Science,* London, Cabinet Office/HMSO.

ASPRONC (1987), 'Symposium Statement: The First International Symposium on New Chemistry — Creation of Frontier Industries Through Chemistry', Tokyo, Association for the Progress of New Chemistry.

Becher, T., Elzinga, A., Kogan, M., and Scott, P. (eds) (1985), *Higher Education and Research Organization,* University of Stockholm Group for the Study of Higher Education and Research Policy (Report No. 34).

Boucher, W.I. (ed.) (1977), *The Study of the Future: An Agenda for Research,* Washington, DC, National Science Foundation.

Bright, J.R. (1986), 'Improving the Industrial Anticipation of Current Scientific Activity', *Technological Forecasting and Social Change,* vol.29, pp.1–12.

Brotchie, J.F. *et al.* (1986), *Spatial Impact of Technological Change,* New York, Croom Helm.

Cabinet Office (1988), *Annual Review of Government-Funded R&D, 1988,* London, HMSO.

Coates, J.F. (1985), 'Foresight in Federal Government Policy Making', *Futures Research Quarterly,* vol.1, pp.29–53.

Dosi, G., Freeman, C., Nelson, R., Silverberg, G. and Soete, L. (1988), *Technical Change and Economic Theory,* London, Pinter Publishers.

Elzinga, A. (1983), *Foresight as Anticipatory Intelligence,* Ottawa, Science Council of Canada.
—— (1985a), 'Technology Forecasting', in Mar *et al.* (1985), pp.309–18.
—— (1985b), 'Mandated Science in the Universities and Picking the Winners: What are the Implications?', in Becher *et al.* (1985), pp.21–49.
—— (1986), 'Foresighting Canada's Emerging Science and Technologies', in Brotchie *et al.* (1986).
—— (1987), 'Politicization of Science, Commercialization and Epistemic Drift', paper presented at the International Conference on Higher Education: Creativity, Legitimation and Systems Transformation, Rosenön, Dalaro, June (to appear in anthology edited by T. Nybohm).
—— (1988), *The Consequences of Evaluation for Academic Research,* Department of Theory of Science, University of Göteborg (Report No.89, Series No.2).
ESRC/NSF (1985), *Rise and Fall of a Priority Field,* Second Joint Symposium of European Science Research Councils and US National Science Foundation, Paris, September, Strasbourg, European Science Foundation.
Foresight Task Force (1983), *Foresight in the Private Sector: How Can Government Use It?,* Report to Committee on Energy and Commerce, US House of Representatives, Washington, DC, US Government Printing Office.
Freeman, C. and Perez, C. (1988), 'Structural Crises of Adjustment, Business Cycles and Investment Behaviour', in Dosi *et al.* (1988), pp.38–66.
Godet, M. (1986), 'Introduction to *la Prospective*', *Futures,* vol.18, pp.134–57.
Hirooka, M. (1986), 'The Role and Development of the Chemical Industry in the Innovation Age', paper presented at Third World Congress of the Society of Chemical Engineering, Tokyo, September, Tokyo, Sumitomo Ltd.
Irvine, J. (1988), *Evaluating Applied Research: Lessons from Japan,* London, Pinter Publishers.
—— and Martin, B.R. (1984), *Foresight in Science: Picking the Winners,* London, Pinter Publishers.
Ishizaka, S. (1986), *Innovation Cycle and Japan's Role,* Tokyo, National Personnel Authority.
Mar, B.W., Newell, W.T. and Saxberg, B.O. (eds) (1985), *Managing High Technology: An Interdisciplinary Perspective,* Amsterdam, Elsevier.
Martino, J.P. (1977), 'Forecasting and its Impact on Policymaking', in Boucher (1977), pp.111–21.
Miles, I. and Irvine, J. (1979), 'Forecasting the Future or Making History', in Irvine, J. *et al., Demystifying Social Statistics,* London, Pluto Press.
Narin, F. and Noma, E. (1986), 'Is Technology Becoming Science?', *Scientometrics,* vol.7, nos.3–6, pp.369–81.
Papon, P. (1988), 'Is a Prospective of Science Feasible?', *Futures,* vol.20, pp.402–9.
Quinn, J.J. (1987), *Establishing New Research Directions,* London, Technical Change Centre (TCCWP-87-001).
Salomon, J.J. (1973), *Science and Politics,* London, Macmillan.
SFS (1973), *To Choose a Future,* Stockholm, Secretariat for Future Studies and Royal Ministry for Foreign Affairs.
Shaw, E.N. (1985), 'Summary of Discussions', in ESRC/NSF (1985), pp.1–8.
Solem, E.(1986), 'Foresight in Government: A Necessary but Rare Component', *Futures Canada,* vol.8, pp.3–7.
STA (1987), *White Paper on Science and Technology 1987; Toward the Internationalization of Japan's Science and Technology (Summary),* Tokyo, Science and Technology Agency.
TNO (1989), *Report on Dutch Prospectives Project* (provisional title), Apeldoorn, TNO Centre for Technology and Policy Studies.
Van Vught, A. (1988), 'Forecasting or Pattern Management', Zoetermeer, Netherlands Ministry of Education and Science.

Wills, G. (1972), *Technological Forecasting*, Harmondsworth, Penguin.
Wissema, J.G. (1981), 'Futures Research: Is It Useful?', *Long Range Planning*, vol.14, pp.29–32.
Ziman, J. (1987), *Science in a Steady State*, London, Science Policy Support Group.

2 The nature and organization of research foresight

1. INTRODUCTION : THE FORESIGHT PROCESS

In Chapter 1, we briefly discussed the nature of research foresight when differentiating it from traditional predictive forecasting. It is, however, now necessary to set out a more detailed conceptual overview of the foresight process in order to provide an analytical structure for the eight country case-studies which follow. This first entails distinguishing several sets of factors or dimensions that can be used to classify the various types of foresight activities commonly carried out. The typology of foresight we shall employ takes into account the following characteristics: (i) the administrative function and location in the R&D system of organizations undertaking prospective analysis; (ii) how broad a range of research is covered; (iii) the objectives and functions of foresight; (iv) the nature of the research under review and its structural characteristics; (v) the relative balance struck between three 'intrinsic tensions' (science-push vs demand-pull, top-down vs bottom-up, and the extent responsibility is devolved to interested parties as opposed to independent 'third parties'); (vi) the time-horizon involved; and (vii) the approaches and techniques used.

A second objective of this chapter is to outline a conceptual model of the foresight process, focussing particularly on activities oriented to setting research priorities but aiming at the same time to shed light on other forms of prospective analysis. This will cover not only the main stages in organizing and executing an exercise but the important implementation phase as well. It will also consider such issues as the resources and expertise required, the types of question which need to be addressed, and who should participate in the various tasks that must be undertaken. The intention, then, is to provide an overall framework for assessing the factors determining success and failure in foresight, especially its impact on the research-policy process.

2. A TYPOLOGY OF RESEARCH FORESIGHT

(a) Organizational characteristics

Probably the simplest way of classifying research foresight relates to the characteristics of the organization involved (e.g. its location in the public or private sector, or its administrative function in the R&D system). The types of inputs required in foresight for determining national research priorities are, for example, not the same as those for planning at the agency level, let alone for articulating a specific programme. Similarly, independent advisory councils responsible for providing critical scientific advice (or for regulating R&D activities) are likely to differ from both the 'producers' and 'users' of research in the issues they address and the approaches to foresight they employ. Seven main categories of organization can be distinguished:

(1) government advisory boards or central agencies involved in the coordination and planning of national science and technology policy;
(2) independent public-sector advisory councils with a broad remit to identify future needs and opportunities for research, as well as potential problems associated with existing policies;
(3) academic funding bodies like research councils whose primary task is to support basic science in higher education institutions;
(4) national academies of science and other professional organizations of the research community;
(5) government departments and mission-oriented agencies financing and/or executing work in strategic research and basic technology;
(6) industrial associations bringing together groups of companies (often from the same technology sector) to discuss matters of common interest or to collaborate in longer-term generic research;
(7) science-based firms in high-technology fields, many of which have been increasing their investment in strategic research.

Given the divergence in research traditions and scientific institutions across different countries, we have opted to use these relatively simple organizational categories in structuring the presentation of our national case-studies. Inevitably, however, there are shortcomings in such a classification scheme. In particular, the boundary between research councils and mission-oriented agencies is not always unambiguous: for example, the principal medical research organizations in France (INSERM) and the United States (the National Institutes of Health) perform a dual function, and could equally well be placed in either category. Similarly, in some cases a government department (like the Canadian Department of Industry, Science and Technology) may have responsibilities both for coordinating national research efforts and for financing its own programme of R&D.

(b) Degree of specificity

A second and arguably more conceptually satisfactory approach to classification is in terms of breadth of coverage. Four levels of specificity can be distinguished:

(1) *Holistic* — foresight concerned with the entire spectrum of fields. Prospective analysis at this level generally aims to gain an overview of possible future directions for the national research effort, and is typically used to coordinate or 'set the scene' for lower-level foresight exercises rather than as a means of providing direct inputs to policy decisions. However, it may also contribute to identifying broad areas like biotechnology which are to be targeted as priorities by government.
(2) *Macro-level* — this relates to foresight focussing on a limited number of research fields. Most funding agencies support a variety of disciplines

and are therefore faced with having to decide on overall priorities. Prospective analysis at this level might, for example, be employed to explore opportunities for interdisciplinary research or to demonstrate the need for a major initiative in one or more identified fields.

(3) *Meso-level* — foresight concerned with a single scientific field or technology sector. It is typically used to determine which research areas have greatest socio-economic potential or intellectual promise prior to selecting the main lines of work for a planned programme.

(4) *Micro-level* — assessments of future prospects at either the level of projects or individual scientific specialties. Systematic foresight of this type is less common in the case of more basic science, although it may well be carried out informally by researchers and programme managers. It becomes increasingly widespread as one moves along the spectrum towards applied research, with many companies, for example, carrying out prospective analysis to set technical goals for projects or to predict the likely impact of R&D on particular products, processes or services (e.g. see Irvine, 1988).

As will be evident, there is often a linkage between the seven types of organization identified above and the specificity of their foresight activities. However, the relationship is by no means straightforward. Bodies responsible for coordinating national science and technology policy (or for determining overall R&D priorities) might be expected to make greatest use of holistic foresight, but they may draw on data from macro- and meso-level studies as well when considering possible themes for targeted programmes. Research councils, ministries and government agencies normally support a range of scientific activities and therefore have to make choices — implicitly if not explicitly — in respect of the relative priorities to be accorded to different fields. For this purpose, macro-level foresight will probably be most appropriate, especially when it can be set in the context of an earlier holistic exercise. Such organizations are also responsible for overseeing individual research fields and programmes, and meso-level foresight therefore has a role in certain types of decision-making, often drawing on surveys of opportunities in particular disciplines by relevant professional scientific institutions. Were individual universities and government laboratories to make greater use of prospective analysis than they do at present, the emphasis would probably be on meso-level and, to a lesser extent, macro-level foresight.

As regards the private sector, most industrial associations are comprised of companies in similar technology areas and their foresight activities consequently tend to be at the meso-level. Micro-level studies would normally be regarded as too commercially sensitive for information to be shared with competitors (they typically focus on specific market opportunities arising from emerging technologies), and are therefore mainly conducted within individual firms. Larger and more diversified companies may additionally undertake meso-level foresight in-house, as well as subscribing to the extensive range of multi-client studies now offered by consultancy organizations specializing in technological and market forecasting.

The foregoing description is, of course, somewhat oversimplified. Independent advisory councils may, for example, carry out foresight at all four levels, while various forms of hybrid organization that have not been considered (such as *ad hoc* committees established jointly by government and industry) sometimes play an important role in reviewing policy for particular fields. It should also be borne in mind that foresight at one level will in many cases both draw upon and feed into prospective analyses at higher and lower levels of specificity.

(c) *Aims, objectives and functions of foresight*

Underlying most research foresight is the belief that science and technology can — and indeed should — make greater contributions to economic, social and cultural welfare. The ultimate objective of foresight is hence to ensure that areas of research likely to yield significant future benefits are identified at the earliest stage possible. The term *socio-economic* is used extensively throughout this volume, often to act as a reminder that the criteria for selecting promising lines of research generally need to go far beyond the direct economic benefits to be derived from exploiting new technologies. Especially where government is involved, such criteria may take into account a variety of other factors, including potential social costs as well as benefits — with 'social' being defined broadly here to cover health and welfare, quality of life, national security (and related geopolitical considerations), environmental protection, political freedom and contributions to culture. The last of these includes societal benefits stemming from curiosity-oriented research in fields like astronomy and particle physics where the pursuit of knowledge 'for its own sake' is normally regarded as the primary goal. Indeed, a central decision in any foresight exercise concerns the relative emphasis to be placed on social and techno-economic criteria. This will vary both among research fields and between the public and private sectors, as well as being influenced by vested interests, organizational ideology, and the wider political and institutional environment. Such non-economic dimensions need to be stressed lest research foresight be dismissed (or, alternatively, be promoted solely) as a device for achieving efficient commercial exploitation of science at the expense of neglecting equally important societal considerations.

Although it is comparatively easy to specify the overall objective of research foresight, this is of limited utility for analytical purposes since it gives only the option of classifying prospective activities in terms of the relative importance accorded to economic and social criteria (something which in practice is difficult to judge since it is rarely made explicit, especially in government research-policy documents). Consequently, it is also worthwhile distinguishing six *intermediate functions* (or more specific aims) of foresight[1] which differ, among other things, in the nature of their linkage with the policy-making process. These are:

(1) *Direction-setting* — this relates to the use of foresight in determining broad guidelines for science policy (e.g. in framing what the Japanese term a national or organizational 'vision' for research), and establishing an agreed agenda to explore further the prospects for a more limited range of future R&D options. Foresight of this type generally focusses on the holistic and macro-levels, and can be further subdivided on the basis of the use-oriented classification scheme for policy research

proposed by Whiteman (1985): (a) *substantive* use refers to analytic information employed 'in formulating the basic outlines of a position on or a solution to a policy problem'; (b) *elaborative* use aims 'to clarify and extend positions or solutions within basic outlines that have already been established'; and (c) *strategic* use[2] denotes the marshalling of analytic information 'to advocate and reaffirm policy positions *after* they have been determined' (Steed, 1987, p.14).

(2) *Determining priorities* — this is arguably the most important objective of foresight and for this reason constitutes the central theme of our country case-studies. It encompasses a variety of activities seeking to identify and select promising options for future research from among a broader range of possibilities, and will often be undertaken in conjunction with launching new policies and programmes (or redirecting existing ones). In many cases, it will be carried out in the context of guidelines formulated by an earlier direction-setting exercise. Foresight aimed at determining priorities can be classified in several ways — for example, in terms of whether the focus is on informing decisions concerning the allocation of research funds, assessing needs for scientific instrumentation or evaluating future requirements for trained researchers.

(3) *Anticipatory intelligence* — the primary aim is to contribute background information on emerging trends in science and technology, in particular concerning developments with major implications for future policy-making (Elzinga, 1983). It may therefore form an integral part of the infrastructure needed to undertake direction-setting and priority-determining exercises, and generally focusses on the holistic and macro-levels since it is difficult to identify significant advances in cross-disciplinary areas when monitoring is conducted only on a field or specialty basis. This category of foresight can be further subdivided into activities oriented towards: (a) providing a prompt indication of potential research opportunities (e.g. through tracking scientific advances reported in specialized journals); (b) giving an early warning of impending threats or difficulties (e.g. the effects of major R&D investments by competitors or changing societal values concerning particular types of research); (c) exploring unlikely but highly significant or disruptive developments — so-called 'wild card' scenarios (Coates, 1985, pp.32-3); and (d) basic science-policy research (e.g. on the growth dynamics of rapidly evolving fields and specialties).

(4) *Consensus-generation* — this refers to the use of systematic analytical and consultative procedures to promote greater agreement among scientists, funding agencies and research users on identified R&D needs or opportunities. A distinction can be made here between foresight activities aiming to develop: (a) *internal* consensus within the relevant research community; and (b) *external* consensus among those financing, executing and using research (e.g. government, universities and industry).[3] Both variants can be carried out at any of the four levels of specificity, although the meso-level (i.e. within a research field or technology sector) is most common in practice. The two may also be conducted in parallel, especially in countries like Japan where cultural traditions require internal and external consensus as a precondition for

government agreeing to finance new programmes (Irvine and Martin, 1984; Irvine, 1988).

(5) *Advocacy* — the deployment of foresight to promote policy decisions in line with the preferences of specific stakeholders in the R&D system. This usually takes one of three forms: (a) advocacy by senior scientists and/or policy-makers explicitly oriented towards mobilizing the research community (e.g. to demonstrate the merits of moving activities from a traditional to a newly emerging field or specialty — i.e. meso- or micro-level foresight); (b) advocacy by researchers which aims to influence decision-makers, in particular by setting out a case for increased funding in terms of either 'what if' or 'what if not' scenarios (e.g. meso-level foresight within a field like medicine); and (c) advocacy by science-policy officials directed at government, the ministry of finance or perhaps the general public, where the intention is to defend or legitimate an agency's research budget (generally macro-level foresight). Advocacy-oriented foresight can undoubtedly have a beneficial impact in drawing attention to promising opportunities or important R&D needs that are not yet widely appreciated. However, the end-result may also be somewhat negative — for example, when a politically powerful scientific community invests heavily in prospective studies to promote its interests at the expense of work in less well-established but intrinsically more promising fields. In cases where advocates are sufficiently astute to enrol members of influential interest-groups, the distinction with consensus-generating foresight can become blurred.

(6) *Communication and education* — three principal functions of this type of foresight can be distinguished: (a) promoting *internal communication*[4] within the research community (e.g. concerning emerging scientific growth-points in an active specialty);[5] (b) *external communication* with industrial and other potential users of research, informing them about promising opportunities associated with developments in specific fields; and (c) *wider education* of the general public, politicians, government officials and others with a less direct, but ultimately no less important, stake in shaping a desirable future for science and technology — in particular, to make available the information required to enfranchize non-experts in debates on priorities for R&D, and more generally to develop the understanding needed to improve the climate of opinion within which research policy is set. Such foresight is typically carried out at the holistic or macro-level, and is often linked with initiatives oriented towards anticipatory intelligence.

The six-fold classification we have outlined covers only the main functions of research foresight, and the examples cited are intended to be no more than illustrative. It should also be noted that various alternatives exist including a further dimension to Whiteman's scheme in which the distinction is made between: (i) *concrete* use of policy research where 'specific conclusions and findings ... are used in promoting action or justifying decisions'; and (ii) *conceptual* use where 'analytic information ... serves to alter the general orientation of policy makers' (Steed, 1987, p.14). Although this cross-cuts several of our foresight functions, a reasonably close parallel can be drawn with our two categories of priority-determination and anticipatory intelligence.

Another possibility is to follow Elzinga (1985a, pp.312–13) in differentiating between *responsive*, *reflective* and *participatory awareness* modes of foresight:

The first is a mode of operation usually associated with the traditional role of commissions, studies and task forces in response to crises and pressures for immediate action. The reflective mode in its most typical form is associated with think tanks and critical reflection in the academic community ... The participatory awareness mode, finally, is the one I associate with an action orientation towards public awareness in a broad sense.

The responsive mode again has certain similarities with our category of foresight for priority-determination, the reflective mode with anticipatory intelligence, and participatory awareness with communication and education. However, despite its evident attractiveness, there are several problems with Elzinga's classification. One is that it is not sufficiently comprehensive — for example, the *proactive* role[6] of foresight in direction-setting, or in advocating new policy initiatives, is not adequately covered by any of the three categories. Another relates to the fact that the first two modes are defined more in terms of organizational responsibility than function (i.e. whether foresight is performed by *ad hoc* commissions or existing think-tanks and academic groups), while the third is of a rather different nature and appears to favour a particular model of politically committed policy analysis which not all might accept.[7] Finally, the notion of purely 'reflective' foresight is something of a philosophical ideal, since even the most independent of think-tanks are generally working in areas where the agenda is set at least in part by a desire to respond, even if indirectly, to current and anticipated policy needs.

(d) Orientation and structural characteristics of research

A further dimension along which foresight can usefully be classified relates to the *orientation* of research — i.e. whether it is curiosity-oriented, strategic or applied (or, alternatively, according to its degree of *finalization* — see e.g. Schafer, 1983). The main focus of prospective exercises in industrial countries has been on opportunities in strategic research, overwhelmingly in natural science and engineering fields but also to a smaller extent in the social sciences (e.g. see Elzinga, 1983; Irvine and Martin, 1984). The application of foresight to assess the prospects for curiosity-oriented research is less common, if only because of the greater intrinsic unpredictability of progress in fundamental science. Nevertheless, possibilities still exist for foresight in such areas, as is shown in the country case-studies which describe several initiatives to identify intellectually promising growth-points within emerging specialties.

At least as important for the foresight process as the orientation of research are its *structural characteristics*. These refer, first, to the *internal disciplinary structure* of the field — that is, whether it involves activities in a *stable* discipline with reasonably well-defined cognitive boundaries (such as the classical fields of physics or chemistry) or in a *complex and dynamically evolving* specialty like chematronics (the new area being formed through the fusion of chemistry, life sciences and electronics). Such hybrid 'fields' have become a focus of foresight not only because of their strategic promise, but also because they are regarded as an especially fertile site for breakthroughs in scientific understanding. As the biologist, Sir Walter Bodmer, has recently observed: 'most areas that are now at the forefront of scientific research are, to a greater or lesser extent, interdisciplinary' (Bodmer, 1988, p.27).

The second structural characteristic which needs to be taken into account is the *external relationship* of the research field in question to wider areas of scientific and technological activity. The nature of this linkage can set limits on the types of foresight that are possible, as well as having implications for the time-horizon considered. At one extreme, developments in fields like biomedical science (e.g. monoclonal antibodies) have on their own spawned entirely new lines of industrial activity. Because discoveries of this sort occur relatively infrequently and in a heterogenous range of disciplines, a priority for foresight in such 'science-push' areas is extensive monitoring of basic academic research (as is often undertaken by pharmaceutical companies). On the other hand, fields like electronics which are characterized by a high degree of integration with complex systems of interrelated technologies (e.g. in the field of communications) tend to follow more predictable development trajectories at least over the medium term.[8] As a result, individual breakthroughs are likely to make a fundamental impact only in the longer term, and to require parallel advances in other elements of the technological system before significant benefits can be realized (Baba, 1988). This means that foresight is generally easier to undertake in technologically integrated strategic fields since the potential influence of scientific advances on the overall system can be determined with greater confidence.

(e) Balance between various 'intrinsic tensions' in foresight

A central issue in foresight concerns the relative emphasis to be placed on *science-push* and *demand-pull* factors when assessing the prospects for research — that is, whether the focus should be on what the Japanese term *seeds* or *needs* (Irvine and Martin, 1984). This will depend in part on the orientation of the work — for example, in curiosity-oriented areas like astronomy greater weight may well be given to promising scientific opportunities, and less to 'external' considerations such as the impact on technology. Conversely, demand-pull factors tend to predominate in more applied research, particularly where there are strong external links to an established technological system. The structural characteristics of the research can also have a bearing on foresight in that areas with an evolving disciplinary structure (especially emerging fusion fields like chematronics) may require increased stress on the science-push dimension.

A second 'intrinsic tension' in terms of which foresight can be classified concerns the balance struck between *top-down* and *bottom-up* approaches. Although there is often considerable scope for choice on this matter, the options may be constrained by the specificity of foresight and the level of organizational involvement, as well as by other factors such as whether a conducive environment exists for cooperation in decision-making among those funding, executing and using research.

The third main 'tension' relates to the extent that responsibility for foresight is allocated to the two principal *interested parties* (scientists working in the area and those directly benefiting from research — e.g. within industry and government) as opposed to more neutral *third parties* (e.g. specialists from other fields or independent advisory councils). On the one hand, the active participation of producers and users of research can be extremely valuable, indeed indispensable, in evaluating science-push and demand-pull opportunities, but it may come at the cost of allowing them considerable leverage in defending established disciplinary or commercial interests. On the other, while independent third parties are less prey to conservative pressures, they may also not be so committed to ensuring that the results of prospective analysis are eventually implemented. The relative stress placed on involvement of interested and third parties is consequently a complex question that needs to be resolved in any prospective exercise, and it can be a key element in categorizing and comparing different approaches to foresight.

(f) Time-horizon

Another important characteristic of foresight is the time-horizon adopted. The convention we shall adopt is to distinguish between the following: (i) *short-term* — dealing with the next one to two years; (ii) *medium-term* — relating to a three- to five-year horizon; and (iii) *long-term* — typically ten years, but ranging from five to as many as thirty years into the future.[9] Because the time-frame is not always clear, it may sometimes be necessary to use the looser terminology of *longer-term* or *shorter-term*. This can also be useful when contrasting the prospects for research areas in which planning periods diverge because of differing orientations, rates of scientific growth (and associated cycles of instrumentational obsolescence) or structural characteristics (e.g. the extent of integration in complex R&D systems requiring large infrastructural investments). The horizon in foresight may also depend on such factors as the lead-time for constructing scientific facilities (being greatest in big-science areas like astronomy), institutional budget-cycles, and even the periodicity of government elections.

(g) Methodological approach

Methodological criteria can be used to classify research foresight along two dimensions — *informal* vs *formal*, and *qualitative* vs *quantitative*. While these may often be related (informal methods tend, for example, to make less use of quantitative techniques), this is by no means always the case. Thus qualitative approaches can span the entire informal-formal spectrum, with more formal techniques including structured interviews, discussions organized around a predetermined set of questions, morphological analysis, Delphi exercises and certain types of scenario analysis.[10] Among the most common informal ways of conducting foresight are through traditional expert committees and unstructured workshop discussions — aptly termed the BOGSAT approach (i.e. Bunch Of Guys Sitting Around A Table) in the United States;[11] conclusions about potential research opportunities or emerging socio-economic needs will in such cases frequently be arrived at on the basis of little more than 'the informed prejudices of wise men' (Irvine *et al.*, 1983, p.27).

Quantitative methods can similarly be classified according to their degree of formality, with simple trend extrapolation falling at one extreme and techniques involving computer-modelling and growth-curve (especially S-curve) analysis towards the other. In between are a variety of hybrid approaches such as: (i) expert committee review in which peer-judgements are complemented by relevant statistical information (e.g. from global monitoring of R&D activities, questionnaire surveys, or bibliometric analyses of publications and citations); (ii) panels or workshops employing explicit criteria to rank proposed research options in terms of their anticipated scientific or socio-economic impact; (iii) the use of modified Delphi methods

Table 2.1 A typology of research foresight: key features and distinguishing characteristics

Characteristics of performing organization	—High-level government advisory boards and central policy bodies —Independent public-sector advisory councils —Academic funding agencies —National academies and professional scientific institutions —Mission-oriented departments and agencies —Industrial associations —Science-based companies
Specificity	—Holistic —Macro-level —Meso-level —Micro-level
Functions	—Direction-setting —Determining priorities —Anticipatory intelligence —Consensus-generation —Advocacy —Communication and education
Orientation and structural characteristics of research	—Curiosity-oriented vs strategic vs applied (or degree of 'finalization') —Complexity and stability of disciplinary structure —External integration with wider scientific and technological systems
Various 'intrinsic tensions'	—Science-push vs demand-pull —Top–down vs bottom–up —Interested parties vs 'third parties'
Time-horizon	—Short-term —Medium-term —Long-term
Methodological approach	—Informal vs formal —Qualitative vs quantitative

to generate quantitative assessments of the timing and likely importance of developments in science; and (iv) scenario analysis combining numerical data with detailed qualitative projections of alternative outcomes of emerging trends and possible counter-trends.

Table 2.1 summarizes the various elements in the typology of foresight we have outlined, highlighting the wide range of possible combinations which exists. However, as stressed earlier, the availability of a suitable framework for classifying and comparing different types of foresight is only one precondition for systematic analysis of the factors determining success and failure. The other requirement is for a conceptual model setting out, in particular, the main issues to be addressed and tasks to be executed in each stage of the foresight process.

3. STRUCTURE OF THE FORESIGHT PROCESS

Any attempt to distinguish conceptually the principal features of a complex social process like research foresight will necessarily have intrinsic limitations. These are likely to be greatest if one attempts to construct a general model covering every type of foresight that might theoretically be encountered (i.e. all permutations of the elements listed in Table 2.1). We shall concentrate, therefore, on outlining a more restricted model relating to priority-determination, a choice reflecting not only the contemporary importance of this type of foresight for science policy but also the fact that it often represents the linch-pin around which other forms of prospective analysis like direction-setting, anticipatory intelligence and consensus-generation are organized.

The schematic representation of foresight for priority-setting shown in Figure 2.1 is based on a synthesis of experience from our previous work (Irvine and Martin, 1984) and the published literature on forecasting and science policy. As can be seen, three main phases are distinguished — *pre-foresight*, *foresight* and *post-foresight* — each of which comprises a number of interlinked elements (e.g. 'strategic analysis'). Before considering these in turn, we should stress that the figure perhaps does not adequately convey the fact that one or more cycles of foresight may be carried out at different levels prior to determining research priorities. Moreover, the emphasis on science-policy processes in the post-foresight stage is not intended to imply that foresight should necessarily be the key determinant in the decision to launch a research initiative. The aim is solely to highlight that foresight does not end with the submission of a report to those commissioning the study, and that continuing prospective analysis may be of use, for example, in steering an ongoing research programme or identifying possible follow-on initiatives. Finally, the figure should *not* be viewed as presenting a normative model against which to judge the case-studies presented later. Nor should it be seen as implying a belief in the possibility of a wholly rational approach to policy-making unconstrained by the political and other factors which ensure that in practice it is much more an *ad hoc* process of 'muddling through'. The intention is rather to provide a *heuristic* model enabling a better understanding to be developed of the nature and role of the different tasks in foresight, how the various elements are interrelated, and the likely effects on the overall

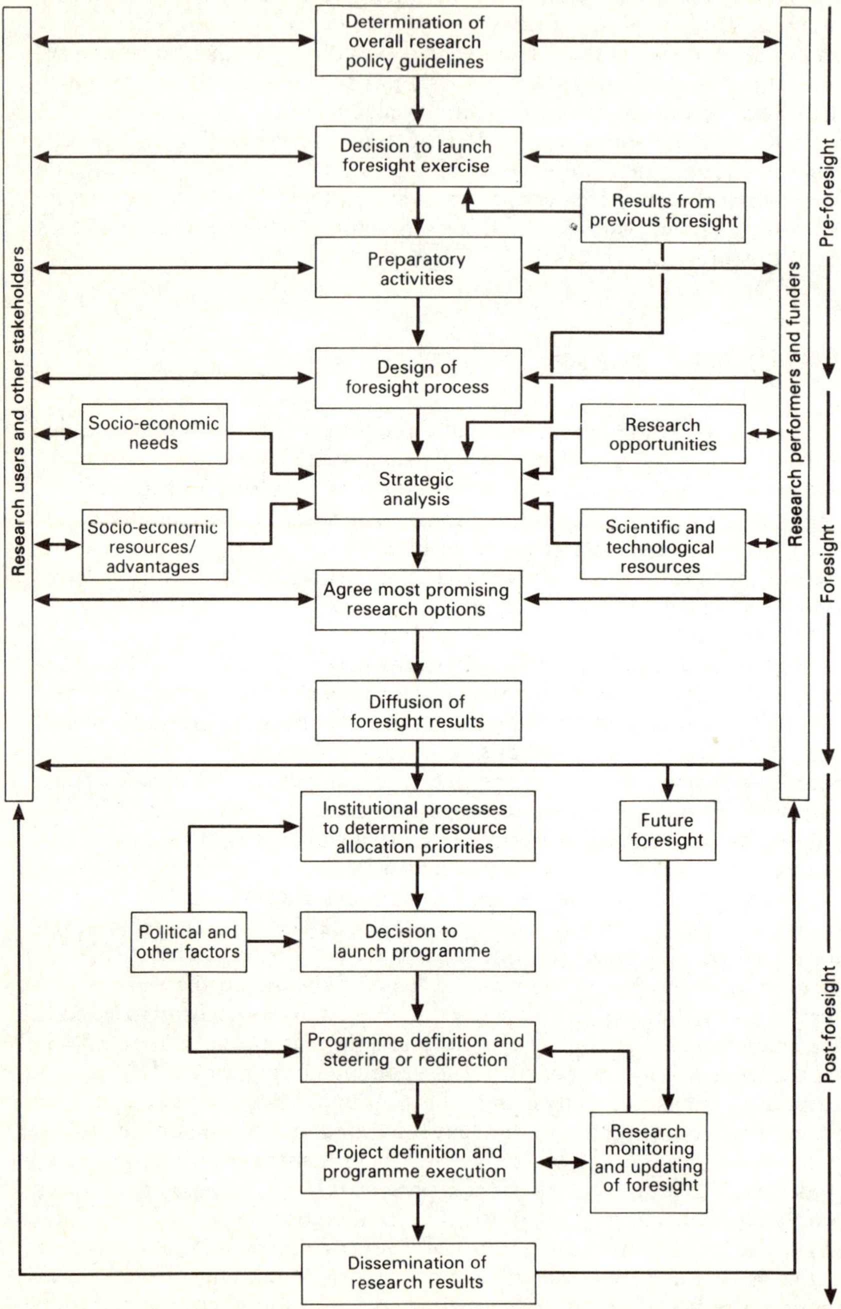

Figure 2.1 *Elements and stages in foresight oriented to priority-setting (including implementation)*

priority-setting process of a failure to execute successfully any of the main steps involved.

(a) Pre-foresight stage

Given that research foresight generally requires considerable investment of resources on the part of organizations and individuals, the first step must be an *explicit decision to initiate the process*. This will often be taken in the context of overall research policy guidelines agreed previously during the wider planning process by the organization(s) concerned (or by the relevant administrative authority). Such guidelines may have been influenced by an earlier direction-setting exercise indicating areas of policy concern or opportunity where options for future research need to be explored in greater detail. Another factor behind the decision may be external pressures from government departments or funding agencies, for example, for systematic foresight techniques to be employed in framing medium-term budgetary priorities. In other cases, the determining factor will be threats and opportunities identified by other types of foresight (e.g. anticipatory intelligence), or a need to draw up more specific plans for a new programme to which funds have been committed in principle (e.g. because of a successful prior initiative focussing on consensus-generation or advocacy).

The second step involves various *preparatory activities*, which include agreeing appropriate terms of reference and determining the resources and administrative support likely to be required. Such decisions will frequently be taken in conjunction with a high-level oversight committee (or steering group), which is given overall responsibility for the exercise. Its two main functions during the pre-foresight phase are: (i) to prepare a strategic plan specifying the principal issues to be addressed, basic guidelines on the approach to be employed, and appropriate operating and reporting procedures; and (ii) establishing an organizational framework for executing the exercise, and in particular for building the links necessary to gain the active involvement and commitment of relevant stakeholders. Figure 2.2 outlines perhaps the most common administrative structure for foresight in which an expert working group is set up to carry out or supervise the various information-gathering and analytical tasks that are required. It is generally serviced by a secretariat shared with the oversight committee, and is responsible for initiating the surveys, specialist panel discussions, workshops and other informal consultative activities necessary to ensure that 'bottom-up' inputs can be combined with the 'top-down' perspective provided by the oversight committee. Also shown in the figure are 'light touch' committees, which provide important informal guidance on critical issues by networking with influential actors in industry, government and the research community.

(b) Main foresight stage

At the core of any exercise aiming to inform decisions on research priorities are four main tasks: (i) detailed design of the foresight process; (ii) strategic analysis of the potential benefits associated with different research options; (iii) agreeing the most promising options; and (iv) disseminating the results to those involved, directly or indirectly, in determining scientific priorities.

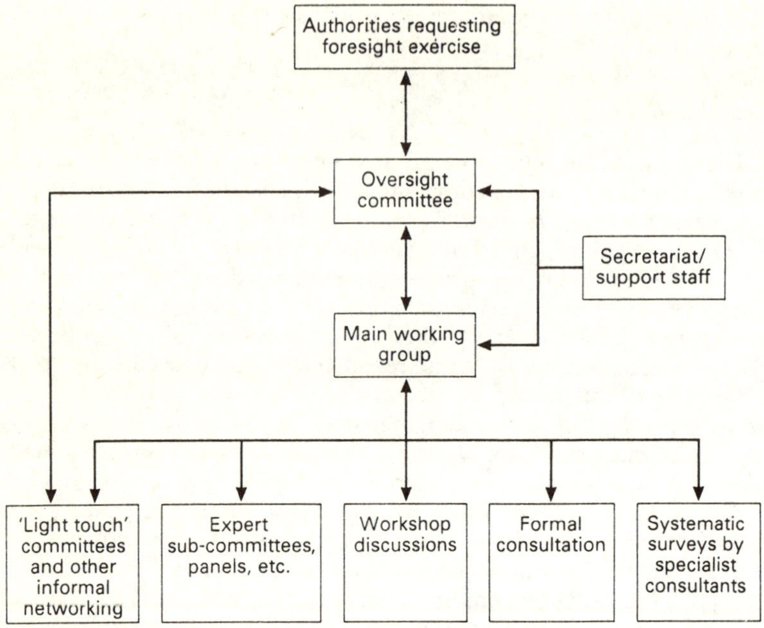

Figure 2.2 Administrative organization of the foresight process

As Figure 2.1 shows, the first of these tasks is to complete the more detailed aspects of *designing the foresight process* begun in the pre-foresight phase. This is primarily the responsibility of the working group and generally involves the following:

(1) further elaboration and clarification of the issues which need to be addressed if foresight is to make an effective contribution to priority-setting;
(2) preparing a study plan and schedule for executing each component of the exercise (the question of timing being especially relevant when foresight is linked to policy formulation within a fixed budgetary cycle);
(3) determining procedures for generating the various types of information likely to be required (e.g. formal and informal consultation, questionnaire surveys or analysis of bibliometric and patent data);
(4) selection of experts to be involved in the exercise (e.g. in study panels or workshop discussions), and appointment of any external consultants or think-tanks required to undertake specialized studies;
(5) establishing procedures for handling the processing and analysis of the large volume of data that is often generated;
(6) ensuring adequate mechanisms exist for coordinating the different foresight tasks, especially those being carried out by external organizations participating in the study;
(7) setting in place a system for checking the quality of data analyses and other information inputs, and for validating any conclusions drawn;

(8) agreeing an administrative structure for consulting and informing funding agencies, research performers and users about progress in the exercise.

The next task, and the one requiring greatest investment of time and resources, is *strategic analysis*. This aims to identify, examine and evaluate the main research options, taking into account their long-term benefits and costs. Because future prospects in most cases depend on a combination of science-push and demand-pull factors, strategic analysis involves examining four sets of criteria: (a) evolving socio-economic needs and threats; (b) emerging research opportunities; (c) comparative advantages and weaknesses associated with economic and other resources; and (d) existing scientific strengths and technological capabilities.[12] These are shown in Figure 2.1 together with a fifth input — the results from other relevant prospective studies, including those at a more macro- or micro-level. However, the emphasis on the different factors in strategic analysis will vary considerably depending on the precise nature of the foresight activity and of the research under scrutiny. This means it is only possible to outline in broad terms the mix of analytical approaches likely to be employed in any specific exercise.

The scope for using formal foresight techniques tends to be greatest in evaluating evolving socio-economic needs for research (e.g. in strategic fields linked to industrial technology or R&D aiming to meet longer-term challenges in areas like health and the physical environment). While the focus of analysis can differ widely (ranging from commercial prospects to broader cultural considerations), it is generally necessary to begin by identifying the underlying factors driving relevant socio-economic trends, followed by an examination of likely scenarios if certain trends dominate or if particular policies are pursued. The results can then be used to determine the potential demands for R&D and the implications for research policy. Analysis of this type may require extensive background statistical information and the adoption of relatively sophisticated forecasting and monitoring or scanning techniques.[13] However, the primary input will consist of informed assessments by research users and other stakeholders on the demand-side factors likely to be important in coming years.

The approach employed in evaluating promising scientific developments and opportunities is in many respects similar to that for socio-economic needs, although formal forecasting may be less common (except perhaps in demographic analyses of research personnel). There has traditionally been a tendency to rely on judgements by the scientific community made during informal committee or workshop discussions, complemented in some cases by information drawn from monitoring trends and developments in R&D (e.g. through networks of overseas scientific attachés or organized foreign visit programmes). Nevertheless, more formal approaches may also be used such as structured expert panel exercises (e.g. to rank the importance of different options for future research) or consensus-based methods like Delphi surveys. In addition, recent developments in information science have made available a range of quantitative techniques for analysing patent and publication data to obtain an early warning of emerging research trends, including opportunities for R&D in specific fields arising from progress in related areas of science or from advances in technology.

The third component of strategic analysis concerns the assessment of comparative advantages and weaknesses associated with available socio-economic resources, this being especially important when considering the prospects for more applied research. As with evaluation of socio-economic needs, the primary source of information is likely to be potential research users and others with the perspective and experience to judge realistically the chances of being able to capitalize on R&D opportunities. The aim of companies will be to identify their relative strengths *vis-à-vis* competitors, while public-sector foresight may focus on macro-level comparisons with other countries. Among the factors typically taken into account in the latter are:

(1) natural advantages or limitations stemming from climate, geography or access to raw materials;
(2) the likely competitiveness of different science-based industrial sectors (e.g. because of the existence of leading international companies or fast-growing domestic markets);
(3) current and projected government policies (especially in such areas as education and training, corporate taxation, industrial standards, health, environmental regulation, and high-technology and defence procurement);
(4) effects of possible changes in government and political structures (e.g. on the environment for future technological innovation);
(5) other socio-cultural constraints or advantages (e.g. language).

The scope for employing formal foresight techniques here is somewhat less than in respect of socio-economic needs, and the same is true for the assessment of scientific and technological resources where the emphasis on supply-side factors means that the principal inputs will come from research performers and funders. The main issues to be considered include:

(1) existing research strengths and weaknesses;
(2) the breadth and quality of scientific and technological skills available now and in the future;
(3) scientific facilities, instrumentation and other capital equipment needs;
(4) the linkages of new research areas with current or planned activities;
(5) the financial resources likely to be available from funding agencies and the opportunity costs for different research options;
(6) the adequacy of the overall scientific infrastructure (e.g. existence of established links between academic or government laboratories and industry).

The final task in strategic analysis is to synthesize what may often be diverse and contradictory sets of information to reach a balanced appraisal of the prospects for each research option in terms of: (i) likely scientific, technological and socio-economic benefits and costs; (ii) the prospects of being able to mobilize the necessary financial, organizational and human resources; and (iii) the broader implications for longer-term policy.

The next step is to *reach agreement on the most promising research options*, and to specify the conditions under which they could best be pursued (including the support mechanisms and institutional resources required). As such, this task represents an extension of strategic analysis but from a more focussed perspective in which the relative merits of different options are explored and contrasted with each other, perhaps in the context of policy guidelines formulated during an earlier direction-setting exercise. It should be emphasized, however, that the results are intended only to be *indicative*, and there are (at least in our model of the foresight process) as yet no direct implications for decisions on resource allocation.

In comparing the prospects for the various options, judgements relating to a wide range of factors must be combined, and the precise criteria and the relative weight accorded to them will depend upon specific circumstances, not least the ideological, political and institutional preferences of the primary audience for the foresight exercise. This said, the general themes addressed will often be similar, and in the case of holistic or macro-level foresight might include the following:[14]

(1) Which broad research fields appear to have the best scientific or socio-economic prospects over the longer-term? What will be the respective benefits and to whom will they accrue?
(2) Which of these fields are likely to prove more risky or expensive to support and develop? Do the problems that are foreseen stem from financial, technical, organizational, socio-cultural or other causes?
(3) What potential exists for building upon existing national scientific strengths in the most promising areas or fields? How do the requirements for specialist expertise match up with the ability and willingness of established R&D organizations to develop relevant programmes, and more generally with the likely supply of trained researchers?
(4) How difficult will it be to capitalize upon expected developments in the various fields? In particular, is research likely to impact on socio-economic sectors related to areas of existing technological or industrial strength? Or are others better placed to exploit the anticipated opportunities?
(5) What implications for research policy arise from the structural characteristics of different fields, especially in respect of pervasiveness of impact on other areas of science and technology, or across several socio-economic sectors? And do potential synergies exist between any of the promising fields which also need to be taken into account?
(6) Given the inherent uncertainties attached to realizing potential benefits in each of the fields, to what extent is it desirable to target resources on a selected number of areas while at the same time providing limited support (i.e. maintaining a 'watching brief' capability) over a broader range of research? How great is the risk that targeting might adversely affect future flexibility in responding to unanticipated scientific or socio-economic threats and opportunities?
(7) Is it possible to be more specific in identifying promising areas or lines of future research and, if so, will this help or hinder subsequent discussions to agree budgetary priorities?
(8) What are the wider organizational and political implications of initiatives in different fields? For example, how closely do they relate to the responsibilities of individual ministries or funding agencies, and what is

the overall degree of 'fit' with government policy in areas like regional socio-economic development? Will collaboration be necessary between agencies, or across government laboratories, universities and industry, and how would the respective roles envisaged change over time (e.g. as work progresses from the strategic research phase to the development of new technology)?

(9) What are the relative financial and other costs of different research options, including the investments in scientific and industrial infrastructure required to ensure potential benefits are exploited effectively? What sets of choices would be possible in the context of different budgetary scenarios? And what might be the implications of deciding to take no action at all in responding to the scientific needs and opportunities identified?

There is often a close relationship between the questions relevant to macro-level foresight and those posed in a meso-level exercise. This is especially the case when a direct link between the two exists (e.g. when meso-level foresight is carried out to explore in greater detail promising opportunities identified during an earlier holistic or macro-level initiative). However, the focus will be rather more specific:[15]

(1) Which areas appear most promising in scientific or technological terms, and how important and pervasive an impact is likely to be made within and outside the field? Are there any active intellectual growth-points in particular subfields, and do the development dynamics so far indicate they will gain in significance rapidly or only over the longer term?

(2) What demands for R&D can be foreseen, and do any lines of research stand out in terms of the nature and extent of their likely socio-economic impact?

(3) Do linked or supporting R&D activities outside the field need to be developed further before the expected scientific or socio-economic benefits can be realized?

(4) In which areas do research groups exist that are capable of developing a leading international position if resources are made available? What level of funding, instrumentation and other support would be required and over how long a period?

(5) Do such groups within universities and government laboratories have the linkages with industry and other research users necessary to ensure that scientific breakthroughs will be exploited? If not, could they be developed by the introduction of new mechanisms for research support like measures to promote cross-sectoral collaboration?

(6) How confident can one be that the areas of research identified as promising will indeed open up major opportunities in the future? For example, is there consensus among scientists in different subfields on the lines of research which should be accorded high funding priority? And to what extent do companies and other research users agree with the scientific community on these priorities?

(7) What would be the resource implications of initiatives in different areas (including related requirements for support of postgraduate training)? And what alternative portfolios of research might be feasible given different budgetary scenarios for a programme in the field?

The final step in the main foresight stage is *diffusion of the results*. While the primary audience in an exercise oriented to determining priorities is normally policy-makers, dissemination of results to research performers, potential users and other stakeholders is also essential (see Figure 2.1). Foresight does not therefore end with the submission of a formal report, an act which may merely signify the beginning of a lengthy procedure in which different mechanisms like conferences, workshops and articles in newspapers and specialist journals are employed to ensure that findings are transmitted effectively to relevant audiences. The effort required to formulate an appropriate dissemination strategy will obviously depend in part on how broad a constituency of interests is involved in deciding priorities for research in the areas covered by the foresight study. Certain follow-up activities can also be required to assist those commissioning the exercise with assimilating the results and translating them into policy. In particular, further clarification may be sought on the assumptions, data and analysis underlying assessments of future prospects, especially in cases where their accuracy or validity is called into question.

(c) Post-foresight stage

The last set of tasks in the foresight process outlined in Figure 2.1 relates to the post-foresight stage[16] — that is, the implementation of findings in policy action aiming to ensure that, as far as possible, the preferred options of today are translated into the realities of tomorrow. The equivalent activities involved during this stage in other types of foresight like consensus-generation and anticipatory intelligence are likely to differ considerably, and any influence on science policy normally involves more diffuse mechanisms. Furthermore, by no means all priority-setting foresight exercises proceed directly to the policy implementation phase. Many at a holistic level, for example, may instead result in follow-up foresight studies at a more micro-level before there is any impact on scientific decision-making.

There will also be considerable variation in the *institutional processes to determine resource allocation priorities* (with differences existing, for example, between organizations solely responsible for funding research and those also engaged in its execution or exploitation). The end-result in most cases will nevertheless be an explicit *decision to launch*, expand or change the direction of *a research programme* (a term used here in a generic sense to cover any initiative, including those relating to training or R&D infrastructure, where more specific choices on particular lines of research remain to be made). The shape of the programme will almost certainly reflect the outcome of compromises negotiated among not only research performers and users but also other stakeholders like trade unions, pressure groups or investment institutions. Decisions are therefore rarely, if ever, made on an entirely 'rational' basis in that a complex set of political and other factors (including bureaucratic considerations) generally mediates the priorities or policies finally agreed, irrespective of whether the link between foresight and programme implementation is direct or indirect. This is explicitly recognized in the model of the foresight process outlined in Figure 2.1.

The subsequent step of *programme definition* is also influenced by a broad array of political and institutional forces depending on the particular environment in which policy decisions are taken. It may well be necessary, for example, to resolve competing claims among agencies concerning responsibility for administering or carrying out research (although in other cases this will have been decided prior to launching the programme). Moreover, even if this results in one agency assuming control, further choices may be needed on the allocation of tasks among different laboratories or divisions, especially in cases where new institutional procedures are proposed. Rip and Hagendijk (1988, p.14) have characterized the type of mechanism involving a single designated organization as *block implementation*, which they distinguish from both *network implementation* (the establishment of collaborative research activities at several institutions) and *allocation implementation* (the more traditional system of inviting proposals from scientists for work in a defined area).[17]

The form of implementation mechanism chosen also affects other aspects of programme definition, especially the organizational arrangements for planning, managing and evaluating research at the project level. Among the tasks which may be involved are:[18]

(1) Further clarification and updating of any prospective analysis undertaken prior to establishing the programme, particularly if there have been delays in securing funding or if scientific breakthroughs have occurred in relevant fields.

(2) Articulation of a hierarchy of broad objectives and detailed goals for the programme, together with a general strategy for tackling them. Rip and Hagendijk (1988, p.11) identify two types of strategy which may be followed:

> If the policy maker takes adaptation as his starting point, and only attempts some modification in the light of his goals, he follows an *accommodation* strategy. If his goals are the starting point, he will be more active in the selection of research and research institutions, in stimulating actors to do what he wants and in creating commitment. Such a strategy can be called *orchestration* of the implementation.[19]

(3) Establishing a suitable management control system for the programme, a key element of which is to make provision for systematic *research monitoring and updating of previous foresight*, along with periodic reassessments of overall strategy. This provides a formal structure for *steering and redirecting* or even terminating research activities in the light of emerging scientific opportunities and changing socio-economic needs.

(4) Setting in place mechanisms for the diffusion of research results and technology transfer, both during the programme and after its eventual completion.

The next step in the post-foresight phase is *project definition and programme execution*. Again, different approaches to planning and management will tend to be associated with each type of implementation procedure. Expert committees and peer-review will play a central role when allocation or network mechanisms are employed, while block implementation will typically result in institutional management having a significant influence on project definition as well as on more general steering of programme activities. Greater reliance may also be placed on formal research evaluation, especially in applied projects where progress can be checked against a series of agreed 'milestones' or technical targets. Such monitoring can then assist with decisions at the programme level as to whether changes are required in the light of updated information on strengths, weaknesses, opportunities and threats provided by ongoing foresight activities.

The final element in the foresight process shown in Figure 2.1 is *dissemination of research results*. Among the tasks involved are: (i) selecting appropriate mechanisms for diffusing or transferring research findings to the various users or potential beneficiaries identified previously; (ii) allocating institutional responsibility for dissemination to those best equipped to transfer relevant knowledge and expertise to the different target audiences; and (iii) establishing follow-up procedures to deal with any problems encountered at a later stage with assimilation or exploitation of results. All these post-foresight tasks are no less central for the foresight process than those undertaken earlier to identify promising research opportunities, since success will ultimately be judged in terms of whether the resultant R&D activities subsequently meet important scientific or socio-economic needs.

4. CONCLUSIONS

Our aim in this chapter has been to set out an overall framework for analysing the lessons which can be drawn for research policy from the wide range of prospective exercises considered in the eight country chapters which follow. This entailed outlining, first, a multi-dimensional typology for classifying different types and functions of foresight, and, second, a conceptual model of the foresight process revealing how the various elements and tasks involved are interrelated. We should reiterate that this model represents an idealized abstraction of one form of research foresight put forward to assist in understanding the nature of the process involved. The need to avoid unnecessary conceptual complexity does mean, however, that the model may suffer from a number of intrinsic limitations. In particular, foresight is rarely such a linear process as we might appear to have suggested, and branching and doubling back to earlier steps may be the rule rather than the exception.[20] Furthermore, in practice each set of tasks is not always carried out in the chronological order indicated — for example, efforts to conduct research may begin before a programme is fully articulated (Rip and Hagendijk, 1988, p.6). Nevertheless, the model should prove helpful in understanding why certain foresight exercises achieve significantly greater impact on priority-setting than others, as well as providing a more general perspective on the way different types of prospective analysis influence the formulation of science policy.

NOTES

1. As with any classification scheme, the boundaries between these categories are often indistinct. Furthermore, it should be noted that individual foresight studies may often serve more than one function.

2. Given the conventional use of the term 'strategic' and the definition employed in this volume (see Chapter 1), a better alternative might be 'legitimatory'. Such a function of foresight is closely related to the notion of foresight as advocacy discussed below.
3. There are many sociologists of science who dismiss this internal/external dichotomy as meaningless. It is put forward here purely as a pragmatic distinction, but one which is nonetheless useful for analysing different forms of foresight.
4. See note 3.
5. This is often linked to the functions of consensus-building and advocacy discussed earlier.
6. Luukkonen-Gronow (1989) has employed a three-fold classification scheme functionally equivalent to Elzinga's in describing the deployment of *post-hoc* evaluation in policy-making — this covers 'instrumental', 'conceptual' and 'persuasive' uses. The term 'instrumental' is perhaps preferable to Elzinga's 'responsive' mode since it encompasses a broader range of foresight.
7. Participatory awareness is not dissimilar to the notion of anticipatory democracy advocated by Toffler (1970, pp.425–40), a concept for which there are few empirical exemplars.
8. For example, the convergence of communications and computing was widely predicted in the late 1970s by electronics companies in Japan and elsewhere.
9. These terms are sometimes used rather differently in the forecasting and research-planning literature.
10. It is not our intention here to provide a comprehensive list of the available methods, together with their respective strengths and weaknesses. The reader is referred instead to standard texts on technological forecasting such as Jantsch (1967), Bright (1968), Lien *et al.* (1968), Arnfield (1969), Ayres (1969), Cetron (1969), Wills *et al.* (1969), Lanford (1972), Martino (1972a; 1972b; 1978), Wills (1972), Fowles (1978), and to the extensive bibliography in Boucher (1977). For an analysis of the relative merits of modelling as an approach to foresight, see Coates (1985, pp.39–40).
11. This acronym was brought to our attention by Dr Christopher Hill of the US Congressional Research Service.
12. The appraisal of Strengths, Weaknesses, Opportunities and Threats is sometimes referred to as SWOT analysis.
13. According to the US Foresight Task Force (1983, p.7), the assessment of socio-economic needs and opportunities can be broken down into four activities: scenario construction, premises quantification, scanning and monitoring. Scanning is defined as the analysis of random signals for possible pattern formation, while monitoring involves the tracking of known variables and, in some cases, the comparison of actual achievements with anticipated performance. However, other possible inputs to this component of strategic analysis include technology assessment and impact analysis.
14. This list of factors is based in part on DITAC (1986, pp.37–8) and Elzinga (1985b, p.43).
15. Many of these questions were identified in Department of Science (1987, Appendix 10).
16. This brief outline of the post-foresight phase draws on the discussion by Laredo (1986) of the processes involved in launching and steering research programmes, as well as the analysis by Rip and Hagendijk (1988) of the implementation of science-policy priorities.
17. In practice, a combination of these mechanisms is often adopted.
18. Several of these tasks relate to the central issues traditionally addressed in science policy such as the management of research programmes and technology transfer. As stressed earlier, their inclusion within the post-foresight phase of our model

should in no way be seen as implying that science policy is subordinate to research foresight.

19. The two types of strategy and the different foresight functions discussed earlier are to a certain extent related. Foresight for communication and education is likely to be most relevant to an accomodation strategy, while initiatives aimed at direction-setting, priority-determination and consensus-generation will normally be more closely linked to the adoption of an orchestration strategy.
20. In particular, many of the important interactions with research performers and users during the 'post-foresight' phase are not shown in Figure 2.1. For example, updating the strategic analysis will involve further inputs from scientists, industrialists and other experts.

REFERENCES

Arnfield, R. (ed.) (1969), *Technological Forecasting*, Edinburgh, Edinburgh University Press.

Ayres, R. U. (1969), *Technological Forecasting and Long Range Planning*, New York, McGraw-Hill.

Baba, Y. (1988), 'Systemic Innovation: Its Nature and How to Benefit from It' (mimeo), Brighton, SPRU, University of Sussex.

Bodmer, W. (1988), 'Can Science Manage?, *New Scientist*, vol. 1629, p.27.

Boucher, W. I. (ed.) (1977), *The Study of the Future: An Agenda for Research*, Washington, DC, National Science Foundation.

Bright, J. R. (ed.) (1968), *Technological Forecasting for Industry and Government*, Englewood Cliffs, NJ, Prentice Hall.

Cetron, M. J. (1969), *Technological Forecasting: A Practical Approach*, New York, Gordon and Breach.

Coates, J. F. (1985), 'Foresight in Federal Government Policymaking', *Futures Research Quarterly*, vol. 1, pp.29–53.

Department of Science (1987), *Research Strength and Economic Benefit: An Examination of Strong Areas of Australian Basic Research and Their Potential for Economic Exploitation*, Canberra, Department of Science.

DITAC (1986), *Selecting Technologies for the Future*, Canberra, Department of Industry, Technology and Commerce.

Elzinga, A. (1983), *Foresight as Anticipatory Intelligence*, Ottawa, Science Council of Canada.

—— (1985a), 'Technology Forecasting', in Mar, B. W. *et al.* (eds), *Managing High Technology: An Interdisciplinary Perspective*, pp.309–18, Amsterdam, Elsevier.

—— (1985b), 'Mandated Science in the Universities and Picking the Winners: What are the Implications?' in Becher, T. *et al.* (eds), *Higher Education and Research Organizations*, pp.21–49, Department of Political Science, University of Stockholm.

Foresight Task Force (1983), *Foresight in the Private Sector: How Can Government Use It?*, Report to Committee on Energy and Commerce, US House of Representatives, 98th Congress, 1st Session, Committee Print 98-B, Washington, DC, US Government Printing Office.

Fowles, J. (1978), *Handbook of Futures Research*, London, Greenwood Press.

Gaillard, J. (ed.) (1986), *Cycle Politique, Programmation, Gestion de la Recherche pour le Développement*, Paris, Institut International d'Administration Publique.

Irvine, J. (1988), *Evaluating Applied Research: Lessons from Japan*, London, Pinter Publishers.

—— and Martin, B. R. (1984), *Foresight in Science: Picking the Winners*, London, Pinter Publishers.

——, —— and Oldham, C. H. G. (1983), *Research Evaluation in British Science*, Brighton, SPRU, University of Sussex.

Jantsch, E. (1967), *Technological Forecasting in Perspective*, Paris, Organisation for Economic Cooperation and Development.

Lanford, H. W. (1972), *Technological Forecasting Methodologies: A Synthesis*, New York, American Management Association.

Laredo, P. (1986), *Le Programme National: Une Modalité Renouvelée de Gestion des Priorités Nationales de Recherches — Reflexions à Propos de l'Experience Française 1982-1985*, in Gaillard (1986).

Lien, A. P., Anton, P. and Duncan, J. W. (1968), *Technological Forecasting: Tools, Techniques, Applications*, New York, American Management Association.

Luukkonen-Gronow, T. (1989), 'The Impact of Evaluation Data on Policy Determination', in Evered, D. and Harnett, S. (eds), *The Evaluation of Scientific Research*, Chichester, Longman, pp.234–46.

Martino, J. P. (1972a), *Technological Forecasting for Decisionmaking*, New York, Elsevier.

—— (ed.) (1972b), *An Introduction to Technological Forecasting*, New York, Gordon and Breach.

—— (1978), 'Technological Forecasting', in Fowles (1978), pp.369–96.

Rip, A. and Hagendijk, R. (1988), *Implementation of Science Policy Priorities*, London, Science Policy Support Group.

Schafer, W. (ed.) (1983), *Finalization in Science: The Social Orientation of Scientific Progress*, Dordrecht, D. Reidel.

Steed, G. (1987), *Alerting Canadians: The Scope and Use of Policy Research by the Science Council of Canada*, mimeo, Ottawa, Science Council of Canada.

Toffler, A. (1970), *Future Shock*, London, Pan.

Whiteman, D. (1985), 'Reaffirming the Importance of Strategic Use', *Knowledge: Creation, Diffusion, Utilization*, vol. 6, pp.203–24.

Wills, G. (1972), *Technological Forecasting*, Harmondsworth, Penguin.

—— Ashton, D. and Taylor, B. (eds) (1969), *Technological Forecasting and Corporate Strategy*, Bradford, Bradford University Press.

3 Research foresight in France

1. INTRODUCTION

The years 1981 to 1986 were a time of great experiment in French science and technology policy. Recognizing the importance of research for innovation, international competitiveness and social development, the Socialist government provided large increases in funding and introduced major administrative changes to public research. Furthermore, it was committed to the formulation of a longer-term strategy for science and technology, and therefore placed considerable emphasis on foresight or *la prospective*.[1]

Central in the new policy were the *programmes mobilisateurs,* which focussed on areas like biotechnology and represented an attempt to repeat the success of earlier technological development programmes in the aeronautical, nuclear and space sectors. Strategic planning and foresight were built into programme management from the outset. Among the organizational changes, the most far-reaching was the creation in 1981 of the Ministry of Research and Technology, which devoted much effort to longer-term planning. In particular, the *schéma d'orientation scientifique et technique* was introduced for monitoring, directing and conducting prospective analysis in relation to the national research programmes. In addition, the Centre de Prospective et d'Évaluation (CPE) and the Centre d'Études des Systèmes et des Technologies Avancées (CESTA) were set up to carry out foresight and evaluation activities.

During this time, the Centre National de la Recherche Scientifique (CNRS) was also committed to developing longer-term policy, and undertook several foresight exercises, both on its own and in conjunction with the Commissariat Général du Plan (CGP). The latter was especially concerned about the potential effects of emerging technologies, and accorded a prominent place to R&D in the Interim National Plan for 1981–4 and the Ninth Plan for 1984–8. Last, in the universities and elsewhere, there was a flowering of theoretical and methodological work on *la prospective* (e.g. Godet, 1985).

Views on how successful these initiatives proved are mixed. Some see the period as a promising start in developing a more strategic approach to research; despite certain problems, the system could have been improved given time. According to others, however, the efforts were largely in vain since research is not susceptible to planning, and a more liberal or less directed approach is required. It is difficult to judge between the two views since the incoming government in 1986 proved less committed to R&D, and especially to longer-term planning. Many of the foresight activities developed during the

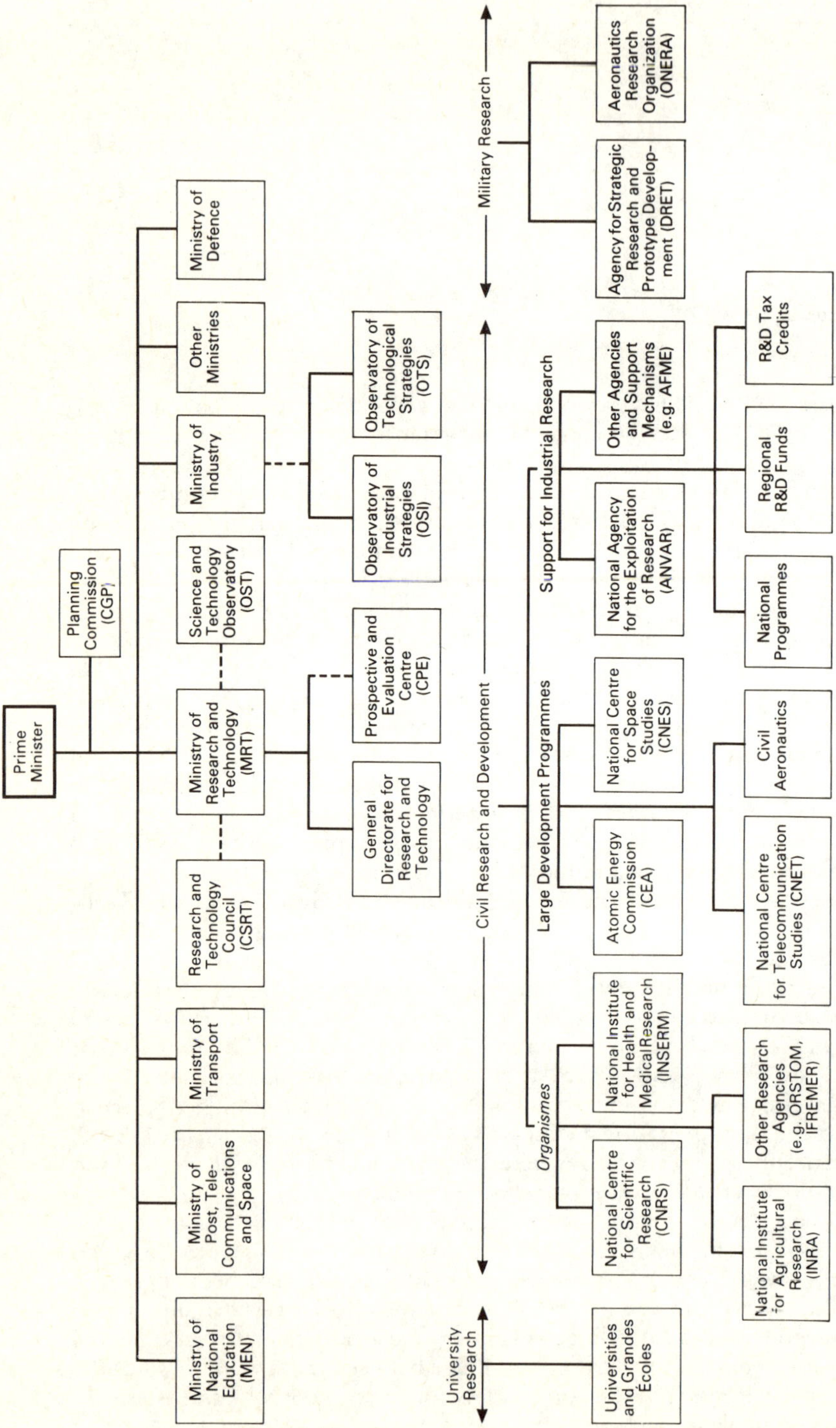

Source: Constructed from government science-policy documents.

Figure 3.1 *Outline of the French R&D system*

previous five years were therefore abandoned until 1988 when the Socialists were re-elected.

Hence, the focus in what follows is primarily on the period 1981–6. We look first at foresight in relation to more basic research in CNRS and other agencies. Next, prospective activities within the Planning Commission, the Ministry of Industry and the Ministry of Research and Technology are analysed. Finally, we examine the strategic planning efforts of five major companies, before concluding by summarizing the main lessons from French experience with foresight.

2. NATIONAL R&D STRATEGY

Any attempt to describe the government-funded R&D system (see Figure 3.1) is made difficult by the complexity of the structure and the fact that it has changed several times over recent years. One permanent feature, however, is the heavy reliance on *organismes* (research agencies). Aside from military R&D (not considered here) and universities (which receive core-funding from the Ministry of National Education[2]), public funds are largely channelled through some thirty agencies. They vary widely in nature, with CNRS traditionally being responsible for most fundamental research.

A second central element is the Ministry of Research. Prior to 1981, the Délégation Générale à la Recherche Scientifique et Technique was responsible for coordinating ministries with R&D responsibilities, but its powers were limited (OECD, 1986, p.62). It was replaced by the Ministry of Research and Technology, which was 'empowered to propose and, in conjunction with the other ministries concerned, to implement government policy in the field of research and technology'.[3] During 1986–8, responsibility for research was subsumed within the Ministry of Research and Higher Education. This represented a downgrading in the importance attached to research, with several R&D responsibilities being reallocated to other ministries and with the Minister of Research no longer reporting directly to the Prime Minister. However, the Ministry of Research and Technology was reconstituted following the 1988 elections.

As regards the development of a national research strategy, there have been four key events — the National Colloquium on Research and Technology in 1981–2, the R&D Act of 1982, the Ninth National Plan for 1984–8, and the Triennial Plan for Science and Technology which replaced the R&D Act in 1985. During the National Colloquium, large numbers of researchers, industrialists, trade unionists, politicians and government administrators examined the likely impact of research on the country's economic and social development. Certain problems were identified with interdisciplinary research areas and technology transfer, the overall view being that *organismes* were insufficiently aware of national needs and new policy instruments were therefore needed. From this stemmed the idea of concentrating scientific activities in strategically important fields through *programmes mobilisateurs* (Barré, 1987, p.32).

It was the passing of the 1982 Act on R&D, however, which made possible the development of a long-term national strategy for science and technology. The Act recognized that three principles are involved in constructing such a strategy. First, longer-term objectives must be set and resources committed over a period of several years. This was achieved through the Act being a 'programming law' (similar to that for military projects) which earmarks resources on a multi-year basis. Second, research policy should be a matter not just for the scientific community but for the nation as a whole, in particular the various 'social partners' in R&D (industry, trade unions and the regions). Third, efforts should be concentrated on priority objectives that would 'mobilize' the research community, and the R&D Act therefore added the *programmes mobilisateurs*[4] to the existing range of support mechanisms (Barré, 1986, p.304).

The Ninth National Plan then integrated much of the R&D Act into the proposals for 1984–8. Of the plan's twelve priority programmes, the third was devoted exclusively to research and innovation, while six others included appreciable R&D components (CGP, 1984a; 1984b). Finally, the Triennial Plan for Science and Technology was established in 1985 following an assessment of the objectives and mechanisms of the R&D Act, and after extensive discussions within the Conseil Supérieur de la Recherche et de la Technologie (the top-level advisory body to the Minister of Research) and in Parliament. This provided the final component in the framework for national strategic planning of research, giving a mandate to the Research Minister to:

> present every year on behalf of the government a report on science and technology activities detailing *the strategic options for national policy* and the level of realization of the objectives set out in the legislation, by presenting evidence of the results achieved in major foreign countries compared with France in international competition. [Revised translation from Barré, 1986, pp.306-7, emphasis added][5]

Hence, by the end of 1985, the concept of a longer-term strategy had been placed at the heart of French research policy.

After the election of the Chirac government, however, strategic planning declined in importance. The uncertain status of the Planning Commission reflected doubts within government about the very notion of a national plan, and meant that CGP was in no position to initiate any foresight activities. The government also dismantled the Ministry of Research and Technology and with it most of the mechanisms for the strategic management of research, with the result that there was no overall national science policy. Individual ministries and agencies were instead left to devise their own policies. Aside from CNRS field studies, there was consequently little research foresight in the two years up to 1988. The new Socialist administration, in contrast, seems determined to continue the initiatives launched by its predecessor in the period before 1986.

3. FORESIGHT IN BASIC RESEARCH

(a) Centre National de la Recherche Scientifique

CNRS finances the entire range of basic research from physical sciences to the humanities. Approximately half its budget of F9,000 million in 1988 was used to support work in its own institutes with most of the remainder going to 'associated laboratories' in universities. CNRS formerly operated on the basis of a rolling three-year *schéma-directeur* (Irvine and Martin, 1984, p.30), but this was replaced in 1986 by an annual strategic plan, the emphasis being more on funding the best groups than identifying promising areas. There are three main inputs to the planning process: (i) prospective studies; (ii) an evaluation of current strengths and weaknesses in French research; and (iii) assessments of individual laboratories. Information relating to these comes from the CNRS National Committee and its forty-nine sections (which play an important role in overall policy coordination), from the agency's seven scientific departments, and from the Directorate of Programming and Budgetary Forecasting.

During the 1980s, CNRS has conducted a wide variety of foresight activities including: (i) macro-level analyses of research priorities; (ii) *ad hoc* studies of individual fields to help develop policy guidelines, or to decide on investments in costly scientific facilities like telescopes; and (iii) international comparative studies.

(i) Macro-level foresight

Initiated in 1984 by Pierre Papon (then Director-General), the *schéma prospectif* (or 'twenty strategic themes') exercise represented an attempt to adopt a more systematic approach to policy-making. Papon recognized that clearer strategic objectives would be essential in making the case to government for budgetary growth, and therefore decided to submit a broad range of promising research areas to prospective analysis. The intention was to use the resulting longer-term perspectives as an input to the rolling three-year *schéma-directeur*, which provided a mechanism for implementing the priorities selected. Since previous foresight studies by CNRS had generally focussed on individual fields, a broader overview was required to define priorities cross-cutting CNRS scientific departments and to help launch new programmes, especially in interdisciplinary areas such as communication science. Another aim, however, was to disseminate information on scientific needs and opportunities to researchers and mobilize their efforts in relation to possible initiatives. It was felt that many French scientists had become too wedded to existing lines of research, and were often slow to recognize the importance of emerging areas.[6*]

Several criteria were employed by the Scientific Directorate in selecting the twenty themes. Some were chosen purely because they represented promising scientific opportunities, while others were more important to technology (for example, fields like biotechnology which were linked to the recently created *programmes mobilisateurs*). In several cases, however, the exercise was used as an opportunity to explore the prospects for international collaboration, or to focus on problem areas such as process engineering where the quality of French work was high but relatively few researchers were involved.

The exercise was primarily carried out by expert panels (often but not always based on existing committees) consisting of CNRS researchers and senior officials, academics and industrialists. The time-horizon was meant to be around five to seven years, but it varied among areas 'from zero upwards'.[*] Although some attempt was made at international comparisons, no formal prospective techniques were employed, nor was much use made of bibliometric statistics or other data-bases. The reports from each prospective study were then synthesized into the final *schéma prospectif* with the assistance of science journalists (CNRS, 1985a).

According to CNRS officials, the exercise was successful both in identifying leading research groups and in assessing the scientific potential of each area. Some of the findings were subsequently implemented in the *schéma-directeur,* resulting in increased support for materials science, new programmes in communication science and work at the interface between chemistry and biology, as well as in the establishment of a centre of excellence in process engineering. Other less direct benefits cited included fostering interdisciplinary cooperation and educating scientists about the benefits of foresight.* On the other hand, given that this was the first such exercise by CNRS, shortcomings were perhaps inevitable. In particular, because the themes were not selected especially systematically, there was some dissatisfaction among scientists in areas not covered. The emphasis on science-push factors also meant that little account was taken of socio-economic needs, except in materials science and in process engineering (where an industrialist chaired the panel). Moreover, in the analysis of scientific opportunities, there were limited inputs from outside CNRS, the small number of foreign experts involved on panels perhaps reflecting a certain parochialism among French scientists with regard to learning from abroad.* As a result, the *schéma prospectif* had little impact outside the CNRS community. Finally, those involved now recognize that insufficient resources were set aside to carry out the level of background research necessary for informed prospective analysis.

Several of these lessons were subsequently taken into account in Prospectives 2005, a large-scale foresight exercise integrating economic and social demands with scientific opportunities. This and the later Europrospective initiative were carried out jointly with the Planning Commission and are described later. However, any possibility of CNRS building upon these efforts was ruled out after the change in administration in 1986 when the *schéma-directeur* system was abolished. Moreover, a reappraisal of the CNRS planning process concluded that a further shortcoming of the *schéma prospectif* was the low degree of legitimacy that foresight has within the research community when carried out by small groups. Instead, it was decided that the National Committee should assume greater responsibility for such exercises.

Previously, individual sections of the National Committee had to produce a *rapport de conjoncture* (i.e. a report on the 'state of the art') every four years, the last occasion being in 1984. They were now required to conduct systematic prospective studies, and to meet at least biennially to analyse national and international trends within their fields. While the main task of sections remained the evaluation of CNRS laboratories, foresight became more important, with the identification of strengths and weaknesses helping to determine which areas should receive increased support. Similarly, whereas laboratories formerly had to produce only a *rapport d'activité,* they were asked to complement this with an analysis of projected future research. This new system of decentralized responsibility for foresight has not, however, been operating long enough to form any judgements on its success, nor on the changes introduced after 1988.

(ii) Field studies

Several field studies were undertaken during the late 1970s in physical science specialties like condensed matter and atomic and molecular physics (CNRS,

1980a; 1980b).[7] These aimed both at generating consensus on future priorities, and at overcoming the tendency among researchers to concentrate solely on their own narrow lines of work. An effort was therefore made from the outset to involve scientists from a variety of fields, as well as those who would ultimately be responsible for research in the areas concerned.* Whereas these field studies concentrated on science-push factors, some of the later ones conducted during the 1980s have given rather greater emphasis to technological and other demand-side considerations. For example, a review of optics (where the French position had deteriorated because scientists were slow to respond to emerging opportunities) was conducted by a committee containing several industrialists, and involved consultations with twenty-five companies (CNRS, 1984, pp.viii–ix). The results formed an input to the *schéma prospectif* exercise, as did those from a survey of opportunities in mathematics. A *comité de rédaction* of no less than fifty-seven researchers was responsible for drafting the mathematics report, which consisted of seventeen chapters on traditional branches of the discipline with little emphasis on cross-field impact and no overview or synthesis chapter (CNRS, 1985b). Furthermore, unlike the optics report which contained sections on prospective analysis, the mathematics review provided little more than a description of the current state of the field. As a result, it was not regarded as particularly useful for policy purposes.*

Another, more successful, study was on the interface between nuclear physics, particle physics and astrophysics. This differed from earlier surveys in focussing specifically on potential interactions among the three fields (as opposed to likely developments within a single discipline). The approach also departed from previous practice in that there was no large *comité de rédaction*. Instead, the first part of the exercise was undertaken by a single person, Jean Andouze, who was invited by the CNRS Director-General to prepare a prospective report while being given complete freedom regarding the approach employed. Having a background in nuclear astrophysics, Andouze was already familiar with work in two of the areas, and this enabled him to prepare relatively quickly a report identifying comparative national strengths and weaknesses (Andouze, 1982). This was used as the basis for a workshop attended by some sixty scientists, in which participants held brain-storming sessions on possible topics of common interest in the three fields. A small cross-disciplinary committee next drew up a short-list of eight promising areas, which were explored in detail by panels chaired by committee members. Again, the approach was informal, with discussions eventually yielding a set of recommendations. These were then considered by the committee and three priority topics identified, to which a fourth was added when the results were later integrated into the *schéma prospectif*. The success of this interdisciplinary foresight exercise was reflected in CNRS's decision to make available an *incitative* budget to stimulate work in what is now recognized as an important new area. French research already shows marked signs of improvement, especially in cosmology and neutrino research.*

An innovative feature of one of the latest field studies (on biology) is that the results were presented not in terms of scientific disciplines but socio-economic problems (Berkaloff *et al.*, 1987). Since the intention was to produce a rolling medium-term plan for biological sciences that would be revised every other year, the emphasis was primarily on developments up to

1990. These were analysed by the survey committee in terms of three dimensions: (a) socio-economic challenges (e.g. utilization of microorganisms); (b) scientific problems (e.g. the genome); and (c) research techniques (e.g. X-ray diffraction). The report was still under review by the National Committee in 1988, so it is too early to assess its policy impact. Nevertheless, certain limitations were already apparent to senior officials, including the fact that it concentrated on science (no industrialists were on the committee), insufficient use was made of quantitative analysis, the international dimension was not taken fully into account (nor indeed were related activities in other agencies), and the recommendations made could have been more precise.*

Related to these studies has been a series of foresight exercises concerning large scientific facilities. A recent example which focussed on astronomy examined whether France should participate in an initiative by the European Southern Observatory to construct a Very Large Telescope, even though this would involve a reduction in existing domestic activities. The lengthy process of designing and commissioning such an instrument required that a time-horizon of fifteen years be considered by the panel of five astronomers conducting the study. A questionnaire was sent to scientific groups asking about equipment, personnel and likely future developments, with larger laboratories also being visited to obtain more detailed information. Another input consisted of brief reviews prepared by individual experts on the state of astronomical research in different countries (including an assessment of national strengths and weaknesses). In addition, publication data were used to provide background information (see CNRS, 1987, pp.255–66). The panel then retreated to a quiet location to prepare its report. A deliberate attempt was made to emulate similar field surveys in the United States (see Chapter 5), with explicit criticisms being made where necessary — something which had been rare in previous French foresight efforts.* One result was the controversial recommendation to close certain observatories. However, subsequent acceptance of the report by the Academy of Science helped legitimize the conclusions.

Within CNRS, the exercise is highly regarded for providing a realistic analysis of both future scientific opportunities and the associated resource and personnel requirements. It made precise policy recommendations and within months its impact was clearly apparent, with the process for closing certain observatories already under way. Two factors are seen as underlying this success, the first of which is that the small size of the astronomy community enabled the panel to involve virtually everyone in reaching a consensus recommendation that the Very Large Telescope should proceed. Second, the magnitude and time-horizon for investment in facilities is such that astronomers have recognized (unlike many of their colleagues in nuclear physics) that survival of their field requires priorities to be agreed and difficult choices made about curtailing existing research activities. As a CNRS official noted, 'the astronomy example shows that a report will be much better when specific decisions *have* to be taken. People are then motivated to take the prospective exercise seriously.'*

(iii) International comparative studies

Several foresight-related studies with an international dimension have been undertaken by the Laboratoire d'Évaluation et de Prospective Internationales

(LEPI). Its predecessor, the Cellule d'Évaluation et de Prospective (CEP), was established in the early 1980s under the CNRS International Division. However, the recognition by the directorate that it was rather 'dangerous' for such a unit to be situated within the administration (where it might be subject to political pressures) resulted in LEPI being constituted as a separate CNRS laboratory.* It remains relatively small, with just five professional staff, and its primary task is to compile and analyse information on international research activities.

Among various data-bases that have been constructed, one (BASEL) has certain potential in relation to foresight, and has been deployed in studies of areas like antibiotics,[8] protein separation and remote sensing. Probably the most detailed study has been a review of developments in chemical reactivity carried out jointly by LEPI and the Service d'Études et de Réalisation des Produits d'Information Avancés (SERPIA). The aim was to explore the merits of three different analytical approaches, the first of which involved a bibliometric survey of *Chemical Abstracts.* This encountered severe problems in 'cleaning up' the publication data, and it remains uncertain whether any reliable results will emerge. SERPIA was responsible for the second component which analysed the French (Pascal) publication data-base using the 'co-word' technique that the group had helped to pioneer (e.g. Callon *et al.*, 1986). Following earlier difficulties with interpreting results, the approach has been modified to yield 'maps' which are simpler to understand and more likely to prove acceptable to scientists. CNRS officials have found particularly interesting the 'strategic maps' which identify national strengths or weaknesses in specific areas, and the degree of centrality and coherence of research activities. This represents a significant breakthrough in the acceptance of this bibliometric technique by policy-makers. And while the approach does not itself yield future projections, it is regarded as offering a tool for initiating debate and negotiations on priorities, and ultimately even generating consensus on likely longer-term trends.*

In the third component of the study, LEPI undertook a multidimensional factor analysis of keywords derived from interviews with fifty leading French chemists identified by the CNRS chemical science and basic physics divisions (Karsky *et al.*, 1988). The exercise was subsequently repeated with British and West German chemists. Interviewees were asked not only about current research, but also about likely trends over the next few years and the longer term. Results from the analysis of the current situation in France are shown in Figure 3.2, the two dimensions of the research 'map' corresponding to each of the main factors analysed. A clear separation is evident between theoretical and experimental work, as is the French emphasis on catalysis and irreversible dynamics. Figure 3.3 depicts trends from the present (T0 — see the left-hand side vertex of the triangle) over the following two to three years (T+1 — the bottom vertex) and after five years (T+2). This indicates a likely shift in experimental work towards more complex chemical molecules and increasing use of lasers, while theoretical research should benefit from the application of computers and mathematical models. The approach is seen as having potential for prospective analysis, although work is still at an early stage and the policy utility remains to be assessed.

The results of such studies have begun to convince a previously sceptical scientific community that quantitative approaches may be able to contribute

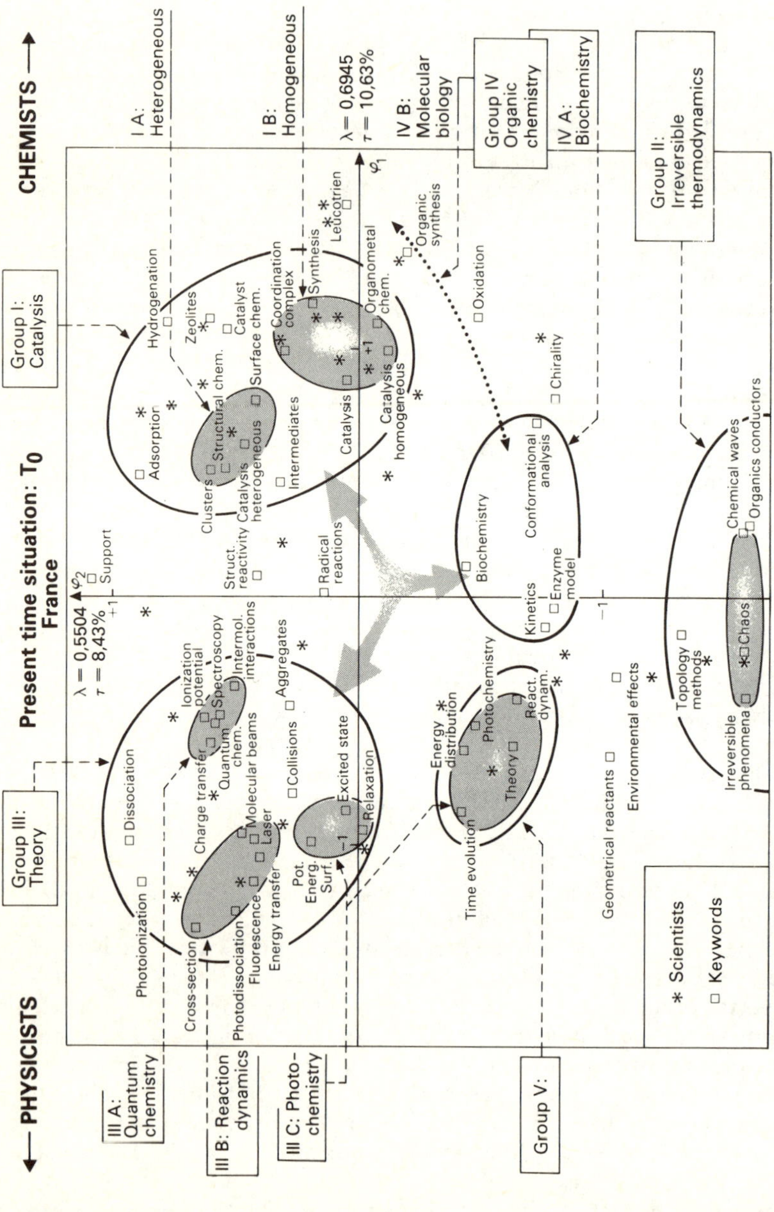

Source: CNRS Laboratoire d'Évaluation et de Prospective Internationales, Paris.

Figure 3.2 Keyword 'map' of French chemical reactivity research

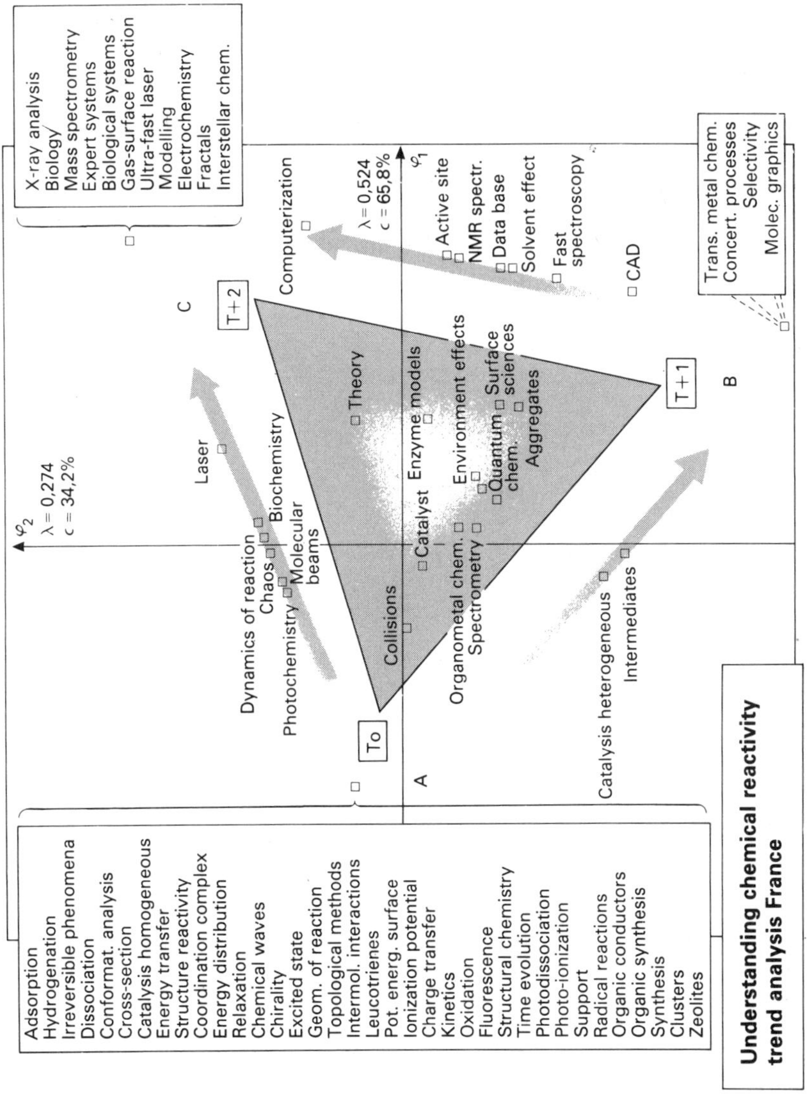

Source: CNRS Laboratoire d'Évaluation et de Prospectives Internationales, Paris.

Figure 3.3 Medium and long-term trends in French chemical reactivity research

to improving the quality of longer-term decision-making, complementing rather than threatening the peer-review process. This is reflected in the establishment by CNRS during 1988 of a Science Policy Indicators Unit (UNIPS) to assist the Directorate of Programming and Budgetary Forecasting with formulating and implementing science policy. It has three main tasks: (i) the preparation of regularly updated science indicators; (ii) conducting evaluations using bibliometric indicators; and (iii) research foresight. The last of these includes responsibility for compiling background information needed for prospective analysis, and providing methodological and logistical support for foresight activities carried out by scientists and other agencies. That such an initiative is now possible stems partly from the progress made in convincing the research community of the advantages of quantitative science-policy methods by involving them in testing and validating different approaches. Nevertheless, a lengthy learning process still remains before scientists fully accept systematic evaluation and foresight as making a legitimate contribution to research policy.[9]

(b) Other agencies

Among other agencies with responsibilities for basic research, the Institut National de la Santé et de la Recherche Médicale (INSERM) finances work in the biomedical and health fields, its budget of F1,500 million in 1987 supporting 4,400 researchers in intramural units and universities. Longer-term planning centres around a prospective report produced every four years by the INSERM Scientific Council. In the first exercise in 1984, field surveys were carried out by the Council's scientific commissions. After consulting with outside experts and in some cases setting up *ad hoc* working groups, each commission produced a report outlining the most promising lines of research. These were then used by the Scientific Council to prepare a *rapport de conjoncture et de prospective* (INSERM, 1984), which listed seven priorities (e.g. molecular genetics and cell biology), together with a qualitative assessment of French research strengths and weaknesses and the implications for the future. Although this report apparently had little effect on the distribution of funds among existing laboratories (where the main criterion remains the quality of research rather than its strategic importance), the priorities identified have played a role in decisions to create new groups.*

The Centre National d'Études Spatiales (CNES) supports space R&D in CNRS and European Space Agency laboratories, as well as in industry and universities. Unlike most other large agencies (its annual budget is over F5,000 million), it has no intramural laboratories, and this limits the leverage exercised by existing institutional interests in strategic planning. In respect of space science, a prospective seminar is held every four years or so to determine the main components of the programme over the next decade. Prior to this, scientific advisory committees meet with researchers to hear their proposals, and prepare prospective reports for specific disciplines like astronomy. Large numbers of researchers, agency staff and representatives from other organizations discuss these reports at the seminar and agree a set of recommendations to be made to CNES, which then bases its longer-term scientific programme on this advice (CNES, 1985). In contrast, priorities for the three-year space technology programme are primarily determined by internal working groups following extensive consultation with industry.

The last of the agencies considered is the Institut National de la Recherche Agronomique (INRA), which is responsible for agricultural research. Its eight scientific directorates each submit an annual set of programme proposals for consideration by the Scientific and Administrative Councils. Although final decisions rest with the Director-General and Scientific Director, most choices are effectively made by divisional directors. Until now, the approach to priority-setting has been rather pragmatic, with only limited use made of consultants (for example, to obtain information on foreign research)[10] and no prospective analysis undertaken of the type carried out by CNRS.[11] Although much INRA research is applied, the main criterion continues to be scientific quality. While attempts are made to obtain some indication of societal needs, INRA has, like CNRS, found it difficult to obtain such information in a systematic form, especially from companies worried about divulging commercially sensitive information during group discussions.[12] However, the fact that INRA's budget increased steadily in real terms during the 1980s (reaching F2,200 million in 1987) has meant that few problems have arisen in relation to initiating work in emerging areas like biotechnology. Whether the same planning approach will prove adequate in future is open to doubt, since any levelling off in budgetary growth will make such decisions dependent on curtailing existing activities.[13]

4. STRATEGIC RESEARCH AND BASIC TECHNOLOGY

(a) *Conseil Supérieur de la Recherche et de la Technologie*

CSRT is the supreme advisory committee on research in France, replacing the former 'committee of sages' when the Socialist government came to power in 1981. Its members include thirty representatives of different sectors (the research community, industry, trade unions and the regions) and ten independent experts. The intention was that CSRT should act as a broad consultative science-policy forum, giving advice to the Minister of Research on the government R&D budget and the annual report to Parliament, the preparation of the National Plan (insofar as it relates to science and technology), and on important emerging areas of research (MRES, 1987, p.4).

Although CSRT is regarded as an interesting experiment in the democratization of national science policy, it has been constrained by three structural problems. First, because it is chaired by the Minister of Research, it is not completely independent. Second, its membership is so large and heterogeneous that there is a tendency to concentrate on institutional aspects of science and technology policy (budgets, mobility, training and so on) rather than becoming involved in specific research initiatives (OECD, 1986, p.65). Consequently, while recognizing the need for a long-term prospective on science and technology (CSRT, 1986, pp.19 and 21), it has made little direct use of foresight. Third, because most CSRT members represent designated sectors, vested interests are never far beneath the surface in discussions. As one official pointed out, it is very difficult to get the CEA representative, for example, to take an open-minded view on the future of energy research. This was contrasted with the predecessor committee where a smaller group of eminent individuals proved better able to discuss strategic policy issues in a less self-interested manner.

(b) Commissariat Général du Plan

Created in 1946, the Planning Commission (CGP) has three tasks: (i) to carry out macro-economic analyses and prepare the national five-year plans; (ii) to act as a tripartite discussion forum for government, industry and trade unions; and (iii) to operate as a think-tank on forecasting. Although CGP experience with prospective analysis has mainly concerned economic planning, some attention has also been given to science and technology issues. For example, an early report entitled *Réflexions pour 1985* (CGP, 1964), drawn up by an expert panel to help prepare for the Fifth National Plan, foresaw the rapid rise of computers (although not microcomputers), but failed to predict the growth of biotechnology and information technology. These omissions were identified by a retrospective assessment of the study carried out twenty years later as background information for a similar prospective exercise on developments up to AD 2000 (Guillaume, 1986). In this, forecasts concerning new technologies featured prominently, together with an assessment of France's standing in relation to other industrial nations (ibid., pp.26–7).[14]

Four foresight studies carried out by CGP merit discussion here, the first of which was commissioned in 1982 from the Observatoire Français des Techniques Avancées (OFTA). The main objectives were to analyse the strengths and weaknesses of French industry, to identify emerging technologies up to 1990 and assess the potential socio-economic impacts, and to formulate proposals as an input to the Ninth National Plan. OFTA focussed on five fields of technology (energy, electronics, materials, biotechnology and manufacturing techniques), and examined their likely effects on different industrial sectors. The resulting report contained general qualitative assessments of the prospects for around fifty lines of technological development.[15] French strengths were seen as lying in areas such as nuclear engineering, aerospace and telecommunications, while weaknesses were forecast in information technology and advanced materials (CGP, 1983, pp.13–14). The report apparently had little impact in CGP or elsewhere, not least because of the limited involvement of research performers and users in the exercise.

A second initiative relates to the Industrial Strategy Groups (GSIs) organized jointly with the Ministry of Industry as part of CGP's remit to provide a tripartite forum in which long-term issues can be discussed.[16] One group was concerned with the diffusion of generic technologies into industry (especially their use by smaller companies), and it set up four subcommittees on microelectronics, data-processing, biotechnology and new materials. Discussions concentrated on the lack of integration between national research policy and industrial strategy rather than on future developments in R&D,[17] although they did play a role in highlighting opportunities in commercially promising fields. The Ministry of Industry and CGP were in 1988 in the process of convening new expert panels, one of which is analysing technological prospectives, a task CGP considers vitally important.[18]

A more successful foresight exercise was Prospectives 2005, which gave rise to particularly fruitful interaction between scientists, administrators and forecasters. This was organized jointly with CNRS at a time when CGP was beginning to recognize the need to link socio-economic planning with that for science and technology. For its part, CNRS had become aware that improved research foresight required better integration of demand-pull factors following experience with the field surveys and the *schéma prospectif*. The benefits of cooperation were therefore clear to both organizations. A mini-symposium to launch the initiative in 1984 attracted good press coverage, and this made it relatively easy to recruit scientists, industrialists, trade unionists and civil servants to serve on an 'orientation committee', which was chaired by a Nobel Laureate. Following presentations by CGP and CNRS officials, the committee agreed a list of seven topics (some of which overlapped with the twenty strategic themes identified earlier by CNRS). Working groups comprising academic and industrial researchers, as well as government officials, analysed the prospects for R&D in each area, taking evidence from other experts and making some use of quantitative analyses of research funding and personnel. Reports produced by the seven groups were presented at a three-day colloquium in 1985 which attracted over 600 participants. This resulted in productive discussions for at least some reports, although other sessions were less successful. Finally, a year later the results were published as a book (CGP/CNRS, 1987).

Although Prospectives 2005 had an impact on the 1986–8 Triennial Research Plan (for example, contributing to the setting up of a materials research programme),[19] the results were not fully implemented because of the change in government and the subsequent replacement of the heads of CGP and CNRS. More important, therefore, was its indirect influence in setting a longer-term national policy agenda, changing the attitudes of trade unions and others towards the links between new technology and employment, and generally helping to pump-prime research foresight.[20*] As for the limitations of the exercise, insufficient use of prospective techniques meant that only very general forecasts of the socio-economic impact for individual technologies could be made. Too much depended on the specialist knowledge of working group members and especially the chairperson — the head of the information technology group was widely regarded as exceptional, whereas others were less creative with the result that their reports were not particularly innovative. In addition, the task of synthesizing the large amount of material proved rather difficult, and insufficient emphasis was given to diffusing the results effectively, the 450-page book being rather academic in tone and inappropriate for the principal audience in government and industry.[*]

Despite these limitations, interest in the results was such that a decision was taken to repeat the exercise at a European level in 1987. A three-day Europrospective colloquium convened jointly with the European Commission brought together 1,000 experts from a wide variety of backgrounds (one third from outside France) to consider the technological, industrial, social and economic dimensions to forecasting. Discussions were divided into twelve sessions, each overseen by an *animateur* in consultation with the central organizing body. Results were presented from recent forecasting efforts by European governments, companies and consultants, followed by a discussion of topics selected beforehand by working groups.

According to participants, although some sessions were successful others produced little that was new. A year later, a team of experts and journalists was still attempting to prepare a synthesis report, and it was, therefore, too soon to judge the longer-term impact. Certainly, the colloquium succeeded in catalysing debate within Europe on the need for collaborative work on the strategic analysis of science and technology. The main criticism, however, is that, as a foresight exercise, it represented little of an advance on Prospectives 2005. Nevertheless, it is planned to repeat the exercise in 1989, and perhaps every two years thereafter. In addition, the European Commission is considering how best to encourage the creation of a network of institutions engaged in research forecasting.

(c) Ministry of Industry

The Ministry of Industry[21] has previously carried out little comprehensive research forecasting.[22] Although individual departments are formally responsible for policy planning in specific areas, in practice this has generally involved no more than short-term economic or industrial forecasting. In the case of information technology, however, a forecast of developments over the next decade was prepared in 1982 and this has since been updated annually. Inputs are primarily made by external contractors (academics and consultants) using trend analysis, expert panels and cross-impact analysis. Technological developments are also monitored through scanning publication and patent data-bases.

In recent years, there have been two initiatives attempting to redress the lack of longer-term foresight. The first involves the creation of an in-house Observatoire des Stratégies Industrielles (OSI), whose role is to analyse the strategies of major corporations around the world as well as national industrial policies (OSI, 1987; 1988). This involves a certain amount of technological forecasting, with relevant information again being provided by outside experts. The second is the establishment of a parallel Observatoire des Technologies Stratégiques (OTS) to monitor the technological strategies of large companies and competitor nations, identify promising fields and examine how industry is preparing to meet the associated challenges. The information compiled on emerging priorities is disseminated both within the Ministry and externally to industry, with growing emphasis on the needs of small firms.

Since 1987, OTS has published a monthly bulletin (*Les Fiches Bleues*) which contains brief reviews of specific industrial technologies.[23] Topics are chosen partly as a result of analyses by consultants of R&D indicators like patents, publications and research funding. Individual reports are generally prepared by OTS staff in association with specialists in the field. In some cases, meetings with industry have been organized to diffuse information on important topics, and these have met with a good response, at least from larger companies. Overall, the OSI and OTS initiatives demonstrate that foresight efforts involving early-warning systems for detecting emerging threats and opportunities can flourish even under a conservative government disinclined to engage in longer-term planning.

(d) Ministry of Research and Technology

In 1988, MRT was reconstituted in a form similar to that existing during 1981–2 and 1984–5. We concentrate here primarily on the latter period, partly because it is associated with a high level of research foresight, and partly because activities within the Ministry of Industry and Research between 1982 and 1984 are described elsewhere (Irvine and Martin, 1984, chapter 3). As today, the responsibilities of the Minister of Research and Technology were to 'propose and implement the government's policy in the field of research and technology'[24] (quoted in OECD, 1986, p.63). Much of the emphasis on foresight stemmed from the priority given to defining strategic research objectives for targeted national programmes (Barré, 1986; 1987). One role of its Mission Scientifique et Technique (MST) was to develop expertise in research evaluation and prospective analysis to help manage national programmes, monitor agencies, and advise other ministries on issues relating to R&D. The three MST divisions (fundamental research, industrial R&D, and social sciences and humanities) were divided into departments, mostly corresponding to national programmes. There was also a small Cellule SOST to assist programme teams with research planning.

The initial step in defining priorities for national programmes can be traced back to the National Colloquium on Research and Technology when some 3,000 people from a variety of backgrounds considered the future for French science and technology. Following a series of regional meetings in late 1981, a four-day conference was held in Paris early the next year.[25] As well as confirming the need for a new policy structure based around targeted support for research, colloquium participants reached consensus on seven strategic themes (e.g. biotechnology). It was also agreed that national programmes should be created as an instrument for mobilizing resources and scientists in solving important socio-economic problems, and as a means of overcoming institutional barriers to interdisciplinary research and technology transfer.

Expert panels (*missions*) each chaired by an eminent president were next set up to carry out strategic assessments of the country's research needs in specific areas. Their main task was to decide whether national programmes were required and, if so, to outline their objectives. In many cases, missions were later transformed into a Groupe d'Évaluation et de Prospective (GEP),[26] the role of which was to conduct the assessment and foresight needed to redirect programme activities in the light of emerging demands and opportunities.

The detailed elaboration of research objectives was carried out by small programme teams overseen by the respective GEPs.[27] Each team, which consisted of MST officials, was then responsible for ensuring that those objectives were met. A key element in programme management was the *schéma d'orientation scientifique et technique* (SOST), introduced by the President of MST, Robert Chabbal, to overcome problems stemming from the complexity of the French R&D system. There were two distinct aspects to the SOST mechanism: (i) it provided a regularly updated information base on programmes, including an evaluation of research achievements, an inventory of resources (both human and equipment), a qualitative analysis of the future evolution of the field, and proposals on priorities; and (ii) it constituted a focus around which dialogue could take place among programme teams, research agencies and the producers and users of R&D, and a 'confrontation procedure' for bringing together science-push and demand-pull factors (Chabbal, 1985, p.i).

There were two phases in the process of preparing the annual SOST. 'Base documents' were first drafted by *ad hoc* groups and synthesized by MST officials (aided by the relevant GEP) to yield a provisional SOST for each programme, outlining strategy and areas where the emphasis should be changed. The proposed SOST was then discussed with research agencies, industry, forecasting experts and, last, with the GEP, and any necessary modifications agreed (ibid., p.ii). The reports for all the national programme areas, together with other related foresight documents (like the CNRS *schéma prospectif*), were finally integrated into an overall SOST (e.g. MRT, 1985b).

While the framework established by MRT represented a pioneering attempt to integrate foresight into the planning and management of research, views on the success of its various components are mixed. As regards the National Colloquium, the wide-ranging and democratic approach to setting overall priorities certainly generated considerable enthusiasm among participants and had a major impact on French research policy. However, it was not particularly systematic — new materials and robotics/automated production were omitted from the list of priorities generated and had to be added later — and the seven original priorities that emerged reflected more the outcome of a political process than a strategic appraisal of national research needs. Another limitation is that the colloquium mechanism cannot be used regularly and thus tends to introduce inflexibility. Once the *programmes mobilisateurs* had been established and the institutional machinery created for managing selected areas, there was no procedure for looking globally at what other fields might merit priority support.

Assessment of the role played by the *missions* shows that few were entirely successful in the task of mobilizing researchers and creating consensus, the problem of persuading people from different sectors to work together often proving insurmountable (e.g. Arvanitis *et al.*, 1986). Furthermore, most failed to undertake systematic strategic analysis to identify priorities *before* programmes were launched (Giri, 1986; Laredo and Rouban, 1985). Even in exceptions like civil engineering where attempts were made to use formal criteria in arriving at a hierarchy of research goals, recommendations were often subsequently ignored by programme teams. This was partly a reflection of the limited effectiveness of many of the evaluation and prospective groups, the policy of involving all relevant government agencies giving rise to such problems as defence of established interests and large unwieldy committees (despite the inadequate representation of industry).[28] Another difficulty was that the GEPs had no resources to commission in-depth prospective analyses, although it is by no means certain that use would have been made of such information had it been available. As one member of the transportation GEP stressed: 'There were several power struggles going on. The Chairman was trying to impose the idea of evaluation and foresight on the group, but there was a lot of resistance. For most members, it was unthinkable to use even very simple bibliometric tools.'* Not surprisingly, therefore, an attempt to stimulate interaction between the teams and the Centre de Prospective et d'Évaluation proved unsuccessful:

> They were in two entirely different worlds: on the one hand, there was the world of the scientists who had come to positions of responsibility for the management of science; on the other were people from a background in academia or administration who were experts in management, or sociology, or economics or strategic analysis. These two worlds not only completely ignore each other, but they also do not recognize that the

other side has knowledge that they do not have ... We need a 'cultural revolution' to get over this.*

Among other factors contributing to the limited influence on programme management of most GEPs was that they met only twice a year, and most members were senior research managers who generally had limited access to the latest technical information (aside from that provided by programme teams). However, the main reason was that the GEPs had no *financial* power — important decisions were prepared beforehand by programme teams more concerned with budgets than with foresight.[29] It was consequently possible for advice to be ignored, resulting in a divergence between formal programme objectives and those actually pursued.[30] This was accentuated by the fact that, as former scientists, many programme team members were used to peer-review but not GEP-style monitoring and strategic analysis.

The SOST mechanism, although it undoubtedly encouraged discussion about longer-term priorities, was also not without limitations. First, little use was made of the systematic prospective techniques then being developed in France (e.g. Godet, 1985), tending to focus on the present and the immediate future. Second, there was an emphasis on science-push factors almost to the exclusion of socio-economic demands. Third, because SOST covered only *existing* programmes, there was no possibility of obtaining a holistic overview — it is significant, for example, that the recently created AIDS programme did not emerge from the SOST procedure. Fourth, little headway was made in integrating agencies like CEA and CNES within a coherent national R&D strategy.

However, the reason why SOST was terminated in 1986 was not so much a result of these shortcomings, but because the stage had been reached where further progress clearly required major institutional restructuring within organizations like CEA. While the SOST mechanism was always resisted by *fonctionnaires* because it was alien to their traditional approach to administration, the agencies over time became increasingly unhappy with the changes suggested in relation to their roles. Consequently, they seized the opportunity to persuade the new government to drop the SOST process. As a senior industrialist pointed out, SOST had unfortunately come to be associated with the belief attributed to the Socialist government that everything can be planned. Consequently, after the 1986 election there was an excessive counter-reaction against such foresight activities.*

As for more general lessons in relation to the use of foresight in the strategic management of research, some observers believe that MRT made a mistake in trying to direct the national programmes itself.* This involved mixing two functions — the elaboration of objectives (together with subsequent monitoring and strategic analysis) and day-to-day management. It often meant, for example, that the head of the programme team in charge of funding and administration was at the same time responsible for preparing material for the GEP. Inevitably, the tendency was to concentrate on the former to the neglect of the latter.* Furthermore, MST officials were reluctant to evaluate their own programmes too critically, with the consequent danger that some grew ever larger and progressively more difficult to control. A better solution might instead have been to place the administration of national programmes outside the Ministry (e.g. in agencies). One would then have needed only a small internal group to initiate new programmes, undertake evaluation and strategic analysis, and, very importantly, curtail activities when necessary.*

Another lesson is that many of the evident limitations with the use of foresight in planning and directing national research efforts stemmed from the fact that the necessary analytical tools were not yet sufficiently developed. For example, a better understanding was needed of what constituted a 'programme' — while some initiatives were sectoral (e.g. transportation), others were based on a specific technology (e.g. new materials) and probably required different mechanisms for programme management. Similarly, more analysis was needed on the role of GEPs if they were to operate effectively.[31]

Nevertheless, after several years of trial and error, by 1986 some progress had undoubtedly been made in establishing mechanisms for carrying out foresight and implementing the results in national programmes. Useful tools were starting to emerge, including the construction of science indicators. This followed high-level moves within MST to develop bibliometric measures as a means of introducing a quantitative dimension to the preparation of SOSTs. Meetings were held with programme teams to explain how such indicators might assist in identifying promising fields, and to persuade them to participate. Also present were staff responsible for the Pascal data-base who, after appropriate keywords had been agreed, compiled relevant publication statistics for each programme area. These were incorporated in the 1985 version of the overall SOST (MRT, 1985b), as well as in the annual MRT report to Parliament (MRT, 1985c) which consequently contained, for the first time, a detailed strategic analysis of French research. Nevertheless, this attempt to employ quantitative techniques had limited impact; as an official centrally involved admitted, much more effort needed to be devoted not only to further development of the indicators but also to presenting the results to programme teams, GEPs and CSRT if their scepticism was to be overcome. However, with the 1986 elections imminent, there was insufficient time to give bibliometric indicators the 'push' required to overcome such institutional resistance.*

The lower priority accorded to R&D and longer-term planning by the new government meant that little attempt was made to create an alternative research policy, foresight activities drifting to a halt rather than being formally scrapped. With the departure of several key science-policy officials, there was no longer the political pressure needed to maintain the SOST procedure in the face of an evident lack of enthusiasm (and even opposition) among agencies and others. Officially, foresight continued to be a responsibility of the Ministry of Research and Higher Education (a division within the General Directorate for Research having the task of *prospective et programmation*), while, in theory at least, the national programmes were still subject to evaluation and strategic analysis by expert groups. However, no prospective studies appear to have been published after 1986,[32] and indeed some GEPs failed to meet during the next two years.* In short, research foresight all but disappeared over this period. Since the recreation of the Ministry of Research and Technology, some progress has, however, been made in reinstating the foresight activities previously abandoned, in particular following the establishment of the Science and Technology Observatory described below.

(e) Public sector think-tanks

The first of the two think-tanks[33] established by MRT to work on strategic issues in science and technology policy — the Centre de Prospective et d'Évaluation (CPE) — began life in 1981 as an in-house unit, although during 1986–8 it became an interdepartmental service group under the Ministries of Research and Industry. A relatively small organization with some thirty staff, its responsibilities include: (i) evaluation of research policies, programmes and agencies; (ii) monitoring and diffusing information on emerging technologies; and (iii) research foresight (CPE, 1987a, p.1). The second of these tasks is organized under a programme on La Veille Technologique, which has represented an important input to national prospective activities. This provides regular digests of information from experts in agencies, research institutes and universities, as well as from scientific attachés abroad (particularly in the United States, Japan, West Germany and Scandinavia). Findings are disseminated through a variety of publications aimed at different audiences including *CPE Flash*, *Liaisons Japon* and *Biofutur Flash*, with in-depth studies being reported in the *CPE Étude* series. These technological monitoring activities are generally well regarded within government and to a lesser extent by industry.*

While the work of the CPE Prospectives Group was, in contrast, severely curtailed under the Chirac administration, it continued to collaborate with the Ministry of Industry in monitoring technological and industrial strategies and helped organize the Europrospective exercise. In earlier years, its main foresight activities centred on the preparation and updating of the influential report, *La Révolution de l'Intelligence* (CPE, 1985). This had its origins in two Technology Consultation exercises which predated CPE, the first being carried out in 1978–9.[34] At the time, inadequacies were becoming apparent in the traditional mechanisms for identifying research priorities through committees and expert discussions. Impressed by the Japanese approach to forecasting, the minister then responsible for research decided to initiate a systematic procedure for framing national R&D policy. Although still relying on the judgement of researchers, the approach adopted was more wide-ranging and thorough than anything previously attempted, with nearly 300 experts being asked in a questionnaire survey to predict future developments in their respective fields. From the responses, which focussed on scientific and technological trends rather than likely socio-economic demands, four major growth areas were identified: (i) new materials; (ii) microelectronics (particularly for robotics and telecommunications); (iii) energy systems and conservation; and (iv) biotechnology. The resulting report (Gaudin *et al.*, 1979) had a significant impact on national thinking concerning the future for science and technology, presenting a broad picture of technological opportunities from which industry, government and the research community could all benefit.

As a result of this generally positive experience, a second Technology Consultation was organized in 1981 under the auspices of the Ministry of Research and Technology. This time, 1,200 experts participated, and a more formal two-stage procedure was adopted whereby results from the first stage were used by small expert groups to compile synthesis reports which were then circulated for comment and revision.[35] In order to obtain the widest possible circulation of the report, the assistance of journalists and the Society of French Engineers and Scientists (ISF) was enlisted, and 20,000 copies of the first edition (CPE, 1983) were distributed.

In subsequently updating the exercise, CPE integrated information from La Veille Technologique as well as findings from recent evaluation and prospective studies. A working group of CPE and ISF officials studied in considerably greater detail the likely effects of technological trends on production processes, lifestyles and human relations. An attempt was also made to broaden the analytical framework by setting the new technologies making up the 'third technological revolution' (biotechnology, microelectronics and advanced materials) in their historical context of growing global economic competition. This resulted in the first version of the report to bear the title, *La Révolution de l'Intelligence* (CPE, 1985). Like the revised edition the following year (Sciences et Techniques, 1986), this was widely distributed. These reports are regarded as especially useful reference documents, being among the first such systematic studies outside Japan. However, while they undoubtedly had a general impact on people's thinking, it is difficult to discern any specific influence on funding decisions. In many respects, the reports are more a review of the state of technology in the mid-1980s than a systematic attempt to look into the future. Formal forecasting techniques were not employed, the approach instead relying largely on informal consultation with little use of quantitative data. Indeed, when CPE tried to adopt systematic prospective methods, this was resisted by civil servants.*

Although much of CPE's work is widely respected, officials voiced several criticisms. One is that, despite its remit, very few CPE studies can accurately be described as prospective exercises.[36] Second, although set up to provide government with specialist expertise in evaluation and foresight techniques, over time it became a relatively independent organization. For example, there was reportedly little enthusiasm for supplying quantitative information and advice on prospective tools to the GEPs.* Moreover, the fact that few CPE staff had previously worked in research institutes or industry contributed to a 'culture gap' which prevented them from linking their work adequately with the needs of decision-makers. Insufficient effort was also given to developing evaluation and prospective techniques and to constructing data-bases on R&D indicators. Because of these problems, as well as the reduced emphasis on research foresight in government, CPE faced major budgetary difficulties in 1988 and its future was most uncertain. Nevertheless, the general view is that CPE has generated useful information not available elsewhere, and that it should be given additional time to prove itself.

The second think-tank — the Centre d'Études des Systèmes et des Technologies Avancées (CESTA) — was created in 1982 as part of the French response to the Versailles Summit. Although remaining an independent public body up to its closure in 1988, it reported through CPE to the Ministry of Research. CESTA was also larger than CPE (with seventy staff), and incorporated the Institut de Prospective des Sciences et de Prévision Technologique (IPSPT) which was charged with undertaking research forecasts for government. A series of twelve programmes was set up under IPSPT to examine the effects of technological change on society (CESTA, 1982), including one (Techno 2000) specifically aiming to assess needs and opportunities in priority areas like telecommunications and biotechnology. In the event, CESTA produced little in the way of systematic prospective analysis,[37] and instead concentrated its efforts on the dissemination and popularization of scientific information, for example, organizing conferences to promote awareness of the impact of new technologies. Without dismissing the importance of this function, officials in government and industry regarded CESTA as having neglected its primary mission of studying the likely impact of emerging fields of science and technology (e.g. MRT, 1985c, p.307). Among the reasons given for this failure were the lack of clear leadership, and staff who reportedly lacked the necessary quantitative and analytical expertise.[38*]

At the time of CESTA's closure, discussions were under way to set up an Observatoire des Sciences et Techniques (OST). The initiative arose from a review of evaluation procedures within national research programmes which examined the possibilities for developing quantitative indicators to help managers monitor progress (Arvanitis *et al.*, 1986). In subsequent discussions, it was argued that prospective analysis requires an improved understanding of the dynamics of the research system and a systematic assessment of the country's relative scientific strengths and weaknesses. The Observatory was thus seen as carrying out several functions: (i) production of an annual science and technology indicators report; (ii) providing detailed input and output data on research to facilitate effective monitoring and evaluation; and (iii) creating a national capacity in formal approaches to research evaluation and foresight (Barré, 1987).

The proposal to establish OST was well received within the Ministry of Research, but a decision was deferred until the new government took office. Following consideration of a report on the role of such an Observatory commissioned from Pierre Papon, the decision was taken to begin operation in 1989. Although it is initially concentrating on producing science and technology indicators, these will at the same time provide the basis for a growing effort on prospective analysis. However, whether officials controlling national programmes and research budgets will prove any more willing to use such information than they were to listen to advice from GEPs remains an open question.

5. RESEARCH FORESIGHT IN INDUSTRY

The five companies covered are primarily active in the electronics, chemicals and materials sectors. They share a relatively decentralized approach to R&D, nearly all development work being carried out within operating groups. However, four have central laboratories where most longer-term research is performed (financed partly through contracts from divisions and partly from corporate funds). In the case of two of the firms, substantial support is also received through government and EEC contracts, primarily for electronics projects. As for tracking external R&D developments, this is generally conducted on a routine basis within operating groups, although for selected priority fields or areas spanning the interests of several divisions some central monitoring may also be undertaken. In each company, foresight plays a certain role in the allocation of corporate research funds. Although this primarily involves informal consultations, committee-based procedures to assess proposals from researchers are also employed in the case of longer-term programmes.

The first of the firms, Thomson-CSF, has recently undergone major restructuring of its business activities, especially within consumer electronics, and this has resulted in a need to concentrate its R&D efforts. Distribution of corporate funds is carried out through a process which has two starting points: (i) longer-term problems or needs specified by operating groups; and (ii) interesting research opportunities identified by laboratory staff. Research group leaders and management are responsible for formulating these proposals into possible projects, and the resulting list is then narrowed down by the R&D Director after taking advice from senior staff and consulting the researchers involved. The whole process is relatively informal, the policy over recent years having been to reduce the paperwork associated with a previous procedure for drawing up an annual Orientation Plan (see Irvine and Martin, 1984, pp.39–41).[39]

In contrast, Compagnie Générale d'Électricité (CGE) employs a relatively systematic approach to research foresight, stemming in part from the government requirement to produce an annual corporate plan when it was publicly owned during 1983–7. In the first of these, operating groups prepared their own plans, the research element mostly focussing on shorter-term R&D. As an input to the second plan, however, the Development Directorate was asked to draw up a longer-term R&D prospective. This was based on a survey of a hundred leading scientists in different fields (mainly from universities, research centres and other companies). It was carried out through open-ended interviews (using a core set of questions), the staff involved discussing their findings with each other before writing them up. Interim results were reported every three months to a working group of experts from within and outside CGE. Some scanning of the research literature and data-bases was also carried out, but the approach was not very comprehensive.[40] After extensive review, the final report was presented to the CGE Board in 1985. Its primary role in relation to operating groups was to set guidelines for the direction of R&D and provide a global overview of developments, especially in neighbouring technology areas. Within the corporate laboratory, the main impact was to broaden the horizons of the five principal research groups, especially in relation to cross-cutting areas like artificial intelligence. The longer-term influence of the report has, however, been limited by the substantial corporate restructuring and subsequent acquisitions which took place after privatization.

Since then, a new Scientific Directorate has been created, one task of which is to eliminate duplication of R&D efforts within the larger and diversified CGE group. Another was to prepare a third prospective review during 1988 of longer-term scientific and technological developments. Again, a relatively comprehensive approach was employed, this time with foreign experts being consulted (reflecting the internationalization of the company). In addition, detailed discussions took place with researchers in corporate and divisional laboratories, and with the R&D directors of operating groups. It was recognized that the main benefits of this process were to encourage staff to think about the longer-term future and to take a broader view in relation to their research interests.

The third of the companies, Rhône-Poulenc, is the exception in organizing virtually all R&D within divisions, although a central Scientific Directorate retains control of a corporate budget of approximately F100 million for strategic research. The process for determining priorities is complex and involves numerous committees: (i) a Scientific Advisory Council (comprising academics and other external specialists), which assists in identifying emerging fields and evaluating the overall R&D programme, as well as providing an interface with the research community; (ii) an internal Research Coordination Committee responsible for budgetary decisions, its members being drawn from the Scientific Directorate, divisional R&D directors and three professors who work closely with the company; (iii) various committees focussing on individual fields (such as organic chemistry or polymers), which supervise R&D programmes and propose initiatives; and (iv) research directorates managing R&D in the operating divisions. In establishing long-term programmes, account is taken of emerging technological needs and opportunities, likely revenue, market position and the company's research strengths relative to competitors. In addition to employing internally generated information, views are often sought from prominent foreign researchers retained as advisers. However, relatively little use is made of bibliometric and patent data-bases, or studies by consultancies.

Besides creating a long-term programme, Rhône-Poulenc utilizes two other mechanisms to support strategic research. First, leading scientists within the company are awarded the title 'associate researcher', which provides access to a small budget to pursue novel lines of research without any restrictions. Second, there is an 'innovation fund' from which R&D staff can obtain support to test the viability of promising ideas. In four years, 150 of the 1,000 topics investigated have subsequently been taken up, some finding direct application, a number being licensed to other firms, and a few leading to the establishment of new research programmes. Although costing only F30 million per annum, these two mechanisms are regarded as essential in maintaining an innovative and flexible approach to R&D.

As regards more formal foresight, Rhône-Poulenc has recently formed a 'creativity group' consisting of researchers who have submitted large numbers of proposals to the innovation fund. This meets periodically to discuss possible initiatives, and the results have been sufficiently encouraging for a decision to be taken to carry out a broader prospectives exercise. This involved bringing together marketing personnel, consultants and forecasting experts with the creativity group to discuss promising research opportunities. A seminar was held in which participants were divided into five groups to discuss research themes identified by the Scientific Directorate. Also reflecting the policy to give greater priority to foresight, a Scientific Direction Group was established during 1988 (its members including a Nobel Laureate and the research director of INSERM) in order 'to lead the prospective reflection essential for preparing for the future' (Rhône-Poulenc, 1988, p.5).

Reorganization of R&D at Pechiney has taken exactly the opposite tack to Rhône-Poulenc, with a corporate laboratory being established in 1988 which receives 20 per cent of its budget from the Board. A rolling three-year plan drawn up annually by a central R&D directorate is used to set priorities for longer-term research. This is based on an analysis of planned divisional R&D programmes to identify areas of common interest or possible synergy which might form the basis of interdivisional research initiatives. Advice is sought from a wide range of experts within the company and outside. This results in a list of up to thirty promising topics which are ranked by the directorate in terms of relevance to corporate strategy, internal research capacity and longer-term commercial prospects. Although an essentially qualitative process, it has nevertheless been found adequate for narrowing down the list to a small number of priority projects which can be financed with the funds available.

The fifth of the companies considered, Saint-Gobain, has traditionally decentralized most R&D to its ten divisions, only operating corporate laboratories for glass and metallurgy (and a joint institute with Bordeaux University for pulp and paper R&D). However, one step towards greater strategic coordination came with the separation of funding for research from that for development. Operating groups now contribute via an overhead payment to a corporate research budget of around F200 million per annum (representing some 20 per cent of total R&D expenditure). Approximately 300 or so proposals are submitted annually by researchers for work with potential longer-term benefits to the company. After discussions with divisions, the Vice-President for Research synthesizes the most promising projects into a coherent programme compatible with overall corporate strategy. This is put before a high-level Research Committee for a final decision. As in Pechiney and Thomson, an attempt is made to keep paperwork and formal procedures to a minimum.

Concern in 1985 that the planning process was failing to provide sufficiently strategic decision-making led to a working group being set up to 'reflect on strategies for technology'. This undertook a comprehensive analysis of relevant technologies, which were classified according to whether the company (i) had an adequate R&D capacity, (ii) needed to improve its competitive position, or (iii) was lagging so badly that R&D in the area should be dropped. The exercise was repeated the following two years but in less detail, and discussions were in progress in 1988 as to whether a more systematic approach should be employed.[41]

6. CONCLUSIONS

The experiments with foresight in France during the 1980s are interesting since, unlike in most other countries, the emphasis was at least as much on using prospective analysis to steer national research programmes as on identifying priorities for new initiatives. Among the main lessons to emerge are the following:

(a) The experiences of CNRS, MRT and other organizations show that foresight typically involves a long learning process. In the case of CNRS, from the early field surveys, through the cross-disciplinary study on nuclear physics, particle physics and astrophysics research, on to the more global *schéma prospectif* exercise, and culminating in Prospectives 2005, one sees an evolution in the degree of sophistication and in the level of success achieved. Similarly, despite several early teething troubles within MRT in applying foresight to the management of national programmes, there was a gradual improvement in the mechanisms for conducting prospective analysis until the change in government in 1986.

(b) A major reason for the long learning process was the resistance to foresight by virtually all sectors — researchers, civil servants, funding agencies and industry. Indeed, large sections of the scientific community remain sceptical, the efforts by proponents of prospective analysis and by forecasting think-tanks like CPE and CESTA to 'sell' foresight not having been particularly successful. Part of the problem involves convincing scientists that they *alone* are not very good at undertaking foresight — that experts on emerging social and economic issues and on prospective methods are also essential. How to integrate such specialists without alienating the scientific community is a dilemma that has yet to be fully resolved. Likewise, many civil servants remain unconvinced of the need for research foresight, or at least that it requires professional expertise (which they themselves generally do not possess).

(c) Among various factors accounting for the opposition of industry to the introduction of foresight, one is the traditional secretiveness of French companies, even in matters relating to longer-term research. Many are apparently unable to distinguish pre-competitive research — where it may be in the interests of firms in a given sector to cooperate — from shorter-term R&D considerations. Consequently, it has been difficult for those engaged in foresight to obtain systematically the views of firms on promising research areas. While it is possible to hold discussions with companies on a bilateral basis, as soon as several are seated around a table they become concerned about giving away information to competitors. Another factor, however, is that during 1981–6 foresight came to be equated by many with socialist planning. Industry therefore welcomed the abolition by the Chirac government of the R&D planning mechanisms established by the previous administration — in most cases before foresight techniques had had a real chance to demonstrate their policy utility.

(d) Given the initial opposition to prospective analysis, it was essential to have a powerful 'product champion' like Robert Chabbal in MRT — once he left, foresight activity ground to a halt. Pierre Papon fulfilled a similar role within CNRS. No such high-level sponsors emerged during the period 1986–8. One reason was the government's antipathy towards longer-term planning, while another was the institutional problem stemming from the decision to subsume research under the Ministry of Higher Education, the lower political profile given to R&D reducing the incentive to engage in foresight.

(e) The examples of the National Colloquium, the missions which created the *programmes mobilisateurs*, the GEPs and Prospectives 2005 show that successful foresight involves balancing the conflicting demands of science-push and demand-pull. Earlier CNRS initiatives like the *schéma prospectif*, in contrast, tended to take insufficient account of socio-economic needs. While foresight must clearly mobilize the research community (as the National Colloquium, for example, did most effectively), scientists should not at the same time be allowed to exert too dominant an influence, which was the case with some of the MRT programme teams.

(f) The experience of MRT illustrates how the foresight process does not end with the establishment of research programmes to implement the priorities identified. Those programmes need to be subject to regular monitoring and strategic analysis in the light of research advances and changing socio-economic demands. Foresight and evaluation should therefore proceed hand in hand. This was explicitly recognized in the remit set for both the Centre de Prospective et d'Évaluation and the Groupes d'Évaluation et de Prospective. Foresight also needs to be viewed as a continuing process rather than just an *ad hoc* activity. Although the National Colloquium was successful in several respects, it was a one-off event which could not easily be repeated. Consequently, the Ministry of Research became locked into a set of priorities, and virtually all subsequent foresight activities centred around those early target areas to the neglect of other emerging fields.

(g) The example of the national programmes also shows how the institutional location of those responsible for prospective analysis is crucial. While foresight needs to be closely linked to day-to-day research management if it is not to be ignored, it should not be so close that there is a reluctance to consider radically different options for an existing programme or to contemplate its abandonment in order to move the resources to new and more urgent priority areas.

(h) Although there can be little doubt that Socialist governments have provided a more conducive climate for foresight during the 1980s, some initiatives were nevertheless embarked upon by the Chirac administration. These included the establishment of the two 'observatories' (OSI and OTS) by the Ministry of Industry aiming to provide a better 'early-warning' system for emerging developments, a form of foresight consistent with the prevailing non-interventionist philosophy of the government. The lesson here is that foresight approaches must generally be compatible with the wider political environment.

(i) Many of the problems with using foresight in the strategic management of the national research effort reflected the relatively primitive state of the analytical tools available, as well as as shortage of personnel with relevant expertise. The decisions to create a Science Policy Indicators Unit within CNRS in 1988, and to establish the Science and Technology Observatory the following year, indicate the determination of the new government to overcome these difficulties.

NOTES

* Refers to material collected during interviews in early 1988, while † denotes information drawn from subsequent correspondence.

1. As noted in Chapter 1, the concept of *la prospective* is similar in many respects to that of foresight — see Godet (1985; 1986; 1987).
2. According to the Director of Research at the Ministry, no research foresight is carried out in relation to the distribution of core-funding to universities and *grandes écoles*.*
3. Article 1 of Decree No. 81-723, 28 July 1981, quoted in OECD (1986, p.62). A year later, it merged with the Ministry of Industry to form a new super-ministry responsible for the country's technological and industrial modernization, but it reverted in 1984 to the Ministry of Research and Technology.

4. *Programmes mobilisateurs* normally focussed on a particular generic (or diffusion) technology, while initiatives concerned with a given economic sector like transportation were generally known as *programmes prioritaires* (Barré, 1986, p.305). The term 'national programmes' is often used to cover both types.
5. The Triennial Plan placed great stress on research evaluation, specifying that this 'should not be limited to the scientific aspects, but should extend to the *strategic*, operational and economic aspects' (ibid., p.307, emphasis added).
6. One example cited was the relative lack of interaction in France between biology and chemistry compared with other countries.*
7. The studies are described in Irvine and Martin (1984, p.31).
8. Among the main conclusions of the antibiotics study was that, during the late 1970s and early 1980s, Britain was perhaps the most innovative country in the field (CEP, 1985).
9. A source of some of the suspicion towards foresight is that 'scientists often confuse *prospective* with *conjoncture* (state of the art) and are unable to evaluate the potential importance of breakthroughs, of new experimental techniques and of shortcomings in established theories. In the background, there is the fear of established groups and disciplines losing influence if new areas are developed.'†
10. A recent example concerns human nutrition, where two scientists prepared an internal report on research in the United States, Canada and Britain, and identified areas which INRA might fruitfully support.
11. INRA officials do, however, contribute to CNRS foresight activities and have on occasions made use of the results.
12. Instead, the Scientific Director and Director-General talk to firms individually, perhaps seeing up to thirty each year.
13. As the Scientific Director observed: 'Part of my job is to decide which new areas to support. But too much is in my head — very little is written down. We need to develop a much more systematic approach to identifying priorities. And we also need a system to follow programmes and see how the money is used. It's a question both of identifying new programmes and of deciding which old programmes should be reduced or stopped ... This is not done very well at present.'*
14. CGP also played a prominent role in the medium-term plans for research and technology involved in the 1982 R&D Act, the Ninth National Plan and the Triennial Research Plan for 1986–8. The last of these explicitly mentions the need to stimulate prospective studies on science (CGP/MRT, 1985, p.57).
15. The only quantitative information was data on current sales, market share and so on.
16. For details of other industrial strategy groups, see Green (1986b, pp.295–6). Certain studies included a substantial research element (e.g. see CGP, 1987, pp.135–85, which contains an analysis of French strengths in various technologies).
17. While the results of these deliberations provided an input to the Ninth Plan, no final report had been published by the time GSIs were abolished in 1986.
18. See Bobe and Senhaji (1987, p.30). Drawn up to help prepare the 1988 government research budget, this CGP report had little impact in the Ministry of Finance, not least because the role of the Commission was then under discussion.
19. It also had an effect on existing national programmes, for example, leading to a partial reorientation of the programme on working conditions. In addition, it helped CNRS defend its budget.*
20. One industrialist also reported that Prospectives 2005 had helped his company decide which research areas to build up and which to cut back.*
21. The title and remit of the Ministry of Industry have changed several times since 1981. From 1986 to 1988, for example, it formed part of the Ministry of Industry, Post and Telecommunications, and Tourism, after which it became the Ministry of

Industry and National Development. In what follows, the term 'Ministry of Industry' is generally used for brevity.

22. While the Ministry has carried out surveys of industrial R&D, these are essentially short-term projections concerned more with economic than technological trends.
23. A recent issue, for example, carried reports on structural ceramics, fibre-optics sensors and desk-top publishing, together with a list of recent patents (OTS, 1987). Topics covered during the year were subsequently published in a collected volume (OTS, 1988).
24. In addition, the Minister 'may be delegated by the Prime Minister to chair the Interdepartmental Committee on Scientific and Technical Research' (quoted in OECD, 1986, p.63).
25. A more detailed description can be found in Irvine and Martin (1984, pp.36–7). See also *Colloque National sur la Recherche et la Technologie* (1983).
26. The GEPs were generally made up of approximately one-third researchers, one-third scientific administrators and one-third industrialists, to ensure that science-push and demand-pull pressures were kept in balance.
27. Another task of programme teams was to assist in servicing the GEPs.
28. For example, only five of the twenty-six members of the GEP for biomedical engineering were from industry (MRES, 1986, Annex No. 0–1).
29. In fairness, however, it should be pointed out that most GEPs had only just begun to operate when disbanded in 1986 and therefore had insufficient time to learn how to operate effectively.
30. 'The scientists [in the programme teams] have had very great power and have been able to manipulate the situation with great success. Where there are no mechanisms to control the relationship between scientists and power, you get periods where scientists and decision-makers are working together "in phase", and others where they are in opposition.'* See also Arvanitis *et al.* (1986).
31. Barré (1987, p.34) outlines several other areas where improved prospective tools are required. These include: the identification of challenges or objectives in the light of the international evolution of science and technology; the location of relevant actors and assessment of their strategies and competence; an audit of French strengths and weaknesses in industry and research; and the identification of technological and institutional blockages.
32. One of the last reports to appear was that on biomedical engineering (MRES, 1986). Published shortly after the Chirac government was elected, it was largely prepared under the former administration as part of the SOST procedure (although the word 'SOST' no longer appears on the cover page).
33. Why it was felt necessary to create *two* such organizations appears to have more to do with institutional politics than any rational division of labour.*
34. This description of the two Technology Consultations is taken from Irvine and Martin (1984, pp.34–5).
35. The findings were published as a series of five reports on major research fields (e.g. MRT, 1982), together with a summary volume.
36. For example, almost all the data in *Une Vision Prospective du Monde* relate to previous years rather than to the future (CPE, 1987b).
37. It was also originally intended that CESTA carry out technological monitoring. When it failed to do so, CPE established its own group to conduct La Veille Technologique.
38. For example, a former senior official commented: 'I was never very impressed by CESTA. The people were not very good — just a lot of hot air — they were not real professionals. They were too involved in following fashions'.*
39. For several years, Thomson also set aside 10 per cent of its corporate research budget to support promising ideas put forward by scientists. Provided the topic fell broadly within the company's interests and the cost was small (around F100,000), authorization was given to carry out a 'quick and dirty' experiment to

test its feasibility. The aim was to formalize 'bootleg' research — that is, to encourage staff to devote approximately 10 per cent of their time to activities outside the official R&D programme. It was accepted that most experiments would fail, but the remainder led to some of the company's most innovative developments.

40. The R&D Directorate has a small technology monitoring and information section.
41. Also visited was Euroconsult, a consultancy firm specializing in corporate strategic planning (Giget, 1987; 1988). In conjunction with the prospectives group at CPE and other experts, Euroconsult has devised a formal methodology for helping companies identify their current technological strengths and weaknesses and assess the longer-term commercial impact of emerging technologies. For proprietary reasons, Euroconsult was reluctant to provide details, but a book summarizing its approach should be published shortly.

REFERENCES

Arvanitis, R., Callon, M. and Latour, B. (1986), *Évaluation des Politiques de la Recherche et de la Technologie — Analyse des Programmes Nationaux,* Paris, Documentation Française.

Andouze, J. (1982), *Rapprochement Possibles entre l'Astronomie, la Physique Nucléaire et la Physique des Particules,* Paris, Centre National de la Recherche Scientifique.

Barré, R. (1986), 'Science and Technology Policy in France — From Planning to Strategy', *Futures,* vol.18, pp.298–308.

—— (1987), 'Prospective et Stratégies pour les Politiques de la Science et de la Technologie: Concepts et Practiques en Émergence' (mimeo), Paris, Conservatoire National des Arts-et-Métiers.

Berkaloff, A., Nacquet, R. and Demaille, J. (1987), *Biologie 1990 — Enjeux et Problématiques,* Paris, Centre National de la Recherche Scientifique.

Bobe, B. and Senhaji, O. (1987), *Réflexions pour une Politique à Long Terme de la Recherche et de Développement Technologique,* Paris, Commissariat Général du Plan.

Boucher, M. (ed.) (1987), *Vers l'An 2000 ... et Après,* Cahiers Français, No. 232, Paris, Documentation Française.

Callon, M., Law, J. and Rip, A. (eds) (1986), *Mapping the Dynamics of Science and Technology,* London, Macmillan.

CEP (1985), *Analyses Bibliometriques: Trois Exemples en Chimie et Biologie,* Paris, CNRS Cellule d'Évaluation et de Prospective.

CESTA (1982), *Centre d'Études des Systèmes et des Technologies Avancées,* Paris, CESTA.

CGP (1964), *Réflexions pour 1985,* Paris, Documentation Française.

—— (1983), *Les Enjeux Technologiques des Années 1985–1990,* Paris, Documentation Française.

—— (1984a), *9e Plan 1984–1988,* Paris, Commissariat Général du Plan.

—— (1984b), *9e Plan de Développement Économique, Social et Culturel 1984–1988,* Paris, Documentation Française.

—— (1987), *Rapport du Groupe de Stratégie Industrielle 'Travaux Publics',* Commissariat Général du Plan, Paris, Documentation Française

—— /CNRS (1987), *Prospectives 2005: Explorations de l'Avenir,* Paris, Economica.

—— /MRT (1985), *Recherche et Développement Technologique: Réflexions pour un Plan Triennial,* Paris, Documentation Française.

Chabbal, R. (1985), 'Présentation de Schéma d'Orientation Scientifique et Technique', forward to MRT (1985b).

CNES (1985), *Seminaire de Prospective: Evolution à Moyen et Long Terme des Programmes Scientifiques Spatiaux*, Paris, Centre National d'Études Spatiales.

CNRS (1980a), *Rapport de Prospective: Physique Atomique et Moleculaire*, Paris, Délégation Générale à la Recherche Scientifique et Technique.

—— (1980b), *Rapport de Prospective: Physique de la Matiére Condensée*, Paris, Délégation Générale à la Recherche Scientifique et Technique.

—— (1984), *Rapport de Prospective en Optique*, Paris, Éditions du Centre National de la Recherche Scientifique.

—— (1985a), *Schéma Prospectif — 20 Thèmes Stratégiques pour le CNRS*, Paris, Centre National de la Recherche Scientifique.

—— (1985b), *Rapport de Prospective en Mathématiques*, Paris, Éditions du Centre National de la Recherche Scientifique.

—— (1987), *Rapport du Groupe de Prospective en Astronomie*, Paris, Centre National de la Recherche Scientifique.

Colloque National sur la Recherche et la Technologie (1983), Paris, Documentation Française.

CPE (1983), *Rapport sur l'État de la Technique*, Paris, Centre de Prospective et d'Évaluation.

—— (1985), 'La Révolution de l'Intelligence: Rapport sur l'État de la Technique', *Sciences et Techniques*, March, Paris, Centre de Prospective et d'Évaluation and Société des Ingénieurs et Scientifiques de France.

—— (1987a), *Le CPE en Bref*, Paris, Centre de Prospective et d'Évaluation.

—— (1987b), *Une Vision Prospective du Monde*, Paris, Centre de Prospective et d'Évaluation.

CSRT (1986), *Rapport Annuel sur l'Évaluation de la Politique Nationale de Recherche et de Développement Technologique*, Paris, Conseil Supérieur de la Recherche et de la Technologie.

Gaudin, T., Portnoff, A-Y. *et al.* (1979), *Premiers Éléments pour un Programme National d'Innovation*, Paris, Documentation Française.

Giget, M. (1987), 'Stratégie: l'Entreprise ne Doit pas Mourir avec son Produit', *Management France* (August/September), pp.4–6.

—— (1988), 'The Bonsai Trees of Japanese Industry', *Futures*, vol.20, pp.147–54.

Giri, J. (1986), *Les Missions d'Orientation de la Recherche Confiées à des Personnalités en 1981 et 1982*, Paris, Centre de Prospective et d'Évaluation.

Godet, M. (1985), *Prospective et Planification Stratégique*, Paris, Economica.

—— (1986), 'Introduction to la Prospective', *Futures*, vol. 18, pp.134–57.

—— (1987), *Scenarios and Strategic Management*, London, Butterworths.

Green, D. (1986a), 'Introduction', *Futures*, vol. 18, pp.115–20.

—— (1986b), 'Strategic Government Planning in France', *Futures*, vol.18, pp.295–7.

Guillaume, H. (1986), *Faire Gagner la France*, Paris, Hachette, Collection Pluriel.

INSERM (1984), *Rapport de Conjoncture et de Prospective*, Paris, Institut National de la Santé et de la Recherche Médicale.

Irvine, J. and Martin, B. R. (1984), *Foresight in Science: Picking the Winners*, London, Pinter Publishers.

Karsky, M., Dore, J-C. and Miguel, J-F. (1988), 'Understanding Chemical Reactivity Tendencies of Research', Paris, Laboratoire d'Évaluation et de Prospective Internationales.

Laredo, P. and Barré, R. (1985), *Dossier Méthodologique sur l'Évaluation — Schéma d'Orientation Scientifique et Technique*, Paris, Ministère de la Recherche et de la Technologie.

—— and Rouban, A. (1985), 'L'Examen des Mécanismes d'Évaluation et d'Inflexion des Programmes', in Laredo and Barré (1985).

MRES (1986), *Rapport de Groupe d'Évaluation et de Prospective en Génie Biomédical*, Paris, Ministère de la Recherche et de l'Enseignement Supérieur.

—— (1987), 'Le Nouveau CSRT', *Recherche Technologie*, 28–9 (June–July), pp.4–6.

MRT (1982), *Le Sol et ses Resources,* Paris, Ministère de la Recherche et de la Technologie.

—— (1985a), *Transports et Equipments — Schéma d'Orientation Scientifique et Technique,* Paris, Ministère de la Recherche et de la Technologie.

—— (1985b), *La Stratégie de la Recherche et du Développement Technologique,* Paris, Documentation Française.

—— (1985c), *Projet de Loi des Finances pour 1986: Rapport Annexe sur l'État de la Recherche et du Développement Technologique, Activités en 1984 et 1985, Perspectives 1986,* Paris, Imprimerie National.

OECD (1985), *La Politique d'Innovation en France,* Paris, Economica.

—— (1986), *Innovation Policy — France,* Paris, Organisation for Economic Co-operation and Development.

OSI (1987), *Cent Acteurs dans la Compétition Mondiale,* Paris, Observatoire des Stratégies Industrielles, Ministère de l'Industrie, des P&T et du Tourisme.

—— (1988), *Stratégies et Compétitivités dans l'Industrie Mondiale,* Paris, CPE-Economica.

OTS (1987), *Observatoire des Technologies Stratégiques,* Paris, Ministère de l'Industrie, des P&T et du Tourisme.

—— (1988), *Douze Technologies d'Aujoud'hui et de Demain,* Paris, Aditech.

Rhône-Poulenc (1988), 'Recherche: Comment "Diriger" les Stratégies', *Présence,* no. 51 (January), Paris, Rhône-Poulenc.

Sciences et Techniques (1986), *La Révolution de l'Intelligence,* Paris, Société des Ingénieurs et Scientifiques de France.

4 Research foresight in the Federal Republic of Germany

1. INTRODUCTION

'Art and science, research and teaching shall be free'; so states the 1949 constitution of the Federal Republic of Germany (quoted in Michaelis, 1981, p.293). As a result, government influence on research planning is weak compared with other nations. Indeed, this is arguably 'the most distinguishing characteristic of science policy ... There is no comparable country in which universities, research-promoting institutions or contract-awarding institutions have greater autonomy in administering R&D funds allocated by the government' (Lübbe, 1974, pp.175–7). Given the limited capacity for centralized decision-making, the political climate is far from conducive to research foresight — particularly under a Christian Democrat administration with its antipathy towards long-term planning and desire to minimize government 'interference'.[1]

Another factor, however, is West Germany's federal structure, the constitution according considerable power to the *Länder* (states). This includes responsibility for financing the universities and much of the research they perform, as well as contributing towards the support of national research agencies and institutes outside the academic sector. Longer-term research planning must therefore meet with the approval of the *Länder*. Yet this can be difficult to achieve, given differing socio-economic interests and the desire of state governments to strengthen research in their regions.[2] Although institutional mechanisms exist for coordinating research across the Federal Republic, their powers are limited. Moreover, any attempt to reduce the growing research 'gap' between richer southern states and the others tends to be resisted by the former.

An outline of the publicly funded R&D system[3] is shown in Figure 4.1. Basic research is carried out primarily in the higher education sector and institutes of the Max-Planck-Gesellschaft (MPG). As in most European countries, there is a dual-support system for universities; core-funding for teaching and research is supplied by the *Länder*,[4] while finance for individual projects and programmes comes from the Deutsche Forschungsgemeinschaft (DFG), the Federal Ministry of Research and Technology (BMFT), and other federal and state ministries. For strategic and applied research, the main source of government support is BMFT which provides core-funds to MPG, the Fraunhofer Gesellschaft (FhG), the National Research Centres and certain 'blue list' institutes.[5] It also has its own research programmes and is responsible for the overall coordination of federal R&D. Under Chancellor Kohl's administration, increased emphasis has been given

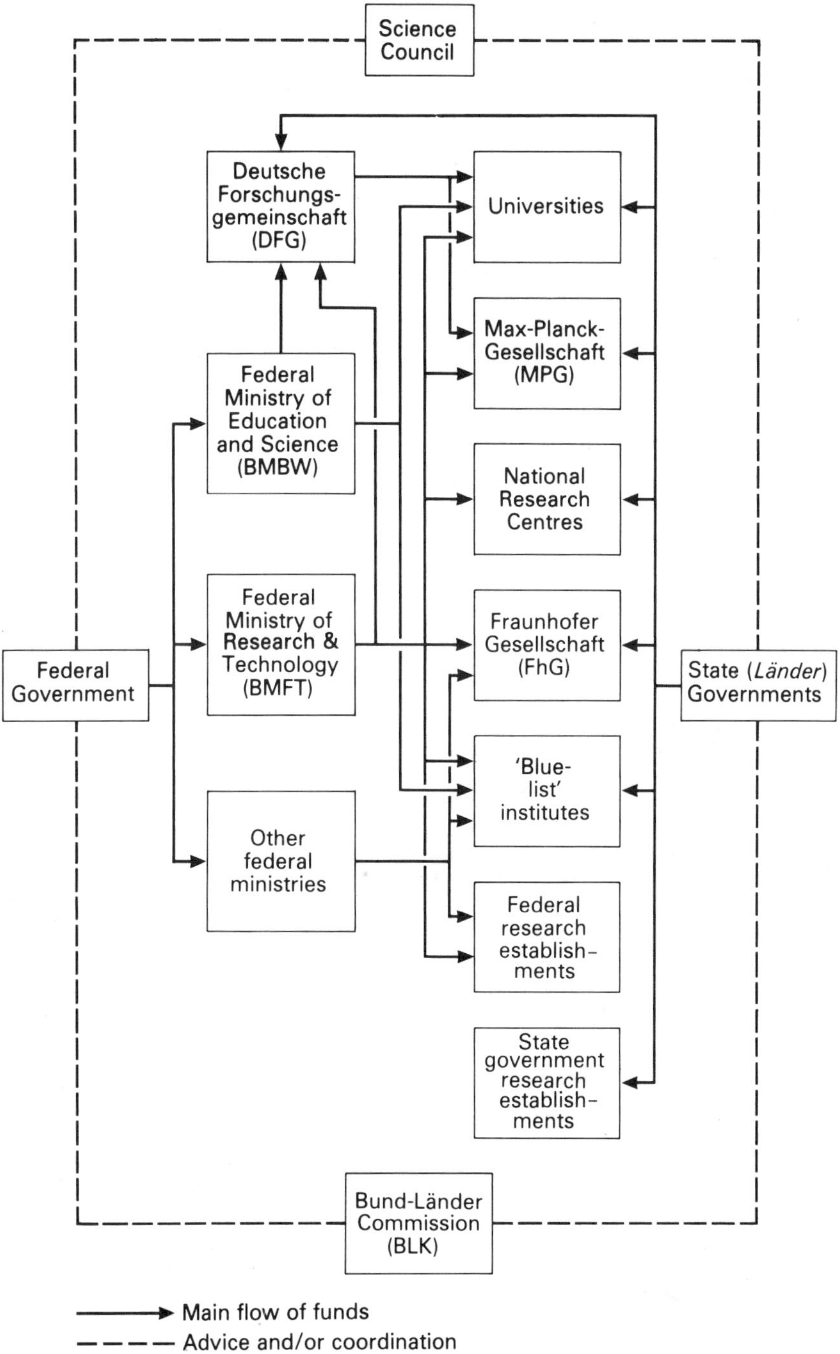

Source: Constructed from official science-policy documents.

Figure 4.1 Outline of the West German R&D system

to indirect mechanisms for supporting R&D, leaving industry and others to identify priority areas. Most government-funded applied R&D is conducted in the National Research Centres and Fraunhofer institutes.[6]

In this chapter, we first look at how decisions on the distribution of resources for basic research are reached in the Science Council, DFG and MPG. This is followed by an analysis of priority-setting in strategic research and basic technology by BMFT, the National Research Centres and FhG. Foresight activities within the state of Baden-Württemberg are also briefly considered. We then examine how longer-term strategy is decided in five large science-based firms, and describe the methodology developed by a major consultancy company for determining industrial R&D priorities, before finally drawing lessons from foresight in West Germany.

2. FORESIGHT IN BASIC RESEARCH

(a) *The Science Council*

The Science Council (Wissenschaftsrat) is the central advisory board for research policy, its role being to coordinate federal and state-funded activities and prepare recommendations on the development of science in universities and elsewhere, taking into account the cultural and socio-economic needs of the nation (von Massow, 1986, p.51). The Council consists of a scientific commission (whose members are appointed for three-year terms by the Federal President) and an administrative commission (with delegates drawn from federal and state ministries), together with a small permanent staff. Although efforts are primarily devoted to monitoring current developments, a recent exercise on university perspectives for the 1990s attempted to look further into the future. This examined the prospects for funding, cooperation between academia and institutes, the need for trained scientists, and the benefits of redirecting university policy to give greater emphasis to research (Wissenschaftsrat, 1988).

Besides reporting on the development of universities and research institutes, the Council also carries out assessments of individual fields. Among the most systematic was a review of clinical research. This found that West Germany's position in the field had seriously deteriorated, largely because university professors were forced to devote too much time to patient care and medical training, and because of inadequate cooperation between clinical and theoretical researchers. One recommendation, therefore, was that a new support mechanism (similar to British Medical Research Council laboratories) was required to free a number of researchers from other duties as well as to improve the links with work in preclinical areas like biochemistry. BMFT consequently agreed to support a programme to be administered by DFG on a competitive basis. No quantitative data on research performance were used in this study, the approach being based on expert review. Despite a certain scepticism towards bibliometric indicators, the Science Council would not be averse to undertaking more systematic foresight, but currently lacks the necessary staff and resources.*

(b) Deutsche Forschungsgemeinschaft

The main role of DFG is to fund investigator-initiated projects across the entire range of scientific, social science and humanities disciplines.[7] These are generally of a basic nature, but include some applied research in areas such as chemistry, engineering, clinical medicine, economics and law. DFG had a budget of DM1,100 million in 1987, provided jointly by the Federal Ministry of Education and Science and by state ministries. It has no laboratories of its own, 90 per cent of its funds going to universities and the remainder to institutes. As the primary source of public support for basic research, DFG sees its function as financing the exploration of novel scientific ideas without any consideration of likely economic or technological impact. In distributing funds, DFG relies on a 'bottom-up' procedure, responding to proposals submitted by scientists.

A large part of DFG activities is funded through the 'normal mechanism', under which researchers seek support for specific projects or fellowships. Applications are judged solely on scientific merit, and no attempt is made at systematic foresight. The only unusual feature is that the 450 peers used to assess proposals are elected by the research community every four years. Current policy is to keep the proportion of funds allocated through this mechanism at around 40 per cent as a safeguard that promising research in any area can obtain support, regardless of the longer-term outcome.*

A second funding mechanism involves 'priority programmes', where researchers from several institutions carry out coordinated work in a given area for up to ten years. Identification of programme topics again emerges through a 'bottom up' process followed by negotiations between researchers and DFG officials. An initiative might begin with a letter from several scientists to the DFG President (e.g. on the need for supercomputers in theoretical chemistry), who discusses the idea with members of the DFG Senate and programme directors. If judged sufficiently promising, the next step may be to arrange a meeting to which other organizations such as BMFT and the National Research Centres are invited, as well as DFG Senators and — above all — scientists active in the field. Discussions tend to be of an open-ended or 'brainstorming' nature and, provided there is agreement on the scientific potential, the researchers responsible are then invited to submit a formal proposal.

Each year, DFG receives up to twenty-five such proposals, and in 1987 was able to fund approximately half. In its deliberations, the Senate (which consists entirely of scientists)[8] assesses the track record of the researchers involved, the quality of the proposal and its cost, and whether DFG is the appropriate funding agency. Decisions are taken largely on the basis of scientific merit,[9] with economic or social factors being considered only in the final stages if the number of high-priority proposals is too great to be met with the funds available. Irrespective of the strategic importance of the area, DFG policy is never to establish a priority programme if the scientific criteria are not first satisfied.[10]

Although formal methods are not used by the Senate in decision-making, consideration was previously given to grading proposals in terms of different criteria (scientific merit, economic potential, and so on). However, this was rejected as too arbitrary and offering no real advantages over the existing open-ended procedure.* Officials admit the process has its drawbacks, in particular the near impossibility of comparing, say, a proposal in mechanical engineering with one in archaeology. There is consequently a tacit agreement in the Senate that none of DFG's main programme areas should be excluded from consideration provided high quality proposals have been put forward. The other main limitation is that the process is not very systematic, with much depending on how persuasive or well prepared a case is made by individual members of Senate. There is a particular problem in assessing the economic impact of engineering proposals since there are only two industrial researchers on the Senate (along with several engineering professors), and they do not have the expertise to judge all fields.

While proposals for priority programmes normally originate within the research community, the initial enquiry can sometimes come from a ministry or politician. In such cases, DFG will convene an expert group to examine the scientific justification for the proposal. In the past, DFG has occasionally tried to implement a priority programme 'from above', but without much success. The policy now is not to initiate a programme unless competent researchers are fully committed to it.*

DFG's third funding mechanism[11] consists of 'special collaborative programmes'. These support interdisciplinary cooperation between scientists within a single university (or a group of neighbouring institutions) for up to fifteen years. Again, the initiative generally comes from researchers, and the decision on DFG support involves a similar peer-review process, although Science Council approval must also be obtained.

Finally, a small but increasing proportion[12] of the DFG budget is tied to supporting targeted fields with funds from BMFT. During the 1950s, DFG accepted earmarked money for restoring the competitiveness of West German universities, but only after lengthy debate about the threat to academic freedom. When in the 1980s BMFT offered DM100 million to DFG over five years for information technology research, arguments raged about jeopardizing autonomy and the imposition of a 'top-down' approach, but again pragmatism finally prevailed over principles, DFG recognizing that BMFT could always distribute the money instead. However, DFG officials are determined that targeted funds should not exceed 10 per cent of overall expenditure since this might give rise to renewed alarm about academic autonomy and flexibility.*

Earmarked finance is currently received for areas as diverse as ocean drilling, empirical social science and cytogenetics. This is officially regarded as additional to — rather than a replacement for — DFG's main budget, but the resources are distributed in the same way (mainly through priority and special collaborative programmes), with similar criteria being applied. In practice, however, some substitution inevitably takes place in that DFG can use its own funds to support other research. DFG was in 1988 considering how to implement recommendations by the Science Council and BMFT on support for clinical research but, as with other earmarked programmes, there were no plans to employ systematic foresight in identifying priorities.

Because DFG support mechanisms depend largely on uncoordinated initiatives from researchers, there is a danger that promising fields may be neglected. Recognizing this, DFG draws up a Grey Plan[13] every four years to provide more systematic background information for decision-making. The most recent represents a distillation of the views of 500 researchers (including elected reviewers, DFG Senators and committee members) on important areas over the next two or three years (DFG, 1987a). Sixteen study groups (each of up to fifteen researchers) were established to prepare chapters on individual fields. The approach varied across disciplines, some groups relying upon telephone calls and informal consultation,[14] while others invited written submissions on the future for their respective fields.

After draft chapters had been revised, they were integrated by DFG officials into an overall report, and an introduction and a section on financial requirements added. The former outlines the limitations of planning and forecasting[15] and senior DFG officials stress that, while the report identifies research trends and likely developments, its predictions may prove to be incorrect or incomplete.* Consequently, although DFG checks whether proposed programmes are in accord with the trends forecast, they will still receive support provided that the traditional scientific criteria are met. In other words, the Grey Plan provides no more than general policy guidelines — it represents a set of perspectives rather than a research masterplan or even a budgetary plan.

Although well regarded within DFG and outside, the Grey Plan is clearly not an exercise in systematic longer-term foresight. First, its perspectives focus almost entirely on science-push considerations, reflecting DFG's view of itself as the custodian of funds for non-directed research. Second, it is more of a stocktaking exercise examining the prevailing state of research, its forecasts confined to the shorter term[16] (except perhaps in the environmental area). Third, although it covers all disciplines, no attempt is made to set priorities among fields.

The only other foresight activities within DFG are occasional field studies. These were more common during the 1950s and early 1960s when a series of reports on strategic areas such as polymers and advanced materials was produced by expert committees, sometimes incorporating findings from questionnaire surveys. They assessed likely trends over the next five to ten years, and provided useful information during the period of university expansion. Such studies have, however, continued, the most recent focussing on astronomy (DFG, 1987b). This was initiated following concern that the institutional locus of astronomical research was moving away from universities and into institutes. Hence, one task of the expert panel convened was to identify projects which could be carried out within academic departments. The resulting report contains 'an impressive catalogue of wishes' regarding equipment needs and estimated costs but little in the way of an attempt to set priorities.*

In short, DFG makes minimal use of foresight, either at a macro- or micro-level. Some officials would perhaps like to do more but are prevented by shortage of staff and by the prevailing ethos that DFG should be a reactive rather than a proactive agency. Others, however, are doubtful whether systematic approaches to identifying promising areas would prove any more effective, and fear the result might instead be to reduce DFG's flexibility in responding to research opportunities.

(c) Max-Planck Gesellschaft

The functions of MPG are: (i) to undertake research in areas of particular scientific importance or potential which require concentration of effort; (ii) to initiate work in emerging areas, especially those outside established academic disciplines or requiring an interdisciplinary perspective; and (iii) to carry out research based on large or specialized facilities (MPG, 1984, p.17). Its fifty or so institutes range in size from twenty to over 1,000 personnel; most are divided into departments of thirty or forty staff, the heads of which constitute the institute directorate. In addition, there are a few independent research groups established for a fixed period. With only a limited number of institutions, MPG cannot cover all basic research and instead seeks 'to concentrate its resources in areas promising special research possibilities' (ibid., p.13), a task made more necessary and at the same time more difficult by the fact that the MPG budget (approximately DM1,000 million in 1987) has remained almost constant in real terms since the early 1970s.

Over the last ten years, MPG has closed an average of two or three departments or (more rarely) institutes per annum, and created one or two new ones. The decision to establish an institute or department depends on a combination of the needs of the research area and the availability of a suitable director. Leadership is extremely important since directors are given complete freedom in selecting staff and pursuing their research interests. The scientific autonomy of institutes is jealously guarded (Dickson, 1986, p.811), although they are regularly assessed by a Scientific Advisory Board. They are also subject to more detailed review every seven years, with a site-visit by an expert panel on which foreign scientists are generally well represented.

MPG's central administrative body is the Scientific Council, which is made up of a section on humanities, another on biology and medicine, and a third on chemistry, physics and technology. Each is responsible for preparing advice on research policy for the MPG Senate, including on the establishment or closure of institutes and departments. Major decisions are often preceded by recommendations from *ad hoc* commissions which seek expert views from outside MPG, especially from abroad. As regards the distribution of funds among sections, the annual budget process starts with the submission of proposals by institutes. These are collated at MPG headquarters and, after intense discussion within the administration, each proposal is scaled down as appropriate. The proposals are next integrated to arrive at a total budget for each section. Thus, there is no attempt to allocate set percentages of the budget to sections, and then to divide up the funds among constituent institutes. Largely because of this procedure, there has been little shift in the overall breakdown during recent years, any changes being more the result of retirement of directors than explicit policy.

The MPG Senate is ultimately responsible for strategic decisions on institutes and departments. To provide policy advice, it set up a Committee for Research Policy and Planning in 1972 with the task of monitoring trends in particular areas and submitting proposals for new initiatives (or the discontinuation of old ones). The Committee is headed by the MPG President, who also chairs the Senate and plays a key role within the organization, although the present incumbent operates a more decentralized style of management than his predecessor, preferring only to become involved in committee deliberations at a relatively late stage.*

Each year, a dozen or so proposals for new institutes are received from researchers within MPG and outside, or from the federal and state governments and the Science Council. In preliminary discussions to narrow down the list, MPG officials first consider such questions as whether gaps exist in the field in West Germany and whether they should best be filled by universities, DFG or MPG. The Senate Planning Committee then decides which of the remaining proposals should be given priority support. As in DFG, a major problem here is the difficulty of comparing initiatives across disciplines. Although one approach would be to employ external criteria relating to the likely technological or social benefits, such considerations rarely feature prominently in the committee's discussions. The next step is to forward the proposal to the relevant MPG section which convenes an *ad hoc* committee of scientists (drawn from Max-Planck institutes, universities and abroad) to recommend whether to set up an institute or department. Given the importance of finding a suitable director, the Senate has, in recent years, often created a fixed-term research group (normally for five years). This makes it possible to initiate new research without establishing a permanent institute and hence reduces the risk of failure.[17]

An example of the above approach concerns the Institute for Polymer Research, the origins of which can be traced back fifteen years to a decision by DFG to launch a special collaborative programme. A subsequent DFG report concluded that an interdisciplinary centre should be created, and this was taken up by the Science Council which recommended the establishment of a Max-Planck institute. An MPG committee then assessed the prospects and identified the broad fields to be covered (the choice of specific topics being left to the directors of the new institute).[18] In 1988, the institute was still in the process of being built up from three departments to an eventual total of six.

Another recent initiative concerns social science. At the time of the decision to close the Starnberg sociology institute in 1981, the humanities section set up a committee to consider how the field should be supported. Its deliberations were extensive, including hearings with international experts, and resulted in two areas being identified as suitable for MPG support — organizational research (*Gesellschaftsforschung*) and cultural anthropology. The latter was judged to be particularly weak in West Germany, and two scientists were asked to prepare a research programme building upon existing strengths in fields like experimental psychology and linguistics. Their proposals were adopted by the committee and subsequently by the MPG section, and a final decision to establish a fixed-term research group was made by the Senate in 1988. The efforts of this social science committee probably represent the closest approach to systematic foresight within MPG.[19] It succeeded in identifying what is widely regarded as a promising line of research at the interface between cultural anthropology and cognitive science.

The converse of identifying a new area in which to establish an institute is the decision to close a Max-Planck department (or institute) when the director retires. It is the task of the relevant MPG section to determine whether — and in what direction — the work of the department should continue. When a vacancy occurs, the remaining directors at the institute (i.e. other departmental heads) can propose a successor. The MPG section then sets up a committee of experts to consider possible candidates and to seek views on them from around the world, eventually putting forward a single name for consideration by the Senate. Deciding whether to close a department is a vital part of MPG planning and of the intellectual renewal of institutes — each new director brings in fresh research ideas without which institutes would stagnate. However, decisions are handled decentrally by MPG sections; the Senate Planning Committee takes note but does not intervene. Only if several department heads are due to retire at around the same time may a Presidential commission be set up to consider an institute's future. However, the current President has made less use of this procedure than his predecessor, preferring to leave such hard choices to individual sections.*

Overall, there is little systematic foresight within MPG, and no mechanism exists for reviewing research across all fields and forecasting longer-term trends. Forward planning is aimed more at locating excellent scientists to become directors than at identifying promising areas. MPG does not keep a central research data-base (e.g. on publications), and has reportedly not even discussed such a possibility.* The virtual absence of formal foresight reflects the belief that the best approach to planning basic research is to choose good advisers and a first-rate President, and to use them to identify leading scientists to become institute directors, who are then permitted the freedom to initiate their own lines of work.

3. STRATEGIC RESEARCH AND BASIC TECHNOLOGY

(a) Federal Ministry of Research and Technology

BMFT accounted for over 50 per cent of federal spending on R&D in 1987, its budget of DM7,600 million dwarfing those of the Ministries of Defence (2,800 million), Economics (1,300 million) and Education and Science (1,000 million — BMFT, 1988, p.357). Its three largest departments are concerned with energy, information technology and aerospace R&D, the other two dealing with promotion of basic research and general administration. BMFT employs several funding mechanisms: (i) grants-in-aid to agencies such as MPG and FhG; (ii) institutional support for the National Research Centres and certain 'blue list' institutes; (iii) subscriptions to international facilities like CERN and ESA; (iv) finance for priority programmes (carried out in institutes, universities and elsewhere); and (v) indirect support for R&D. BMFT funding of academic research grew slowly up to 1985, but increased considerably thereafter to reach DM500 million in 1987, a figure regarded as disproportionately high by many academics since it is equivalent to half the funds disbursed by DFG. * Among areas receiving support are emerging chemistry and biology specialties, with several molecular biology laboratories being funded and programmes on AIDS, neuroscience and genome sequencing having been launched since 1986.

In line with the philosophy of decentralized decision-making, BMFT's three main departments each has a longer-term planning group. Periodically, these may set up committees to review developments within particular fields (e.g. computer science) or to identify priorities in strategic areas like materials research (see PERA/OTIS, 1989). Also, there is a Planning Division reporting directly to the Minister. This has been a source of tension since departments maintain that they have a better understanding of developments in their own areas than it has.* Within the Planning Division, medium- to long-term issues are handled by two sections responsible for 'technology assessment and future tasks', and 'research policy and planning'.

Since 1984, decisions on the overall distribution of expenditure have been made biennially at the so-called 'corridor' discussions. Involved are the heads of BMFT departments, sections, and planning and financial divisions. Trends in the funding of energy, IT and aerospace research as a percentage of the BMFT budget are analysed, and comparisons made between the shapes of the financial 'corridors' for these fields. The aim is to force officials to recognize that they are engaged in a 'zero-sum game', and their plans should be seen in relation to those for other disciplines. From this process emerges some consensus on longer-term priorities. A budget ceiling is set for each field, within which officials then have to fit their planned research commitments.

As regards the inception of priority programmes, ideas for initiatives often originate from scientists in national centres or companies receiving support. An official will then arrange for industrialists and researchers to be brought together to discuss what BMFT might contribute. Next, a formal proposal is prepared setting out the objectives, time-scale and funding requirements. This is then reviewed by the Planning Division, which may suggest improvements or alternatively make a case against support. Comments are submitted with the proposal to the Minister for a final decision.

The process can be illustrated by the example of information technology. The initial IT programme for 1985-8 had its origins in a report prepared by the Director of the Max-Planck Institute for Solid-State Physics after informal discussions and meetings. Towards the end of the four-year period, the question arose of priorities for a possible follow-up programme. A series of workshops on the theme of 'IT 2000' was therefore organized by BMFT to establish a broad policy framework. As an input, four expert committees were set up to study the prospects for microelectronics, telecommunications, information and data-processing, and consumer electronics. These assessed the relative strength of West German industry, identified major technological trends, and estimated the level of government R&D funds required. However, in the public debate which followed release of the IT 2000 report, what had begun as an initiative to help BMFT set priorities came to be seen more as an IT lobby requesting large sums of public money. The proposals were widely regarded as unrealistic, senior ministry officials being especially disappointed with the outcome.* In particular, the view of Planning Division staff is that there are few IT sectors where any real progress has been made in international terms over the last fifteen years despite billions of marks having been invested by government. Their assessment is that West German industry is in general not sufficiently strong to develop new products and set market standards, and indeed is probably in a weaker position now than a decade ago.* Needless to say, such views are not shared by most of those within the IT department, especially the microelectronics group.†

Aside from such activities at a departmental level, the other main foresight efforts of BMFT concern work on 'early-warning' systems to pick up emerging research developments as promptly as possible, and thus enable policies to be formulated on the provision of appropriate infrastructural support for R&D. Such an approach is more consistent with the prevailing government philosophy that strategic decisions should as far as possible be left to the 'market'.

A series of early-warning studies has been commissioned by the Planning Division since 1985, one of which was carried out by the Institute for Economic Research (IFO) at Munich. This aimed to explore the emergence of promising fields through an analysis of rapidly growing areas of patenting, without directly resorting to the judgement of technical specialists. Use was made of a large patent data-base covering all major countries and patent classes for the period 1979–85. To avoid inconsistencies in intellectual property laws and practices, only patents registered in more than one country were included (with care being taken to avoid double-counting) in a computer analysis of 150 fields with growth rates above a set threshold.[20] Results comparing the leading industrial countries were presented to BMFT and outside experts in 1987.[21] Subsequent evaluation has shown, however, that the attempt to avoid problems stemming from the technical limitations of the data-base was not fully successful.[22] Thirty active fields of patenting identified by IFO were subject to an independent analysis based on consultation with experts, and this revealed that only half reflected genuine technological trends. The remainder were statistical artefacts, resulting either from stochastic processes or from conflating two or more patent symbols (defined by similar keywords) in the automated search procedure[23]* Although IFO disputes the extent of these problems, further work is planned to improve the method by attaching abstracts to clusters and then inviting researchers to validate the labelling. BMFT had yet to decide in 1988 whether to fund a follow-up project.

A second early-warning study was less exploratory, and concerned an attempt by CHI Research to identify West German technological strengths using US patent data. Two main indicators were employed: first, technological *activity* was assessed by comparing the percentage of West German-invented patents in different classes with those for other industrial nations; second, a measure of technological *quality* was derived by analysing the number of citations in US-registered patents to previous German-invented patents, again in comparison with foreign competitors (Narin and Olivastro, 1987, p.2). Among the findings were that the nation's patent share is relatively high in traditional manufacturing, textiles and chemicals sectors, while it is much lower in electronics, computers and communications, with an appreciable decline evident in pharmaceuticals over recent years. Likewise, highly cited West German patents tend to be concentrated in traditional technologies such as vehicle engines, process manufacturing and bulk chemicals rather than in newer science-based areas like electronics and pharmaceuticals (ibid., pp.4–7).

Presentations of the results to government officials, industrialists and researchers were reportedly well received.* Within BMFT, the study is seen as having been useful, the regular meetings with CHI Research providing an opportunity to develop an understanding of the strengths, weaknesses and policy utility of patent analysis. More specifically, BMFT Planning Division staff regard the results as having helped clarify their views on the prospects for West German industry to compete in certain technology areas like mass-produced integrated circuits. One basic limitation of the approach, however, is that it was impossible to establish whether, for example, the decline in pharmaceutical patenting in the United States reflects a real decrease in technological output or merely a change in market strategy towards North America by major companies.

Four of the early-warning studies have been carried out by the Fraunhofer Institute for Systems and Innovation Research (ISI) as part of a wider programme to develop research indicators (based on bibliometrics, patent data and technometrics) for use in forecasting as well as in its work on innovation policy. The aim of the first study was to explore the use of econometric indicators (notably, an index of 'revealed comparative advantage' based on balance of trade in high technology products) for assessing West German competitiveness, and hence identifying areas where research resources should be concentrated. One finding was that the United States has been dominant in *very* highly R&D-intensive areas (e.g. aerospace and computers), while Japan leads in sectors which are slightly less R&D-intensive (e.g. consumer electronics), in both cases followed by West Germany. From this, it was concluded that it would be inadvisable for the country to switch its longer-term research efforts to attempt to compete in high-technology areas (Grupp and Legler, 1987; Grupp and Schmoch, 1987).

Evaluation of the results showed that econometric indicators are too historical to provide a useful early warning of emerging technologies, nor do they reveal the extent to which R&D has contributed to commercial success. A second study therefore aimed to measure the dynamics of technological growth directly without recourse to econometric data. The approach was based on 'technometric' indicators derived from detailed specifications of technological innovations in six strategic fields (e.g. lasers and solar cells), with comparisons drawn of the relative contributions made by West Germany, Japan and the United States (Grupp and Hohmeyer, 1986; 1987; Grupp *et al.*, 1987a; Grupp and Schmoch, 1987, p.5). Although enabling ISI to produce a convincing demonstration that investment in R&D is correlated with technological performance and subsequent economic benefits (e.g. Grupp and Hohmeyer, 1988), the study again suggested that the use of technometrics as an early-warning indicator is somewhat limited since it can only be applied to existing products, or at best to prototypes. Furthermore, the cost of obtaining information on technical specifications from firms is rather high (typically requiring several person-months of effort per field). However, one interesting result which caused concern was the finding that just a single professor in the Federal Republic was working on biosensors and he was receiving funds from a Japanese company. This contributed to the BMFT decision to launch a programme in the field.

The third ISI study explored the use of patent statistics in providing an early warning of emerging technologies. Because patents can reflect market strategy as well as technological output, it was necessary to distinguish between the two effects by comparing the trends evident in US, European and West German data-bases. Another task was to devise a reliable automated search procedure involving a combination of patent classification symbols and keywords taken from abstracts (Grupp, 1988, pp.14–18).[24] Provided the analyst is familiar with differing national patent practices, it is then possible to detect technological developments at a relatively early stage (Grupp and Schmoch, 1987, p.9). Data on laser patents, for example, reveal rapid growth in the United States over 1975–85 in crystal-growing for solid-state lasers, but little change in West German activity (ibid.; Schmoch *et al.*, 1988, pp.167–73). Follow-up work has concentrated on comparing patent indicators with bibliometric, co-citation, technometric and R&D funding data. The results show interesting correlations among several indicators — for example, between publication and patent data (but not with most of the co-citation statistics produced by Bielefeld University (discussed below).[25] An important conclusion, therefore, is that as wide a range of indicators as possible should always be used.

More recently, BMFT has authorized a further three-year study, involving both ISI and CHI, to investigate the links between basic research and industrial R&D using a combination of American and European patent data.[26] The range of indicators has also been broadened to include research cited by patents and scientific publications produced by industrial authors. BMFT hopes the study will transfer CHI expertise to the Federal Republic, and also provide a more reliable means of identifying emerging technologies.*

Taken together, the ISI studies for BMFT have enabled a group to be built up which has proved its ability in terms of both generating income from contracts and carrying out pioneering work in the area of quantitative research policy. Moreover, of the foresight approaches explored, technometric indicators have proved useful in pointing to 'gaps' in the nation's research efforts,[27] while patent indicators may provide an early warning of rapidly growing technologies and hence make a useful contribution to strategic planning. The results have also been well received within BMFT; true, they have not always been acted upon, but it would be unrealistic to expect such findings to represent more than just one input among many to decision-making. In addition, several larger companies regard the work as promising, and at least one reportedly has plans to set up a group to carry out similar studies.*

Finally, mention should be made of a closely related project funded by BMFT's social science section which was undertaken by the Science Studies Unit at the University of Bielefeld. Although initially aiming to develop techniques for providing an early warning of national scientific weaknesses,[28] the research has been more methodologically oriented than the other studies, exploring two different bibliometric approaches. The first focussed on publication and citation data for the period 1973–82 contained within the CHI/NSF Science Literature Indicators Data-Base, and the results provided interesting comparisons of West German scientific performance with other industrial countries (Weingart *et al.*, 1988). The larger component of the study focussed on co-citation analysis[29] using data for 1984 purchased from the Institute for Scientific Information. The first step was to develop suitable software for generating clusters of co-cited papers — i.e. 'maps' of specialties. Analysis of these revealed evidence of national strengths in chemistry, physics and engineering, for example, and of weaknesses in biology and geosciences (Weingart *et al.*, 1988), in line with some of the results obtained from the CHI/NSF data-base although a direct comparison of the two methods proved difficult.

Within BMFT, the co-citation work is viewed as a promising approach to mapping science, but one that is still rather 'academic' and too unproven to be used for policy purposes.[30] Furthermore, officials at other agencies were generally sceptical or even rather critical.* On the positive side, it was recognized that the Bielefeld group was still in the process of developing expertise in the technique, and that its initial efforts should be judged more in terms of establishing a domestic capability in macro-level science indicators.[31] Among the criticisms, one concerned the process of labelling co-citation clusters, which was judged by the scientific experts consulted to be inadequate. Second, it was felt that the level of methodological innovation achieved was small in relation to the several person-years of effort invested. Third, the group was criticized for failing to appreciate that acceptance of the results would require more intensive participation of the scientific community. Last, officials in BMFT and elsewhere expressed regret that greater emphasis was not given in both components of the project to analysing the implications of the findings for science policy.*

As regards the overall impact of the early-warning studies, the approaches investigated are generally regarded as exploratory rather than yielding reliable policy-relevant results. Certainly, science and technology indicators still have some way to go in terms of development and validation before they achieve the same public acceptance as economic statistics. Yet while the BMFT Planning Division would like to engage in more research foresight, it has limited powers compared with the three main departments which are also competing directly for funds. In such an environment, there is limited scope for prospective studies which ideally need to focus on fields cross-cutting the interests of these departments.

(b) The National Research Centres

The thirteen National Research Centres had a combined budget of DM2,600 million in 1987, 90 per cent provided by BMFT and the rest by state governments (BMFT, 1988, p.85). Their task is to conduct R&D in fields requiring interdisciplinary cooperation and the concentration of personnel, funds and equipment. The centres cover the spectrum from curiosity-oriented science (in high-energy physics, for example), through strategic research to development of industrial technology (AGF, 1985a, pp.2–3). They are legally independent bodies with considerable autonomy in relation to determining research priorities (von Massow, 1986, p.36).

The centres have cooperated in two foresight initiatives. In the first, undertaken in 1974, they established an Applied Systems-Analysis Programme to strengthen and coordinate work in the field and adapt it to research policy-making (Geimer and Geimer, 1978, p.55). Later, the Applied Systems-Analysis Group was created to extend this activity and identify 'white spots' (or growth points) in science and technology over the next ten to twenty years. The group failed entirely in this task, ending up doing traditional technology assessment and being disbanded in 1982 (AGF, 1983, p.6). Among the lessons drawn from this experience was that forecasting should not be carried out by professional analysts in isolation from scientists in the areas concerned (Irvine and Martin, 1984, pp.54–5).

The other exercise took place in 1985 and represented an attempt to generate research perspectives for the next decade. This was carried out over a six-month period through a series of meetings of laboratory directors, with inputs made by committees in each centre which studied the future for their main field of activity. It resulted in a widely circulated report identifying a number of major trends (such as a decrease in nuclear energy research), but not in a particularly systematic manner (AGF, 1985b).[32] It did, however, point to the need for a better early warning of emerging research needs, and a committee was subsequently set up to explore the use of various indicators like publications and patents. Work was still at an early stage in 1988.

(c) The Fraunhofer Gesellschaft

The primary task of the Fraunhofer-Gesellschaft (FhG) is to carry out a broad range of strategic and applied research for government and industry.[33] It operates a network of thirty-five institutes and R&D groups, and in 1987 had a budget of DM600 million. Institutes derive most of their revenue from contracts but also receive a basic subsidy from FhG corresponding to a fixed proportion of their external income. They enjoy appreciable autonomy provided they meet 70-80 per cent of costs from commissioned research. The annual budget-setting process is based upon financial plans prepared by each institute in which targets for contract income are agreed with FhG, the remainder to be made up with the basic subsidy. Thus, there is no central coordination of priorities for the distribution of FhG funds with the result that there has been little demand for research foresight.

Nor has there been much of a role for systematic foresight in more strategic decision-making. Changes in research direction can come about in three ways: through an existing institute moving into a new area; through closure of an institute; or through the creation of a group or institute. Over the last decade, for example, approximately one institute a year has been set up by the FhG Senate after consideration of detailed proposals on needs and opportunities (FhG, 1986, p.5). A key role is played by members of the planning department who monitor the market for R&D, mainly through informal mechanisms such as attending conferences and discussions with researchers. The main criterion is whether the demand for contract research in a given field will support an institute in two or three years' time, and any foresight is therefore concerned only with the shorter term. Limited experiments have been conducted with co-citation and patent analysis, but it was found that 'these did not offer a real solution — they did not give us better results than we get by monitoring things ourselves'.*

Among the areas in which initiatives have been taken is high-temperature ceramics. This arose from the recognition that the narrowness and poor integration of West German materials research with industry provided potentially important opportunities for work in the field. Over several years, the planning department reviewed the level of existing expertise in different areas of materials science and considered whether suitable candidates with a commercial orientation might be found to direct a new institute. After extensive discussions with experts, the choice finally focussed on high-temperature ceramics. The planning department did experiment with quantitative indicators in this instance, but only after the area had been identified through the conventional informal approach.

Another example is provided by computer graphics. Here, a group was established around a promising team at Darmstadt University, whose leader had already demonstrated an ability to obtain grants and contracts, and ensure effective technology transfer to industry.[34] After consultations with scientists and Fraunhofer institutes working in related areas, the Hesse state government was approached and agreed to provide support for more basic research. In this case, FhG decided to set up a group of around forty researchers for an initial five-year period, after which, subject to a review, it will be upgraded into a full-scale institute. Possible initiatives in surface technology and high-temperature superconductors have also recently been considered by the FhG planning department, but, after discussions in West Germany and abroad, it was decided that it would be premature to create a new institution. In surface technology, the market for contract research, although growing, is still highly specialized and diffuse, while the high-temperature superconductor field was judged to be at too basic a stage of development.

Overall, FhG's approach towards establishing institutes or groups is based on informal consultations with researchers and experts in industry. The latter are generally conducted on an individual basis to overcome concern about revealing 'secrets' to commercial competitors. As regards the choice of experts to consult, again no very systematic approach is used. And even though demand for contract research is such an important criterion, no formal assessment of future markets is attempted,[35] the central office lacking the capacity to carry out more detailed strategic analyses.

(d) *Foresight by the Länder*

During the 1980s, the *Länder* have become increasingly involved in establishing their own research policies (von Massow, 1986, p.56). Baden-Württemberg, the richest state, has probably been the most active, its Ministry of Science and Cultural Affairs supporting R&D through a variety of mechanisms including priority programmes. The Prime Minister has been strongly committed to the role of research in providing the infrastructure to stimulate economic progress and, together with his Cabinet, personally directs the state's science policy. There has therefore been the political power to implement institutional changes combined with a willingness to invest in longer-term research.

Two foresight initiatives have been important. In the first, the Research Commission Baden-Württemberg examined the future for strategic research in the region. The Commission, which consisted mainly of industrialists together with a few scientists and government officials, assessed current research, analysed major problems, and attempted to identify scientific areas linked to the technological needs of the state. Its recommendations on priorities (FKBW, 1982) reportedly had a considerable impact on policy,*and it was therefore decided four years later to carry out a similar exercise. Following discussions between the Prime Minister and university rectors, the Research Commission Baden-Württemberg 2000 was set up to determine priorities for academic research over the next decade (MWK, 1987; 1988). Although the group included just one industrialist (the other members being scientists, this time drawn from outside the state), the longer-term R&D needs of local industry, especially smaller companies in sectors such as mechanical engineering, were given particular attention. The Commission's approach to deciding what scientific and technological infrastructure should be provided through universities relied largely on consultative discussions with experts carried out by subcommittees, and no systematic foresight methods were employed. Given the Prime Minister's close interest (he attended the Commission's first session), the resulting report is again likely to prove influential.

4. RESEARCH FORESIGHT IN INDUSTRY

Of the five science-based companies covered,[36] two are in the electrical and electronics sector (Siemens and AEG) and three in chemicals (BASF, Bayer and Hoechst). In general, strategic research is conducted in central laboratories financed by a mixture of contracts from divisions and corporate funds (and, in the two electrical companies, partly through BMFT and EEC programmes). Decisions on such work generally involve an element of foresight, with a central research strategy group often playing a coordinating role. Experiments with formal approaches to foresight developed by consultancies have met with mixed success.

Despite having gone through a period of stagnation, Siemens devotes a high proportion of turnover to R&D, the figure having risen from 9 per cent to 12 per cent since the mid-1980s. Out of the total R&D budget of DM6,200 million in 1986–7, 14 per cent was devoted to longer-term research. Decisions on corporate R&D policy used to be made on an incremental basis with little regard to overall strategy. However, it was recognized that clearer priorities were needed, particularly as product life-cycles were becoming shorter (they are currently as little as two or three years in microelectronics), while the lead-time to develop new technologies has been lengthening (a typical figure is ten years). Hence, the corporate technology planning department is now responsible for the increasingly important role of coordinating preparation of the company's longer-term research strategy.

The annual planning procedure begins with research group leaders proposing projects, which are then integrated and scaled down by their department heads after discussions with business divisions and others concerning the R&D requirements for future products. The resulting set of proposals is submitted to the next level of laboratory directors, where the process is repeated. Finally, negotiations in committees (which include representatives of corporate and product divisions) lead to an agreed research budget for the next two years. This short-term budgetary plan is, however, situated within a five-year strategy based on a qualitative analysis of projected technological developments, which is in turn informed by a more speculative ten-year outlook.[37] The medium- and long-term plans are both updated annually using a procedure which is eclectic rather than formalized. It draws principally upon individual discussions and expert panels, together with some trend and S-curve analysis. In one recent case, scenario analysis was also carried out, but this is unusual and other techniques like Delphi are not employed. While prospective studies of fields are carried out, they are organized on an individual rather than a global basis, although over time most areas are covered.

As in other firms, the emphasis in strategic planning is on integrating science-push and market-pull factors (such as the relevance and pervasiveness of novel technologies). R&D fields are categorized in terms of whether they are likely to yield key technologies with an impact on a wide range of products and their degree of maturity. Inputs are made by staff responsible for product groups about the potential of existing technology which still remains to be exploited, when this will be exhausted and what might replace it. Other information, for example on market trends, comes from multi-client surveys. More specific studies by consultants have, however, been found to yield little useful information despite their considerable cost. Hence, as in much of industry, the principal source of background knowledge remains the company's own monitoring of external R&D, both through data-bases on patents and scientific literature,[38] but principally through researchers' personal contacts with their counterparts in other companies and with academics.

Although AEG is less than a quarter the size of Siemens and devotes only 8 per cent of turnover to R&D, until recently it suffered similar problems with coordination of research. After the takeover by Daimler–Benz in 1986, the new R&D Director created a structure in which the 300 or so previously disparate projects were grouped around seven or eight focal points (e.g. digital communication), each with a rolling five-year programme to ensure steady progression from basic research through applied R&D and into product development. Overall strategy is overseen by a small central group which aims to combine analyses of technological trends (to identify lines of development and hence potential focal points for research) with assessments of the R&D requirements of business divisions. A common mechanism for achieving consensus among researchers and divisions on technological trends and scientific needs is the expert workshop. Use is also made of questionnaire surveys (conducted by the research strategy group) and interviews — for example, to obtain information from divisions on market developments, emerging technologies, the strengths of competitors, and the opportunities and risks associated with introducing new products and processes.

Once a focal point has been agreed, the next task is to construct a five-year programme. This involves a two-stage consultative procedure in which discussions on overall strategy are first held between research departments and senior divisional managers (particularly from production, marketing and development). Questions relating to the execution of each programme and the constituent projects are then addressed in a lengthy process during which department heads within central research institutes must secure 70 per cent of their funding from divisions.

As regards more systematic foresight, attempts have been made in a few fields like energy to look further than five years ahead. Delphi methods are not employed on the grounds that they involve too much effort and have not proved especially successful in relation to research. Trend analysis, however, is sometimes carried out, with expert groups examining the future development of technologies. Yet, as the head of research strategy pointed out, there is little point in identifying opportunities unless there is a commitment on the part of individuals to 'champion' new technologies. This requires having 'a vision of what the work will realize' and an ability to see projects in terms of an overall programme.* Part of the effort to improve the strategic planning of R&D since 1986 has therefore been aimed at fostering an organizational culture among researchers which encourages the emergence of such champions.

Foresight in the three chemical companies draws upon work by the Chemical Industry Association (VCI), which is also responsible for distributing the Chemical Industry Fund to leading university scientists. Around DM15 million (raised through a sales levy) is allocated annually by a committee of industrialists and academics on the advice of small expert groups covering the main chemistry specialties. Each of these grades all university professors in the field, making use of personal knowledge as well as information provided by VCI (e.g. on publications).[39] Professors above a certain threshold are awarded some funds and those at the top appreciably more. The philosophy of the chemical industry is that basic research should primarily be conducted in universities and applied work in their own laboratories (rather than by FhG), and the fund is one way of ensuring effective links between the two. Leading academics often participate in workshops convened periodically by VCI to discuss future research opportunities — for example, in advanced materials (FCI, 1986). The latter was followed by a meeting two years later in which small panels of university scientists and industrialists prepared overviews of developments in four fields (e.g. biologically active materials), with particular emphasis on identifying commercially important 'deficit' areas where gaps existed in basic research. These were then combined into a brief report presented to the annual VCI meeting.*

The first of the chemical companies, BASF, has a relatively centralized approach to R&D, 50 per cent of which is carried out in its corporate laboratory. Longer-range and more general research financed from central funds accounts for around 15 per cent of total R&D, with priorities being set through a combination of bottom-up and top-down procedures. The former involves a process in which lists of proposals from scientists are gradually narrowed down through a series of discussions. They are then integrated into programmes by research division heads who, together with the director of R&D, determine the distribution of funds within the corporate laboratory. Decisions are based within the framework of an annually updated three-year plan prepared for the BASF Board by a planning group which coordinates discussions between business divisions and the central laboratory. Priorities take into account a range of criteria — R&D opportunities, market trends, production requirements and costs, government regulations and environmental issues. This involves the planning group in market-oriented analyses (for example, of product quality), with inputs often being made by *ad hoc* expert panels. The approach is informal, with panels free to determine the best procedure for conducting their investigations. Occasionally, Delphi techniques may be employed, and in some cases outside consultants are also commissioned to assess the scientific and market opportunities associated with particular fields. The proposed three-year plan is finally submitted to the BASF Board, which agrees an overall research budget and the main priorities, taking into account wider corporate strategy.

Bayer is in contrast rather more 'science-driven' — all its main business areas have stemmed from research advances, with products developed in the last fifteen years representing 40 per cent of sales (Bayer, 1987a, p.9). R&D is relatively decentralized, the corporate research division accounting for only 10 per cent of total expenditure (with two-thirds of its funds coming from divisions). Considerable emphasis is placed on a bottom-up approach in line with experience that the company's most important innovations (such as polyurethane) have resulted from unforeseen discoveries made by individuals.[40] Consequently, Bayer believes the best policy is to grant researchers a high level of freedom within an environment in which unplanned advances can be exploited. Decisions on funding are made primarily on the basis of expert judgement. Bayer has experimented with formal approaches but found they reveal little that is new, and fail to predict important breakthroughs. It has therefore concluded that the extra effort involved is not worthwhile.

In terms of the organization of research, Hoechst has more similarities to Bayer than to BASF, although over 30 per cent of R&D is conducted at the main site. Most longer-term work is undertaken by the corporate research department which accounts for 11 per cent of total R&D. Just over half its finance comes from a central fund administered by a Long-Term R&D Committee (consisting of four members of the Hoechst Board). New projects can emerge either from the bottom-up or from the top-down, but in neither case is the process very formal.[41] Although recognizing the danger of lack of overall strategy associated with this type of approach, the company has, like Bayer, found formal approaches to be of limited use in the past, at least for longer-term research where progress depends upon talented individuals and often unpredictable discoveries.

In 1987, concern about the absence of a coherent strategy resulted in Hoechst establishing a committee to identify promising longer-term research. After informally discussing current strengths and know-how, it selected four fields for scrutiny (e.g. electronic materials). Task-forces were set up to analyse areas where Hoechst was less active and needed to expand its efforts. These addressed such issues as: (i) the likely implications for the company in 1995 if no major changes were made to the existing research portfolio; (ii) priorities for new initiatives and the costs involved; and (iii) areas of existing research which might be cut back or terminated. After much debate, it was agreed to use external consultants to assist one task-force. However, the experience proved somewhat disappointing in respect of the specialist expertise offered, although the consultants did provide a methodology for structuring discussions (which admittedly some task-force members found rather cumbersome), enabling thirty topics to be analysed fully in terms of technological and market criteria. While many researchers, especially older ones, remain sceptical about this experiment, others believe that the adoption of a systematic approach did help identify clearer strategic priorities.*

Among the main advocates of more formal approaches to research planning in West Germany has been Arthur D. Little, which sees itself as combining the technical skills of Battelle with the management expertise offered by the Boston Group. Among the questions it helps clients address are the following: how important is technology to the firm's business? Are resources being invested in technologies with strategic potential? Is the firm competitive in those technologies? Should R&D be more centralized or decentralized? Do appropriate structures exist for planning and managing R&D? And is there effective dialogue between researchers and business managers (ADL, n.d.a)? The approach that has been developed since 1987 for the 'strategic management of technology' (ADL, n.d.b) is relatively labour-intensive, a typical study costing up to DM1 million. It is consequently suitable only for larger companies.

The first step is to identify the novel technologies likely to affect the client's business, together with the underlying research areas. After scanning databases and technical journals , brainstorming sessions are held with a dozen or so experts from within the company (including research directors, product development heads, and product and sales managers) and outside (from universities and research institutes) to provide a more focussed analysis of emerging technologies. At the same time, an assessment is undertaken of the research requirements of likely innovations. In this way, a matrix can be constructed relating the cross-impacts of individual technologies on different products and processes.[42]

Identified technologies are then classified into four categories according to their potential impact on competitiveness: (i) base technologies (essential for a given product or process but widespread and easily accessed); (ii) key technologies (which provide a commercial advantage through product or process differentiation or through improved economics); (iii) pacing technologies (not yet widely applied but having the potential to change the basis of competition in the sector and therefore of high research priority); and (iv) emerging technologies (often still at a basic stage and with uncertain prospects, but with the promise of developing into pacing, and, perhaps subsequently, into key technologies — ibid.).[43]

The client's standing in each technology is next evaluated relative to competitors, using various criteria (for example, current patenting position and the availability of good researchers). A profile of corporate strengths is thus arrived at and investment priorities for pacing technologies identified. Besides assessing the potential for each technology, an attempt is also made to assess possible barriers (e.g. the need for major capital investment) in order to gauge the rate of development and implementation, and to estimate the likelihood of success.

In parallel with this segmentation of technologies, market-related factors (such as price and performance) are likewise broken down and classified, and the client's position relative to competitors ranked using various criteria. (Such analysis may reveal that the strongest challenge will come not from the most technologically advanced rival but from one with the strongest market position.) By combining the technology and market segmentation results, the consultants and their client can identify the fields on which longer-term work most needs to be concentrated and hence arrive at an optimized research portfolio.[44] While certain scepticism about the Arthur D. Little approach exists among industrialists,* the fact that it has been employed in some thirty studies for companies (and governments) over the last five years suggests at least some success in assisting firms with strategic planning of R&D.[45]

5. CONCLUSIONS

Following concern a few years ago about a 'gap' between the mechanisms for supporting basic science and those for applied R&D, several developments have recently served to ensure that strategic research receives higher priority. First, academics have become generally more sympathetic towards industrial needs, in part because of the limited availability of government funds for fundamental science during the 1980s. Second, DFG has received a small but increasing proportion of its budget in the form of earmarked funds for national priority areas such as information technology. Third, MPG has likewise extended its coverage to include longer-term strategic research (e.g. polymer science and computer science). Fourth, BMFT has been giving greater emphasis to fields like biotechnology. Finally, several National Research Centres have been restructuring, cutting back in nuclear energy R&D, for example, and thereby freeing resources to be deployed in emerging areas.

Although all the main scientific organizations now see themselves as partly responsible for 'strategic research' (a term still not normally used in West Germany), the level of formal foresight carried out remains small compared to other countries. This can be explained in part through wider political factors, including the considerable post-war autonomy of basic research, the country's federal structure, and the Christian Democrat government's unwillingness to engage in any activity which might be construed as centralized long-term planning. However, a number of more specific factors have also contributed, from which certain conclusions can be drawn:

(a) Many of the lingering doubts towards foresight among agency officials stem from the limited success of experiments in forecasting and priority-setting in the 1970s by BMFT, the National Research Centres and others. That those exercises failed was not least because they did not fully involve academic and industrial scientists, who therefore remained unconvinced of the merits of formal approaches to research planning. Some of the early-warning studies have likewise made little progress in this respect, although the ISI work, with its stress on participation by researchers and companies, is starting to have an impact.

(b) During the 1980s, there has been a switch in attention from traditional extrapolative forecasting to the development of early-warning systems for detecting emerging research opportunities and changing R&D needs. The latter are more compatible with current federal government philosophy towards planning, an illustration of how foresight must take into account the wider political context in which it is conducted.

(c) The need to balance the conflicting requirements of 'science-push' and 'demand-pull' in foresight is clearly recognized by West German companies in the mechanisms they employ for determining longer-term research priorities. The limited use of systematic foresight by DFG and MPG, in contrast, is primarily attributable to their overriding concern with fostering scientific opportunities rather than trying to relate them to socio-economic demands, although their high degree of autonomy has also been important.[46] Consequently, within these two agencies the emphasis has been on responding to proposals from leading scientists rather than identifying potentially important fields.

(d) Although foresight has tended to focus on selecting promising research areas, the converse is to identify less important activities where the scale of effort can be cut back—a task which is essential under conditions of level funding. While decisions of the former type are less widespread than in other countries, this is by no means true of the latter, with MPG and FhG periodically closing down research departments or even entire institutes. This form of foresight is seen as crucial if funding agencies are to preserve their flexibility to respond to emerging opportunities and if institutes are to maintain their intellectual vitality.

(e) With the possible exception of the Prime Minister of Baden-Württemberg, few product champions for foresight have emerged in the public sector. The promising work by ISI on early-warning indicators would benefit from high-level support within BMFT to help overcome the deep-rooted scepticism and hostility of researchers and companies, but even the original sponsor of the initiative has since moved to another position. Likewise, a product champion has still to appear for the prospective studies undertaken by the National Research Centres. Yet the vital role of such individuals is fully recognized in companies such as AEG.

(f) The wide range of skills and high level of technical expertise required for foresight have perhaps not been available to the same extent as in other countries, the structure of the universities having done little to encourage the development of professional competence in the quantitative science-policy field. However, since 1984 the various early-warning studies have helped transfer expertise from abroad and stimulated the emergence of a national centre of excellence at ISI.

(g) The most systematic foresight studies have been undertaken by Arthur D. Little for companies and government agencies. Although the methodology used appears to offer certain advantages in identifying strategic research priorities, it also entails considerable effort and expense, and many in industry and elsewhere remain to be convinced that the costs are worthwhile. Furthermore, foresight is seen as more relevant to some fields than others. Among chemical companies, Bayer in particular argues that foresight is of limited utility in pharmaceuticals research because major innovations have previously stemmed from unanticipated discoveries.

If we look to the future, two further factors seem likely to constrain the acceptance of foresight in research planning. The first might be described as a certain pragmatism on the part of the German people. For many, the attempt to look into the future on a systematic basis represents something that is too 'philosophical' or 'academic' to be worth expending much effort. Second, companies in West Germany, as elsewhere in Europe, have traditionally been rather secretive, even in relation to their longer-term research objectives. This is especially true in the important chemicals sector. In the information and communication industry, in contrast, domestic and EEC programmes like ESPRIT have started to have an effect in persuading firms that benefits can be derived from cooperating on precompetitive research, and that this first requires arriving at a measure of consensus on long-term priorities. Certainly, there is still a long way to go in changing cultural attitudes within industry towards collaborating with competitors on matters of mutual interest. However, if this trend continues, it is not inconceivable that foresight may in future come to be accepted as having at least some role to play in both the public and private sectors.

NOTES

* Refers to material collected during interviews in 1987 and early 1988, while † denotes information drawn from subsequent correspondence.

1. Every four years, the federal government presents Parliament with a report on research policy. The seventh (in 1984) stressed the intention to support individual initiatives, with the government maintaining a low profile (von Massow, 1986, p.53).
2. 'Where resources are limited, particularly in the sphere of the new technologies, great competition inevitably develops between the Länder' (ibid., p.56).
3. An excellent but somewhat dated description of West German science can be found in Geimer and Geimer (1978). For a more recent account, see von Massow (1986).
4. Capital investment is funded jointly by state governments and the Federal Ministry of Education and Science. With the exception of the efforts by Baden-Württemberg, little research foresight is carried out in relation to university funding. However, if student numbers decline in the future, universities may be forced to specialize more (Neffe, 1986a; 1986b), in which case the need for systematic prospective analysis could increase.
5. Funded jointly by federal and state governments, the 'blue list' institutes carry out a mixture of basic and applied research. They are overseen by the Bund-Länder-Commission (BLK), and are also subject to review by the Science Council. However, BLK has a coordinating role only — there is no central policy for the institutes and no attempt at research foresight, so these institutes are not considered further here.
6. A number of other institutes are supported by individual federal ministries and state governments (von Massow, 1986, pp.40–1).
7. Other roles of DFG are to further collaborative work, support postgraduate training, advise the government on scientific matters, and administer West Germany's research interests abroad.
8. The term 'scientist' is used in its broad sense to include scholars in the humanities and social sciences.
9. DFG officials pointed out that Georg Bednorz, who shared the 1987 Nobel Prize for Physics, had previously studied at Münster University within the solid-state chemistry priority programme. The programme was supported because of its scientific merit, no-one foreseeing its contribution to high-temperature superconductors.
10. For example, a recent assessment of AIDS research revealed a shortage of good scientists. It was therefore decided to set up scholarships for training researchers from fields such as virology and cell biology instead of creating a priority programme.
11. Other mechanisms include core-funding for research groups and the operation of central facilities (DFG, n.d., pp.22 and 26).
12. The figure grew from DM17 million in 1984 to 75 million in 1988.
13. Details of earlier Grey Plans can be found in Irvine and Martin (1984, pp.47–8). The term 'Grey Plan' (e.g. DFG, 1983) has, however, been replaced in the latest report by 'Perspectives' (DFG, 1987a).
14. The heads of DFG review groups have a good overview of their field and already possess most of the information needed to prepare such reports.*
15. At the time the first draft was prepared, high-temperature superconductors had just been discovered. DFG therefore added a few sentences at the proof stage, and officials cited this as an example of the unpredictability of science.*
16. The fifth Grey Plan attempted a more systematic approach to foresight but this proved unsuccessful (see Irvine and Martin, 1984, p.47).
17. It is also claimed that such a mechanism provides a means for improving relations with universities, in that groups with a limited lifetime must seek close cooperation with neighbouring academic institutions (MPG, 1984, p.20). This is important since there is some evidence of tension between Max-Planck institutes and universities. Although there is meant to be approximate parity between the

two sectors, long-standing rivalries have been fuelled by differences in resources, opportunities and status. Increased teaching loads, budgetary restrictions and democratization of decision-making within universities have made for a less conducive research environment (Dickson, 1986, p.811).

18. See Irvine and Martin (1984, pp.50–1) for a description of the origins of the Polymer Research Institute.
19. The head of the committee was also involved in the US field study on social science research (see Chapter 5).
20. For details, see Faust (1987a; 1987b).
21. Several industrial experts who attended were reportedly rather critical.*
22. Because titles for patent applications from countries not using Roman characters are transliterated by computer (the rules for which are not unambiguous), keyword searches are especially difficult for Japan, the Soviet Union and Korea. The resulting information is consequently somewhat unreliable.*
23. One observer explained this in terms of the researchers' lack of technical knowledge, resulting in unreliable labelling of patent clusters. In some cases, the automated search procedure led to single fields being split into two or more parts, while in others different areas were conflated.*
24. In 1987, the Federal Ministry of Economics funded ISI to develop further the patent-search procedure and make it available to smaller companies (Grupp and Schmoch, 1987, pp. 10–11). After sixty studies (on average costing a few thousand marks) had been carried out, the service was transferred to the Patent Interpretation Office of the Bavarian Institute for Trade and Industry.
25. The only significant correlation was for solar generators (Schmoch *et al.*, 1988, p.355).
26. The current disadvantage of European patent data is that the computerized information solely covers references made to *prior* patents and papers. There is no data-bank comparable to that compiled by CHI Research for US patents where the 'inverted' matrix of citations received by patents from *subsequent* patents is stored. In a follow-on study, an inverted citation data-base is to be produced for selected fields.*
27. A second shorter technometric study was commissioned by BMFT in 1988.
28. There has also been some overlap between the Bielefeld study and other early-warning work, especially at ISI.
29. Details on co-citation methodology can be found in Small and Griffiths (1974), Griffiths *et al.* (1974) and Franklin and Johnston (1988).
30. Although funds for the second phase of the project were initially sought from the BMFT Planning Division, the methodology was not regarded as sufficiently established to be used for policy purposes, and support was instead again provided by the Social Science Section.
31. Another positive observation was that the work had identified productive West German scientists, but this was tempered with doubts as to whether such an elaborate methodology as co-citation analysis was really necessary to arrive at these results.*
32. No use was made of quantitative data, the report being the product of open-ended discussions and successive revisions of drafts.
33. It also conducts research for the Ministry of Defence, and provides technical support to government and industry.
34. He had, in fact, already set up his own company.
35. Occasionally, studies are commissioned from consultants, but these are mostly very small. In addition, ISI reports regularly to FhG Central Office.
36. The need for a selective approach means that the findings reported here may not be entirely representative of the situation in industry.
37. These procedures may soon change because of corporate restructuring.†

38. Publication data-bases have even yielded interesting information on Siemens' own research, for example, revealing unsuspected duplication of effort among its 41,000 R&D staff.*
39. On one occasion, citation data were provided, but it was concluded that these added little to the evaluation.*
40. See the list of innovations in Bayer (1987b, pp.22–6).
41. For example, after senior staff from one corporate research group attended a meeting on high-temperature superconductivity, they decided to launch an initiative in the area.
42. The large volume of analysis is carried out by computer, which enables results to be shown to workshop participants almost instantaneously and hence speeds up the process of arriving at consensus.
43. To take the example of desk-top publishing, base technologies include TV monitors and microprocessors (which will be developed in other industrial sectors and then applied in workstations), while applications software, CAD and automated assembly techniques currently represent key technologies. Pacing technologies, in contrast, include video storage and voice recognition, the latter of which could (by removing the need for keyboards) have a revolutionary impact on the way workstations are used and hence on markets.
44. The applications of such foresight exercises are not confined to R&D, but may also relate to broader corporate strategy. For example, a position of both technological and market strength is probably required if a 'leadership' strategy is to be successful. If, however, the company is strong in technology but has only an average market position, a 'niche' strategy is likely to prove more effective, while if its market position is weak a joint venture may represent the best option.
45. For further details on the approach, see ADL (n.d.c).
46. 'Value for money' has yet to surface as a major issue in relation to West German research policy.

REFERENCES

ADL (n.d.a), *Arthur D. Little Europe*, Brussels, Arthur D. Little.

—— (n.d.b), *Strategic Management of Technology*, Cambridge, Massachusetts, Arthur D. Little.

—— (n.d.c), *Managing Innovation*, Cambridge, Massachusetts, Arthur D. Little.

AGF (1983), *Arbeiten zur Systemanalyse und Technologiefolgenabschätzung in der AGF*, Bonn, Arbeitsgemeinschaft der Grossforschungseinrichtungen.

—— (1985a), *Programme Survey*, Bonn, Arbeitsgemeinschaft der Grossforschungseinrichtungen.

—— (1985b), *Zur Thematischen Orientierung der Grossforschung in den 80er und 90er Jahren*, Bonn, Arbeitsgemeinschaft der Grossforschungseinrichtungen.

Bayer (1987a), *Bayer Annual Report 86*, Leverkusen, Bayer.

—— (1987b), *Bayer — Names, Figures, Facts*, Leverkusen, Bayer.

BMFT (1988), *Bundesbericht Forschung 1988*, Bonn, Bundesminister für Forschung und Technologie.

DFG (1983), *Aufgaben und Finanzierung 1983 bis 1986: Grauer Plan VII*, Bonn, Deutsche Forschungsgemeinschaft.

—— (1987a), *Perspektiven der Forschung und Ihrer Förderung: Aufgaben und Finanzierung VIII, 1987 bis 1990,* Bonn, Deutsche Forschungsgemeinschaft.

—— (1987b), *Denkschrift Astronomie*, Bonn, Deutsche Forschungsgemeinschaft.

—— (n.d.), *Deutsche Forschungsgemeinschaft: Organisation and Functions*, Bonn, Deutsche Forschungsgemeinschaft.

Dickson D., (1986), 'Germany's 75 Years of Free Enterprise Science', *Science*, vol. 234, pp. 811–12.

Faust, K. (1987a), *Früherkennung Technischer Entwicklungen auf der Basis von Patentdaten*, Munich, Institut für Wirtschaftsforschung.

—— (1987b), *Early Identification of Technological Advances on the Basis of Patent Data*, Munich, Institut für Wirschaftsforschung.

FCI (1986), *Schriftenreihe des Fonds der Chemischen Industrie*, Heft 27, Frankfurt, Fonds der Chemischen Industrie.

FhG (1986), *Jahresbericht 1985*, Munich, Fraunhofer Gesellschaft.

FKBW (1982), *Abschlussbericht*, Stuttgart, Forschungskommission Baden-Württemberg.

Franklin, J. J. and Johnston, R. (1988), 'Co-Citation Bibliometric Modelling as a Tool for S&T Policy and R&D Management: Issues, Applications and Developments', in van Raan (1988).

Geimer, R. and Geimer, H. (1978), *Science in the Federal Republic of Germany: Organisation and Promotion*, Bonn, Deutscher Akademischer Austauschdienst.

Griffith, B. C., Small, H. G., Stonehill, J. A. and Dey, S. (1974), 'The Structure of Scientific Literatures II: Toward a Macro- and Microstructure for Science', *Science Studies*, vol. 4, pp. 339–65.

Grupp, H. (ed.) (1987), *Problems of Measuring Technological Change*, Cologne, Verlag TÜV Rheinland.

—— (1988), 'Innovation Dynamics in OECD Countries by Technometric Analysis', paper presented at Workshop on Innovation at the Crossroads Between Science and Technology, Haifa-Jerusalem.

—— and Hohmeyer, O. (1986), 'A Technometric Model for the Assessment of Technological Standards and their Application to Selected Technology-Intensive Products', *Technological Forecasting and Social Change*, vol. 30, pp. 123-37.

—— and —— (1987), 'The ISI Approach to Technometrics', in Grupp (1987), pp. 123–37.

—— and —— (1988), 'Technological Standards for Research-Intensive Product Groups and International Competitiveness', in van Raan (1988).

—— Hohmeyer, O., Kollert, R. and Legler, H. (1987a), *Technometrie — Die Bemessung des Technisch-Wirtschaftlichen Leistungsstandes*, Cologne, Verlag TÜV Rheinland.

—— Hohmeyer, O. and Schmoch, U. (1987b), 'Indikatoren zurFrüherkennung von Entwicklungslinien im Entwicklungs- und Innovationsprozess', *FhG-Berichte*, vol. 1, pp. 24-7.

——and Legler, H. (1987), *Spitzentechnik, Gebrauchstechnik, Innovationpotential und Preise*, Cologne, Verlag TÜV Rheinland.

—— and Schmoch, U, (1987), 'Early Indicators for Measuring International Technology Transfer', paper presented at Workshop on Indicators of International Technology Transfer, Evanston.

Irvine, J. and Martin, B. R. (1984), *Foresight in Science: Picking the Winners*, London, Pinter Publishers.

Lübbe, H. (1974), 'Science Policy in the Federal Republic of Germany', *Research Policy*, vol. 3, pp. 172–80.

Michaelis, A. R. (1981), 'The Recovery of Science in Germany', *Interdisciplinary Science Reviews*, vol. 6, pp. 283-311.

MPG (1984), *The Max-Planck Gesellschaft and its Institutes: Portrait of a Research Organization*, Munich, Max-Planck Gesellschaft.

MWK (1987), 'Forschung Baden-Württemberg 2000', in *Gründe und Hintergründe*, pp. 6–9, Stuttgart, Ministerium für Wissenschaft und Kunst, Baden-Württemberg.

—— (1988), 'Perspektiven für die Forschung in Baden-Württemberg', *Mitteilungen über Wissenschaft und Kunst*, vol. 1, p. 13, Stuttgart, Ministerium für Wissenschaft und Kunst, Baden-Württemberg.

Narin, F. and Olivastro, D. (1987), *Identifying Areas of Strength and Excellence in FRG Technology*, report to BMFT, Haddon Heights, NJ, CHI Research.

Neffe, J. (1986a), 'Ethologist Takes the Helm', *Nature*, vol. 319, p.166.

—— (1986b), 'Student Squeeze in West Germany', *Nature*, vol. 324, p.200.

PERA/OTIS (1989), *BMFT Takes Stock of Achievements in Materials Research*, London, Overseas Technical Information Service.

Schmoch, U., Grupp, H., Mannsbart, W. and Schwitalla, B. (1988), *Technikprognosen mit Patentindikatoren*, Cologne, Verlag TÜV Rheinland.

Small, H. G. and Griffiths, B. C. (1974), 'The Structure of Scientific Literatures I: Identifying and Graphing Specialties', *Science Studies*, vol. 4, pp. 17–40.

Van Raan, A. F. J. (ed.) (1988), *Handbook of Quantitative Studies of Science and Technology*, Amsterdam, Elsevier.

Von Massow, V. (1986), *Organization and Promotion of Science in the Federal Republic of Germany*, Bonn, Inter Nationes.

Weingart, P., Sehringer, R. and Winterhager, M. (1988), 'Bibliometric Indicators for Assessing Strengths and Weaknesses of West German Science', in van Raan (1988).

Wissenschaftsrat (1988), *Empfehlungen des Wissenschaftsrates zu den Perspektiven der Hochschulen in den 90er Jahren*, Cologne, Wissenschaftsrat.

5 Research foresight in the United States

1. INTRODUCTION

Any analysis of foresight in the United States must begin by stressing the widespread scepticism if not antipathy existing towards centralized government planning. One result is that there is 'no coherent policy or integrated plan' for federal research (GAO, 1987, p.15). Instead, a philosophy of pluralism prevails, reflecting the view that choices concerning the distribution of public funds are intrinsically difficult and mistakes can best be avoided by having several centres of decision-making. Certainly, US basic research is a diverse enterprise, the parts of which operate largely independently, with funds being disbursed by federal agencies and departments having very different missions and organizational structures (Teich, 1986a).[1] Although the Office of Science and Technology Policy (OSTP) is theoretically in a position to coordinate policy, its influence has in practice been limited. Furthermore, even in the annual budget-setting process, neither in the Office of Management and Budget (OMB) nor in Congress do any procedures exist by which federal research priorities can be determined:

> The organization of ... OMB, of the budget itself, and of the congressional appropriations committees creates certain obstacles to the transfer of funds among scientific fields and projects; obstacles that inhibit making explicit trade-offs among projects and fields of science. [Baldwin and Hill, 1988, p.13]

Consequently, the 'shape' of the overall federal research effort only emerges through an adversarial process involving funding agencies, the Executive Office, Congress and the scientific community.

After outlining the structure of federal R&D, we first describe the efforts at national priority-setting by OSTP and lower-level foresight initiatives by the Office of Technology Assessment (OTA). Next, foresight relating to basic research is considered including exercises by the National Academy complex,[2] the National Science Foundation (NSF) and the National Institutes of Health (NIH). This is followed by an examination of foresight focussing on strategic research and basic technology, particularly in the Departments of Energy, Commerce and Defense. Finally, after looking at the activities of several consultancy firms,[3] we analyse the conclusions to be drawn from US experience with foresight.

2. STRUCTURE AND ORGANIZATION OF FEDERAL R&D

The structure of the federal R&D system is summarized in Figure 5.1. The top

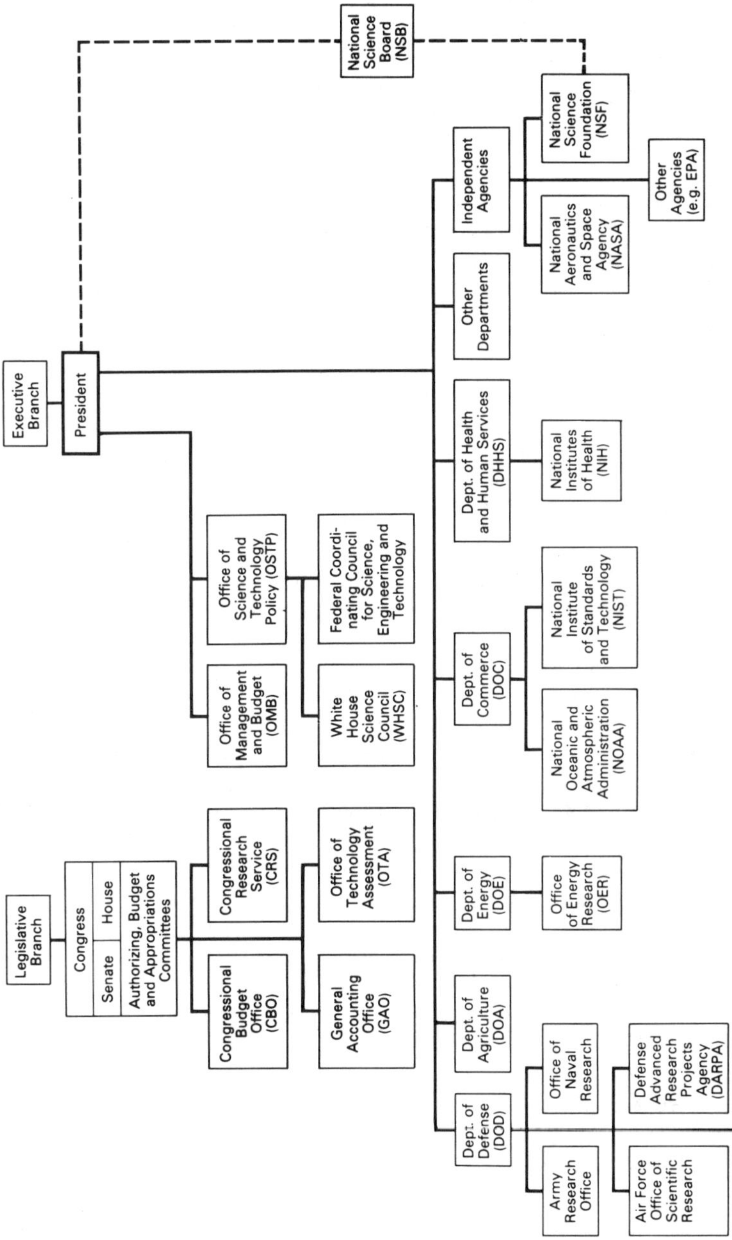

Source: Constructed from various federal science-policy documents, especially GAO (1987).

Figure 5.1 Outline of the US federal R&D system

six funding agencies — the Departments of Health and Human Services, Defense, Energy and Agriculture, and NSF and the National Aeronautics and Space Administration — accounted for 96 per cent of the $56,000 million federal R&D budget in 1987.[4] NSF mainly funds basic work in natural science, engineering and social science, together with some generic research not supported by other agencies. Its research budget ($1,500 million in 1987 — NSF, 1987a, pp.24–5) is, however, less than one third that of NIH ($5,600 million) which finances biomedical and health sciences in universities as well as its own laboratories. The other large agencies also support basic research, providing grants to academic scientists and financing seventeen federally funded R&D centres (FFRDCs) in cooperation with universities, and several others jointly with companies.

Within the Executive Office, the two bodies in a position to shape federal research policy are OMB and OSTP. The main function of the former is to prepare an annual budget on behalf of the President. Because this is done on an agency-by-agency basis, R&D is not treated as a whole during the budget-setting process,[5] agencies with research responsibilities coming under six different OMB divisions (Baldwin and Hill, 1988, p.13). Only after agency programmes have been decided does OMB, as an accounting exercise, carry out a cross-cutting 'special analysis' of R&D commitments. This identifies any trends and policy shifts, and provides 'the nearest equivalent that exists in our system to an administration science policy' (Teich, 1986a, pp.30–1).

OSTP was established by Congress under the National Science and Technology Policy, Organization and Priorities Act of 1976. Although one of its functions should be to help OMB in reviewing agency research budgets, in practice it 'does not have the authority or the responsibility in the budget-making process that was intended' (GAO, 1987, pp.45–6). Because OSTP's role is to generate initiatives while OMB's is to control federal expenditure, this has in recent years given rise to tension between the two as budgets have become tighter, resulting in decreased collaboration (Teich, 1986a, p.34).

The function of the Federal Coordinating Council for Science, Engineering and Technology, which was also set up in 1976, is to assist OSTP with policy issues affecting more than one agency. Most of its work is carried out through inter-agency panels, some of which (for example, those on geosciences and advanced computer research) have engaged in a limited amount of foresight. The White House Science Council also offers guidance to the Presidential Science Advisor on changing research-policy needs.[6] In addition, the remit of the National Science Board (NSB) includes not only providing policy advice and direction to NSF, but also assisting in the formulation of federal science policy.[7] Historically, it has concentrated its planning efforts on the former, although a 1981 report called for a greater involvement in framing national policy (GAO, 1987, pp.47–8), a recommendation recently repeated by a task-force reviewing the prospects for NSF in the 1990s (Goodwin, 1988, pp.49–50).

Like the Executive Branch, Congress does not have a 'federal R&D budget', research coming under numerous House and Senate committees. The review and approval of agency submissions involves separate budget, appropriations and authorizing committees.[8] In the second and third of these, responsibility for research is divided among several subcommittees, and science and technology are forced to compete with other items like housing.

In each case, the research 'cake' is sliced up differently and there is little coordination among the relevant subcommittees.[9] With research policy being formulated in two chambers and three sets of committees, 'Congress must be understood as a complex set of interacting power centres ... [It] is fragmented by nature — and by design — in the way it treats issues, including basic research funding' (Teich, 1986a, pp.37 and 47). Coupled with the constraints imposed by the annual budget cycle, this limits the scope for oversight and strategic planning, although Congressional hearings and special inquiries[10] do address emerging longer-range issues. Unless Congress moves to a multi-year authorization cycle, as a recent report (GAO, 1987, p.33) suggested, the possibilities for global research foresight will remain limited.[11]

In overseeing federal research, Congress is assisted by four bodies — OTA, the General Accounting Office (GAO), the Congressional Research Service (CRS) and the Congressional Budget Office (CBO) (see Figure 5.1). CBO prepares occasional reports on federal R&D support including a recent assessment of the proposed Superconducting Super Collider (CBO, 1988). GAO, besides its primary responsibility for evaluating agency programmes, also performs government-wide R&D policy evaluations (GAO, 1987, p.51), but has not engaged in research foresight. OTA's role is to provide Congress with an early indication of the potential impact of new technologies. Finally, CRS carries out analyses and prepares Issue Briefs, several which have related to research. However, because of the Congressional preoccupation with the annual budget cycle, these tend to focus on important issues for the following year (for example, AIDS and high-temperature superconductors).[12]

There are various other components of the US research system which, although not part of government, nevertheless play a vital role in federal science policy. A major contribution is made by the National Academy complex through numerous studies conducted by its operating unit, the National Research Council (NRC). Previously, these studies were somewhat limited in scope, since most were performed in response to requests from federal sponsors. However, the academies have now begun to obtain discretionary funds from other sources, which can be used to initiate studies independently (GAO, 1987, p.52). Another key organization is the American Association for the Advancement of Science (AAAS).[13] Because of the complex federal budget process, AAAS produces an annual analysis of R&D funding broken down by agency, discipline and performing institution. It also convenes a colloquium to discuss the budget and associated policy issues. Only after this and the OMB 'special analysis', do government research priorities for the coming year become clear. In other words, US R&D 'policy' is largely arrived at *after the fact* through aggregating the policies of individual agencies, although certain broad guidelines (for example, the favourable treatment of basic research under the Reagan administration) are of course set at the top.

3. NATIONAL PRIORITY-SETTING

(a) *The Five-Year Outlook*

OSTP is officially responsible for the coordination, oversight and planning of science and technology (GAO, 1987, p.41). One of the initial tasks set by

Congress in 1976 was the preparation of an annual Five-Year Outlook on Science and Technology 'to identify ... situations and conditions which warrant special attention within the next five years to which American science and technology can contribute'. The aim was to force the Executive to take the question of longer-term research priorities more seriously and to provide a regular opportunity to cross-examine the President's Science Advisor. However, OSTP was never enthusiastic about such a task (not least because of staff constraints at the time) and in 1977 delegated it to the National Science Foundation.

NSF was similarly unenthusiastic, recognizing that it was in a 'no-win' situation — a thorough critical analysis might offend OSTP while a politically acceptable report would risk forfeiting the respect of the scientific community (Hahn, 1986, p.19). After lengthy negotiations, it was agreed that NSF would subcontract most of the work to the National Academies, the Social Science Research Council and the AAAS. The first two Outlooks (NSF, 1980a; 1981a) were produced in this way, and, perhaps as a result, both were too lengthy and 'academic' for most politicians or administrators to assimilate.[14] Furthermore, despite the Congressional mandate to set priorities, OSTP had effectively sidestepped this request, abdicating responsibility to NSF which lacked the authority to coordinate research spending and priorities across agencies. The first two Outlooks consequently had virtually no impact either in Congress or elsewhere.*

Faced with the evident lack of interest, Congress abolished the requirement for a separate Outlook in 1982, instead requesting OSTP to devote one chapter of what shortly became the Biennial Science and Technology Report[15] to the subject. Two Outlooks have since been prepared for OSTP by the National Academies' Committee on Science, Engineering and Public Policy (COSEPUP). Recognizing the unsuitability of the first two reports for Congressional needs, COSEPUP realized that it had to make the third briefer and more focussed.[16] After canvassing Academy members, the Committee chose ten topics (for example, surface science) and commissioned articles from academic and industrial experts. These were subject to thorough review, with two being rejected. An introductory chapter outlining general policy issues was also prepared, emphasizing that the disappearance of traditional disciplinary boundaries was making it increasingly difficult to determine which areas of science should have priority. As a result, the 'highest priority ... [should be] balanced strength and an insistence on the highest quality. We cannot choose selective excellence in science' (COSEPUP, 1983). In other words, the desire of Congress and others for sharper priorities was seen as neither desirable nor realistic. The fourth Outlook (COSEPUP, 1985a) was even briefer. Published as a chapter in the 1983–4 Biennial Science and Technology Report (OSTP, 1986), it consisted of abstracts of the fourteen Research Briefings conducted in 1983 and 1984 (see below), with a short discussion of selected issues in science and technology — for example, cooperative work across disciplines.[17]

How successful were the Outlooks? On the positive side, they represent one of the few attempts to develop a more integrated approach to science policy in the United States. As one official involved put it: 'In concept, the Outlook is a very interesting idea. They could have been very useful if the government reacted on the basis of analytic thinking. But they don't! The documents were virtually unused within NSF and even more so elsewhere'.* An inherent problem was the contradiction between the interests of those preparing the Outlooks, who sought to focus on scientific opportunities, and Congress which was more concerned with policy and priorities. In hearings before the House Committee on Science and Technology (1984), the Outlooks were much criticized.[18] In the view of OSTP, they were of little use to the Executive[19] and because of the lengthy review procedure the information they contained was bland and dated (Ling, 1984). A survey of members of Congress and others found there was a need for better summarizing and a clearer sense of priorities (Cort and Cort, 1984). As one witness pointed out, 'the average Member of the House spends eleven minutes a day reading, [so] I suspect that the first and second reports represented an undigestible mass of information' (Rettig, 1984, p.1).

The basic fault lay with the 1976 Act which was not sufficiently specific in relation to the Outlook and failed to provide OSTP with the motivation to do its job properly. Consequently, 'it is not surprising that OSTP ... sloughed its responsibility on to NSF and ... others' (ibid., p.2). Yet this gave rise to an extended and tenuous relationship between the question-asker (Congress), the question-and-answer manager (NSF on behalf of OSTP) and the question-answerer (the National Academies and other contractors): 'This latter group ... have no real understanding of what the Congress wants and, perforce, must impose their own sense of your intention on their work' (ibid., p.3). Hence, the Outlook never fulfilled its promise of offering a mechanism for regular, longer-term analysis of national research issues and identification of priorities. It had a negligible impact on federal decision-making despite successive attempts to to make it more policy relevant, and no further Outlooks are therefore anticipated.

(b) The Research Briefings

Criticism of the Outlooks and of the state of longer-term research planning more generally began to grow at the beginning of the 1980s (e.g. Staats, 1980; GAO, 1981). The Science Advisor and Director of OSTP, George Keyworth, realized that increased support for basic research depended upon the scientific community agreeing priorities in areas likely to have the greatest impact on industry and defence (Keyworth, 1983). He therefore launched a series of Research Briefings on fields judged to have particular promise. A total of thirty-six were prepared by COSEPUP and the National Research Council (NRC) between 1982 and 1987 (COSEPUP, 1985b; 1985c; 1986; 1988).

The approach adopted was rather *ad hoc* or even circular in that a field, in order to be considered in a Briefing, had first to be recognized as a 'priority'. The National Academies were initially asked to assist in setting priorities among scientific fields but refused, arguing this was the responsibility of OSTP and other government agencies. They agreed instead to identify (but not prioritize) areas having potentially high 'leverage' *within* individual fields. After discussions in the National Academy complex in 1982 on scientific needs and opportunities, COSEPUP prepared a list of thirty possible topics. Keyworth then reduced the number to twelve, of which COSEPUP took the top seven. A similar approach was adopted the following year, but in 1984 and 1985 OSTP adopted a more proactive role, adding several of its own topics.

However, in the next two years the procedure reverted to the Science Advisor selecting from a COSEPUP list.

Briefings were prepared by *ad hoc* panels serviced by NRC staff. These typically consisted of a dozen experts, large enough to ensure representativeness and credibility without being too unwieldy. Membership tended to be broader than that for the NRC field surveys discussed later, with most panels including several industrialists. At its first meeting, each panel considered background material and was briefed by experts and government officials. Discussions then concentrated on identifying promising research opportunities. As a guideline, panels were advised to devote 60 per cent of their attention to science and 40 per cent to potential applications—the latter being sufficient to ensure the interest of government without dominating the science-push aspect. Next, NRC officials or selected panel members prepared a first draft of 'no more than twenty pages' for discussion and revision. Once agreed, the draft was reviewed by COSEPUP. Dress rehearsals of the Briefings were also held before the Committee which on several occasions helped improve their quality.[20] Presentations were then made to OSTP and NSF, and subsequently to other agencies.[21] The resulting reports were widely circulated within government, industry and the scientific community.

According to some officials, the initiative represented the nearest approach to wide-ranging research foresight in the United States.* Certainly, it provided concise reviews of opportunities in key fields accessible to those without in-depth expertise but who nevertheless needed an idea of likely developments. Several succeeded in drawing attention to emerging areas and in building a sense of community among researchers, while others improved the dialogue between scientists and policy-makers. In a few cases, Briefings had an immediate impact on funding (for example, in the hazardous chemicals field). To this extent, the initiative can be judged reasonably successful, especially in the light of the modest expense (an average of approximately $25,000 each, excluding some staff costs).

Other interviewees, in contrast, were more critical, pointing out that the excitement generated by the early Briefings soon faded. As the list of subjects covered grew, so they became more narrow. Whereas those in 1982 included mathematics and materials science, the three chosen for 1987 were 'order, chaos and patterns', 'biological control in managed ecosystems' and 'chemical processing for information storage and handling' (COSEPUP, 1988), with high-temperature superconductivity only being added later. This narrowing in focus meant that the end-product perhaps became too sophisticated for a lay audience in Congress and elsewhere. Furthermore, the Briefings lost their 'product champion' as Keyworth's enthusiasm diminished and his interest moved on to the space station and SDI, a trend which continued under his successor.[22] However, the main criticism is that, like most US foresight efforts, the Briefings shied away from the task of identifying inter-field priorities. When introducing the 1984 report, for example, the Chairman of COSEPUP explicitly emphasized that:

> presentation of these nine briefings ... does not imply that other topics are considered to be of lesser importance to the progress of US science and technology. *No priority setting among fields ... is implied* ... [Instead, briefing panels concentrated on] identifying those research areas *within* the field likely to yield the highest *scientific*

dividends as a result of incremental federal investment. [COSEPUP, 1984, pp. v–vi, emphasis added]

Consequently, it is debatable whether OSTP was correct in assuming that the Briefings fulfilled the requirement for foresight and priority-setting set out in the 1976 Act which envisaged a more central role for the Executive in long-term research planning (Teich, 1986a, p.31). With the demise of the Five-Year Outlook and the increasingly specialized field-based approach of the Briefings ruling out any possibility of inter-field comparisons, OSTP never succeeded in establishing a structure for agreeing overall priorities (GAO, 1987, p.46; Knezo, 1988, p.33).[23]

In recent years, OSTP's task has been made more difficult by a general decline in its authority. As part of the Executive Office, its influence depends on the relationship between its Director and the President. During the late 1970s, Frank Press established close working links with President Carter, White House staff and OMB, and this is credited in part for the support given to basic research (GAO, 1987, p.43). Keyworth was also successful in this respect during the first Reagan administration and thus able to launch initiatives like the Engineering Research Centers. However, during Reagan's second term, OSTP's position was downgraded and the Director's access to the President curtailed (Knezo, 1988, p.44). Suffering from weak leadership, it now has limited influence within the Executive Office and with agency heads (GAO, 1987, p.43). Yet, at the same time, OSTP has become identified with certain specific interests of the President such as SDI, and has as a result lost credibility with the research community (Knezo, 1988, p.49). If anything, it is therefore now in less of a position to engage in holistic foresight or to set priorities for federal R&D than ten years ago.

(c) Office of Technology Assessment

The task of OTA is to respond to requests from Congress for 'objective analyses of major public policy issues related to scientific and technological change' (OTA, 1987, p.3). Its 100 professional staff (drawn from social sciences as well as scientific disciplines) are responsible for planning and executing studies. However, they make extensive use of expertise in industry, universities and public interest groups, with panels helping to define studies, monitor progress and review draft reports (OTA, 1987, pp.3 and 8). The two main approaches employed are case-studies and workshops, but limited use is sometimes made of computer-modelling and scenario-building (Procter, 1987, pp.7–9). Although most OTA exercises are primarily concerned with technology assessment, they frequently contain a research foresight element. In addition, a recent review of R&D evaluation techniques included consideration of the role of forecasting and strategic planning in government decision-making (OTA, 1986b, pp.69–72).

To illustrate the OTA approach, let us consider a report on new materials which was rather more oriented to research foresight than normal. In 1985, the Senate and House of Representatives requested an analysis of the military and commercial opportunities opening up in the field and recommendations on R&D priorities (OTA, 1986a). The study was carried out by an OTA team aided by an external advisory panel, and drew on the results of four workshops, each attended by five to eight experts. Such workshops have been described as:

> an intense, activist form of Delphi exercise in which all the expert opinions are compared and analysed and mutually understood. No consensus is necessarily sought but a thorough understanding of the basis of agreement or disagreement ... Workshops can be used to force experts to try to understand the future. [Procter, 1987, pp.7–8]

Besides fieldwork undertaken by in-house staff, twelve contractors were commissioned to help define the issues. In line with normal practice, OTA officials prepared the draft report, the panel giving advice on interpreting data and suggesting where to obtain further information. As well as analysing opportunities in structural ceramics and polymer matrix composites over the next twenty-five years, the report identified three categories of R&D priorities: (i) 'very important' (for example, impact resistance); (ii) 'important' (for example, tribology); and (iii) 'desirable' (OTA, 1986a). After extensive review, the report was submitted to OTA's Congressional Board.

Several features distinguish this report from most other US foresight studies. First, it squarely addressed the issue of longer-term priorities. Second, it achieved a good balance between science-push and demand-pull factors, largely because industrialists were well represented within the advisory panel and among workshop participants, contractors and reviewers. Third, the analytical basis was comparatively strong, reflecting the widely respected professionalism of OTA staff; for example, projections were made of when new technologies (such as bioceramics) would be implemented and the likely size of markets (ibid., pp.24–5).[24] The materials study is, however, somewhat unusual even for OTA which, given the preoccupation of Congress with annual budget-setting, is not always able to devote so much attention to the issue of longer-term priorities.

4. FORESIGHT IN BASIC RESEARCH

(a) The National Academy complex

The National Academy of Sciences (NAS), National Academy of Engineering (NAE) and Institute of Medicine (IOM) are first and foremost honorific bodies. However, they also supply scientific and technical advice to government, mainly through the National Research Council (NRC), which provides a mechanism for calling upon academy members and other experts to take part in the 300 or so studies carried out each year. Separate from NRC but again reporting to the three academies is COSEPUP, which in 1981 replaced the former NAS Committee on Science and Public Policy (COSPUP).[25] The Committee's remit is to respond to requests from OSTP, the National Science Board, agencies and Congress for information and advice relating to the health of US science and technology (COSEPUP, 1987).

The NRC studies involving most research foresight are the so-called 'field surveys' — assessments of individual disciplines and the promising areas within them. Beginning in the early 1960s, a dozen or so reports on various physical sciences were produced by COSPUP in the period up to 1974.[26] Since NRC assumed responsibility, however, coverage has been extended to engineering, life sciences and even social sciences. Reports published in the 1980s include those on astronomy (Field,[27] 1982), materials research facilities (Seitz and Eastman, 1984), mathematics (David, 1984), chemistry (Pimentel, 1985), physics (Brinkman, 1986), engineering (Puckett, 1987), behavioural and social sciences (Gerstein *et al.*, 1988) and chemical engineering (Amundson, 1988), while those on biology and on materials science and engineering had still to appear in mid-1988.[28] The approach employed can be illustrated by focussing on chemistry (widely regarded as one of the more successful surveys), engineering (with its more immediate industrial relevance), and behavioural and social sciences (because exercises in this area are rare).

In chemistry, it was clear by the early 1980s that the field had undergone a revolution in techniques, instrumentation and capabilities since the Westheimer survey in 1965. After a planning group had reported on the need for a new study and the likely cost, NRC obtained funding from a consortium of agencies, foundations and companies, and set up a committee in 1982 under the chairmanship of George Pimentel. An attempt was made to ensure that its twenty-six members were representative in terms of sectors (academia, industry and government), geographic regions and chemistry subdisciplines, but the main criterion was scientific merit.[29] The committee was charged with examining: (i) contemporary research frontiers; (ii) opportunities for chemistry to address societal needs; and (iii) the resources needed to exploit those opportunities (Pimentel, 1985).

After surveying several hundred chemists on potential topics to address, committee members organized themselves into five task-forces. These were deliberately *not* constituted on the basis of traditional subdisciplines, partly because such boundaries are irrelevant to policy-makers and partly because it might precipitate disputes among specialties or between strategic and more fundamental research. Instead, the (unnamed) task-forces were allocated responsibility for a number of topics identified in the survey. Each commissioned half a dozen or so eminent scientists to prepare brief reviews of intellectually exciting opportunities, the potential impact of research, and the resources and instrumentation required. These were then synthesized into a draft report, which was circulated for critical review.[30] Where gaps were identified, experts were commissioned to provide additional material. The report finally appeared three years after the study began and at a cost of over $600,000. A further year or more was devoted to dissemination, with illustrious chemists appointed to lead regionally-based efforts to ensure that the findings were diffused to politicians, industrialists and university presidents.[31]

No use was made in the survey of formal foresight techniques, the only empirical data reported being economic statistics employed to demonstrate the importance of the chemical sector. Although the three central chapters of the report are structured around the different types of impact made by chemistry (for example, on energy and materials), no quantitative assessments were made of likely effects, and the prospective analysis was confined to general research trends over the next decade. Unlike many other field surveys, the Pimentel report did identify a small number of priority areas like chemical reactivity and catalysis. However, in common with most it concluded that federal support should be doubled over the next three or four years (ibid., pp.14–15). While the attempt to set priorities was widely commended, the recommendations on funding were regarded as politically unrealistic. *

A broadly similar approach was adopted in the engineering survey, which aimed to identify research priorities with 'a strong potential to improve the industrial competitiveness, quality of life, and/or national security of the United States' (Puckett, 1987, p.25). Commissioned by NSF, it was carried out by a board of fifteen, aided by seven panels. Industrialists were well represented, making up a third of board and panel members. The panels were again not established along disciplinary lines but on the basis of engineering systems (for example, manufacturing systems). Altogether, they identified about eighty promising topics. The highest priorities from each panel's list were then presented to the board for discussion. Fourteen 'research areas of current opportunity' were finally selected, including three of 'especially high priority' — complex system software, advanced engineering materials and manufacturing systems integration — although the board refused to rank the remainder (ibid., p.26). Officials praised the study for emulating the Pimentel report in identifying priorities (although some felt that the most novel ideas had been 'squeezed out' during the exhaustive review process).* Scepticism was, however, again voiced about the funding recommendations—for example, that NSF should increase the annual budget of its Engineering Directorate from $150 million to $400 million by 1990 (ibid., p.6).

The survey on behavioural and social sciences was conducted by an NRC committee which had already produced a report assessing 'the value, significance and social utility of basic research' in the area (Adams *et al.*, 1982, p.v), and was in the process of preparing a second summarizing achievements over the past fifty years (Smelser and Gerstein, 1986). In 1983, concern about cuts in federal funding of social science led NSF to request the committee to extend its terms of reference and help define future priorities. The first step was to conduct a survey of 2,400 researchers to identify possible topics and participants. From the replies, the committee selected thirty-one cross-cutting themes (for example, internationalization). Interdisciplinary working groups were convened to assess research opportunities and needs, preparing concise overviews of specific areas (Gerstein *et al.*, 1988, p.xiii). The final report drew heavily upon this material and reviewers' comments, but was largely drafted by the co-chairmen of the committee and NRC staff. The study took four years to complete and cost over $800,000.

Yet again, the report recommended a large increase in funding, in this case by $240 million per annum over the following three to four years — similar in magnitude to total federal expenditure on basic social-science research (ibid., p.249). Furthermore, although it listed some areas of interest (with the implication that those not mentioned were less important), no clear priorities were identified. Little attention was given to future socio-economic considerations, the emphasis instead being on setting out the intellectual opportunities for curiosity-driven research.[32] It was claimed that the diverse range of disciplines involved made it difficult if not impossible to reach consensus on priorities — or even on the criteria to be employed.*

Although we shall not examine the other field surveys in detail, it is worth noting how success in identifying research priorities has affected subsequent impact on policy. The astronomy study is acknowledged to have been especially influential, having set out clear criteria for scientific choice and ranked in order a list of potential programmes (Field, 1982, pp.14–17). One observer commented that this had involved 'blood on the floor' sessions, but the end-result was that astronomers largely succeeded in obtaining what they wanted from government. Likewise, the materials research panel ranked nine new or improved facilities, although the task was found 'difficult and painful' (Seitz and Eastman, 1984, pp.iv and ix). As shown below, the Department of Energy was consequently more ready to implement the recommendations.

A marked contrast is provided by the physics report. Despite the introduction promising 'a brief discussion of priority setting ... in Chapter 3' (Brinkman, 1986, p.4), none appears, the authors choosing to ignore this aspect of their remit just like earlier physics surveys (Pake, 1966; Bromley, 1972).[33] This made the DOE's task in taking up the recommendations much harder than for materials research facilities. Physicists can, however, claim that their report at least set out future research opportunities, something which not all surveys have managed. The mathematics report, for example, instead concentrated on arguing that the subject provides the foundation for all other sciences yet receives less than half the federal funds required (David, 1984).[34] Perhaps the most extreme case of failing to confront the priority question, however, is a study of geosciences which merely produced a description of *all* earth science, with no discrimination as to currently important areas, let alone those offering the most exciting future opportunities.* Regarded as probably the least successful field survey, it had no policy impact and appears not even to have been published.

On the positive side, the field surveys have clearly served several purposes. First, they have provided an opportunity for researchers to think systematically about prospects for their field and to see developments in narrow specialties in a wider context. Second, by involving large sections of the community, they have fostered a sense of identity and common purpose. Third, they have in some cases forced scientists to confront policy issues and wrestle with priorities, with those not shying away from the 'painful' challenge of arriving at consensus often being rewarded with increased funds. Even when priorities have not been explicit, lists of promising opportunities have still proved useful to agency officials in defending programmes against the threat of cuts. Last, some, like the Pimentel report, have provided a platform from which to educate Congress, industry and the academic community on the potential benefits from research, helping to develop a constituency to lobby for resources.

Yet, within agencies and elsewhere in government, NRC field surveys are seen as suffering from the same limitations as those conducted earlier by COSPUP. First, many are regarded as little more than special pleading; what starts out as an attempt to define longer-term policy and priorities is all too often 'hijacked' by the research community to advocate a doubling of its budget over the next few years. Second, the approach involved is rather cumbersome, time-consuming (partly because of the heavy reliance on volunteers) and costly. Third, reports are often long, academic and inaccessible to outsiders, although NRC officials claim the situation has now improved.†

A fourth criticism concerns the emphasis on science-push to the virtual exclusion of demand-pull considerations. In the case of physics, for example, the survey committee contained just one industrialist despite the longer-term commercial potential of many specialties in the field. Related to this is the fact that formal foresight or strategic analysis techniques have generally been ignored. Instead, studies have relied on the informed but nonetheless subjective judgements of experts.

Finally, most surveys have avoided identifying explicit priorities.[35] No consideration has been given to what might happen should federal research funding cease to grow, let alone decline, perhaps because of a fear of generating self-fulfilling prophecies, but also because of the institutional problem associated with the studies being carried out under the auspices of the National Academy complex. On the one hand, the National Research Council is at least partly beholden to the Academies — honorific societies concerned about their reputation among researchers. On the other, NRC must respect its charter in responding to government requests for science-policy advice. In the inevitable tension between the two sets of demands, the latter is more likely to suffer.[36] The result is that even those surveys which have identified priority areas meriting increased support have shied away from pointing to declining areas where commitments might be reduced, since this would obviously offend sections of the research community.[37]

(b) National Science Foundation

NSF is the only federal agency with the primary mission of financing basic science, although since 1968 it has also been charged with certain responsibilities for applied research (NSF, 1987b). Organized into directorates and divisions on the basis of disciplines, it covers most areas of natural science, engineering and social science. One third of the programme officers are 'rotators' on secondment from universities, national laboratories or companies. Although forbidden to operate its own laboratories, NSF supports several central facilities managed by consortia of universities. However, most funds are allocated through peer-reviewed grants in a process which:

> involves both 'bottom up' and 'top down' decision-making. From the bottom, demands of various areas are weighed against one another by NSF management, with oversight by [NSB], OMB and Congress. From the top, administration and Congressional policy priorities are factored into funding decisions at the macro-level, as, for example, several years ago, when the White House moved to ... de-emphasiz[e] the social and behavioral sciences. [Teich, 1986a, p.8]

In practice, the distribution of resources among directorates has changed little over time, the policy being more one of incremental adjustment of historical shares in gradual response to proposal pressure rather than attempting to set major new priorities (Healy, 1987, p.9). The exception is when additional funds are made available through initiatives like the Science and Technology Centers programme discussed below. Being primarily concerned with basic research, NSF sees its mission as funding good work in *all* areas rather than just in priority fields.[38]

Although divisions informally attempt to anticipate where research is heading, NSF staff are mostly sceptical about formal foresight techniques, perhaps reflecting previous experience.[39] In 1977, three approaches to identifying research areas with long-term promise were explored in studies commissioned from consultancies. However, the experiment proved singularly unsuccessful (Irvine and Martin, 1984, pp.76–83). Two years later, the first in an annual series of Status of Science Reviews was produced in response to a request from the National Science Board for information to be used in long-range planning (NSB, 1979). The report consisted of separate sections prepared by NSF divisions summarizing the judgements of programme directors on current directions and opportunities. It contained nothing in the way of synthesis, even though an explicit concern of the Board was to draw comparisons among fields. Five further Reviews appeared subsequently, but again none contained any overview (e.g. NSB, 1983). Although initially regarded as a useful statistical compilation, the Reviews became progressively more fragmented and in 1984 NSB decided to abandon the exercise.

Three years later, however, following President Reagan's acceptance of the proposal to double NSF's budget over a five-year period, interest in longer-term planning was revived[40] and a strategic plan was drawn up for 1988-92 showing how the additional funds might be deployed. NSF directorates prepared short summaries of current expenditure and longer-term goals, which were then combined in an overall Long Range Plan (NSB, 1987). Despite recognizing that 'priorities among disciplines ... will have to be set' (ibid., p.8), the plan contained only a long list of 'illustrative' research opportunities rather than an ordered set of priorities.[41] Given the prevailing federal budget constraints, the impact of the exercise is likely to be limited.

The traditional planning philosophy of NSF is also reflected in the two major initiatives embarked upon since 1983 — the Engineering Research Centers (ERCs) and Science and Technology Centers (STCs). Both presented opportunities for employing foresight to identify areas of national importance in which centres might work, but in neither case has this occurred. The ERC programme is designed to stimulate multidisciplinary engineering research in fields relevant to future industrial competitiveness,[42] the initiative reportedly stemming from a meeting in the Presidential Science Advisor's office to discuss the Research Briefing on computing in design and manufacturing (NRC, 1986, pp.v–vi). Concern about the state of academic engineering (particularly the limitations imposed by traditional departmental structures) led to the proposal to establish cross-disciplinary centres focussing on the next generation of technological advances, and integrating research with education. Such centres would also actively involve companies in planning, research and training, and raise substantial external funds from industry and other sources.

Following recommendations by the National Academy of Engineering (NAE, 1983), it was decided not to preselect fields. The first programme announcement by NSF merely listed seven illustrative examples identified by NAE (such as computer-integrated manufacturing — see NSF, 1984, p.1), the intention being to leave the choice of areas to researchers. However, almost all the initial proposals fell within these fields, so no guidance was given in the second and third years.[43] Topics were suggested in the next year to fill 'gaps' in the range of existing centres, but these were subsequently dropped. NSF has concluded from experience that the teams of academic and industrial engineers who submit proposals are closer to the technological frontier and hence in a better position to identify priorities. Consequently, instead of attempting to target areas, NSF has fallen back on its traditional reactive mode and reliance on peer-review.[44]

The establishment of STCs was first proposed in the Packard–Bromley report (1986, p.5) on the health of universities, and was taken up by NSF following the promise to double its budget. The National Academy of Sciences was invited to provide advice on the nature and management of the new centres. An expert panel was convened which 'attempted to identify areas ... that seem most promising in relation to economic competitiveness' (NAS, 1987, p.viii). However, given the difficulty of forecasting more revolutionary advances, the panel 'quite deliberately chose not to prejudge which areas might be particularly relevant to economic competitiveness' (ibid., p.ix), a course which NSF has begun to follow when the necessary funds were eventually made available. As with ERCs, therefore, a responsive approach is being employed rather than a proactive strategy of preselecting priority fields on the basis of foresight.

(c) *National Institutes of Health*

The mission of NIH is to carry out biomedical research aimed at improving the nation's health (NIH, 1987, p.2), and its budget represents nearly 10 per cent of federal R&D. Although the heads of the twelve component institutes (and various other units) report to the NIH Director, they enjoy significant autonomy in respect of both budgets and research policy. A typical institute supports both intramural and extramural programmes, the latter determined primarily by unsolicited proposals which are evaluated in terms of scientific quality and mission relevance (Teich, 1986a, pp.3–5).

The allocation of funds among institutes is largely determined by Congress and reportedly depends more on the influence of powerful health lobbies and the concerns of members of Congress (such as their worries about cancer and ageing) than on any systematic assessment of national health needs.* While the Director's Office coordinates the annual budget-setting process[45] and can modify proposals *prior* to Congressional decisions, it has no power subsequently to redistribute funds among institutes. Hence, although acting as a catalyst in establishing shorter-term NIH priorities (Moskowitz, 1988, p.2) and overseeing cross-cutting policy issues, it would be wrong to think of the agency as having a single centre of strategic planning.[46] Certainly, the Director's Office makes no attempt to engage in longer-term research foresight and staff appear to have little sense of what are likely to emerge as the main NIH-wide priorities over the next decade. Indeed, when asked to predict what would be the most important new areas in ten years' time, a senior official replied, 'I don't find that a helpful question'.* Responsibility for longer-term planning and priority-setting instead rests mainly with individual institutes, which undertake most of the research foresight within NIH. In what follows, we consider the efforts of two institutes, the National Heart, Lung and Blood Institute (NHLBI) and the National Eye Institute (NEI).[47]

As early as 1972, Congress mandated the predecessor of NHLBI to develop a national research plan. Over 200 medical and scientific advisers participated in an exercise to assess emerging health problems and identify research opportunities. They divided the field into twenty 'categories' (for example, hypertension) and for each a panel examined the current state-of-the-art, short-term needs (where the basic knowledge was already in place) and longer-term goals (requiring new knowledge). Panels were provided with briefing documents (for example, recent scientific reviews and quantitative material prepared by economists, biostatisticians and demographers). From each panel's discussions, some consensus on priorities emerged, enabling a draft report to be prepared. Panel recommendations were then combined to yield a five-volume report setting out a plan for the next decade (NHLI, 1973). This was later updated annually, with more detailed reviews every five years.

The research plan forms one input to the annual budget-setting process, which begins with officials in the three NHLBI divisions analysing the structure of existing funding. Advisory groups then assess whether changes should be made in the light of new needs in each of the twenty categories employed in planning, and agree their recommended lists of priorities. Next, at what was described by one interviewee as a 'tooth and claw' meeting, the lists for each division are narrowed down and combined. Finally, after meeting with divisional heads, the institute director produces an overall set of priorities, taking into account such criteria as timeliness, scientific merit, feasibility and resource requirements. The time-horizon has, however, now been reduced to five years following a review of the planning process a decade after the initial exercise (NHLBI, 1984). In other respects, the process was judged reasonably successful in that appreciable progress had been made in most categories, with bibliometric data suggesting that targeted areas had advanced faster than others.

The National Eye Institute was similarly mandated to produce long-range plans, and to date these have covered the periods 1976–9, 1978–82 and 1983–7.[48] The first was prepared by the National Advisory Eye Council together with senior institute staff and leading researchers, identifying twenty-four areas of need and opportunity (NAEC, 1976). Encouraged by the favourable response it received, the Council embarked on a second and much larger planning effort in which 150 experts participated (NAEC, 1977). The third exercise in 1982 was yet more comprehensive, involving some 350 people. Aided by NEI staff, six advisory panels reviewed the current state of programmes and identified future opportunities, drawing upon the advice of clinicians, researchers and lay groups (such as the blind). Two sets of research areas were distinguished — the 'program base' where current funding levels should be maintained, and 'program development priorities' warranting new or increased support (NAEC, 1983). Panel reports were then submitted to the Advisory Council, which integrated the recommendations and converted them into budgetary estimates.

The planning process in both institutes consequently relies on expert consensus rather than formal foresight techniques, although some use is made of statistics on morbidity and treatment costs. In addition to helping set longer-term priorities,[49] the plans serve several secondary purposes: (i) fostering communication among institute staff and advisers; (ii) diffusing information within the community and inviting proposals in identified areas; and (iii) providing a promotional tool for the wider public, thereby ensuring that groups lobbying Congress put across a consistent message and one which, as a result, is more likely to be well received.

Besides these institute-wide activities, NIH also engages in more specific foresight studies, for example, on diabetes and atherosclerosis (Moskowitz, 1981, pp.269–75). In addition, there is a requirement that 1 per cent of its budget be devoted to evaluation, and this has financed studies by organizations like CHI Research attempting to identify leading-edge research meriting increased support (Healy, 1987, p.21). The reaction of the scientific community to such assessments has, however, been rather mixed.* Thus far, NIH has not engaged in systematic monitoring of overseas research, perhaps because of its previous dominance of many biomedical specialties.

Overall, NIH's approach to longer-term planning can be characterized as a responsive bottom-up process: research on AIDS, for example, began within an existing intramural programme rather than as a targeted priority. The limited reliance on foresight reflects both the relative generosity of Congress over recent years (rendering the need for priorities less immediate) and the autonomy of institutes. Although the division of NIH into disease-based institutes may be essential for the Congressional budget-setting process (Teich, 1986a, p.6), it has appreciable disadvantages from the point of view of research and strategic planning. Constrained by existing institute boundaries and having a director with circumscribed powers, NIH inevitably faces problems in relation to new interdisciplinary initiatives like sequencing of the human genome.[50] Unless the director is granted greater powers and especially the discretion to shift resources between institutes — as advocated in a report by the Institute of Medicine (1984, p.5) — and unless NIH moves towards multi-year budgeting (thereby reducing its vulnerability to short-term political factors), the scope for effective development of foresight in setting overall research priorities seems likely to remain limited.

5. STRATEGIC RESEARCH AND BASIC TECHNOLOGY

(a) Department of Energy

Around 20 per cent of the $4,900 million spent on R&D by the Department of Energy (DOE) in 1987 was devoted to basic research, two-thirds of which financed work in multi-programme national laboratories. Although applied R&D is organized within divisions based on fuel form (fossil, nuclear, etc.) and subject to little overall coordination, basic research is mostly supported through a central Office of Energy Research (OER). Its major programmes include high-energy and nuclear physics, fusion, health and environmental research, and other 'basic energy sciences' necessary to fulfil the agency's mission. An Energy Research Advisory Board (ERAB) comprising academic and industrial experts provides guidance on overall policy, while external advisory committees assist in setting research priorities at the programme level, with peer-review being employed to evaluate individual proposals (Teich, 1986a, pp.10–11).

Since 1980, DOE has undertaken several foresight exercises initiated partly in response to political pressure. The agency's large array of expensive programmes attracted the attention of the incoming Reagan administration, in particular those involving more applied R&D which showed little evidence of recent progress and which clearly required restructuring. Efforts were therefore made to identify programme areas likely to be more (and less) strategically important over the next fifteen to twenty years. After a preliminary study by consultants proved unsuccessful, ERAB embarked on a major foresight exercise in 1981 which, using explicit criteria and a modified Delphi procedure, succeeded in generating clear priorities (ERAB, 1981; Irvine and Martin, 1984, pp.84–5). However, its impact was mixed. Although certain priority fields (like materials science) received increased support, other recommendations were ignored. For example, the attempt to curtail R&D on magnetohydrodynamics was overturned by the Senate, with the result that large sums continued to be spent on development activities at the expense of basic research, exactly the opposite of what ERAB intended. This illustrates how even the most rigorous foresight study can represent no more than one input to decision-making in a situation where wider political considerations may override 'rational' attempts to set priorities.

In 1984, the Secretary of Energy asked ERAB to conduct a similar Long-Term R&D Strategy exercise, examining energy needs over the next twenty to forty years and identifying 'those areas of energy R&D for which Federal investments are most necessary and appropriate' (ERAB, 1985b, Appendix A). Responsible for the study was a committee of twenty-four, nearly half drawn from industry. Four panels were set up to analyse energy supply, demand, research and 'infrastructure' (including management, organization and personnel). To provide a quantitative baseline for the forecasts, an energy consumption/price projection was prepared with upside and downside variations.

The main activity involved was a study-week in which ERAB members attempted to determine R&D priorities, as far as possible in quantitative terms using a voting procedure. However, unlike in 1981, the prioritization process broke down,[51] partly because discussions were not as effectively chaired but also because participants had apparently not read the briefing documents sufficiently thoroughly beforehand to enable a common set of basic assumptions to be agreed.* Time constraints unfortunately precluded making a second attempt, so the study committee decided to drop the idea of formulating priorities. (Since this meant abandoning the effort to develop a 'strategy', the name of the exercise also had to be changed to *Guidelines* for Long-Term Civilian R&D.) The final report proposed three broad national goals (one of which was to maintain a strong science base) but, as regards specific research areas, it merely reproduced a list of all the fields currently supported by DOE (ibid., p.4). No attempt was made to distinguish priorities, even though this had been part of the study's terms of reference.[52]

Perhaps unsurprisingly, the exercise's impact seems to have been limited. Although referred to on Capitol Hill and used for advocacy purposes by groups within DOE, the results were not employed in any systematic way in the budget-setting process.* Nevertheless, DOE continues to retain some faith in foresight, and in 1988 ERAB was in the final stages of a study to forecast R&D needs up to the end of the century. The approach involved brainstorming by three expert panels, but this time groups were forced by their chairmen to narrow down the list of issues identified to the most important three or four on the basis of an agreed set of criteria. (The procedure of gradually whittling down a list proved easier to operationalize than ranking items in order, as the 1985 exercise had attempted.) Following intensive debate at a plenary session, the lists were further reduced to one or two items per panel. The results are to be published in a report entitled *R&D Initiatives for Energy Competitiveness*.

Other foresight activities within DOE have related to the NRC field surveys. Given their politically unrealistic recommendations on research support, DOE decided to establish panels to review the three reports of most direct relevance — on materials research facilities, chemistry and physics. In the first, the success of the Seitz-Eastman survey in arriving at priorities enabled the panel to endorse the report's broad conclusions. However, some of the more detailed proposals were revised in line with the pragmatic principle that only one new facility should be allocated to each DOE laboratory, thus securing the support of as many laboratories as possible, together with their local Senators and Congressional delegations. The panel also recognized that not all the Seitz-Eastman recommendations could be implemented simultaneously with the resources likely to be available, and therefore developed a phased plan for constructing the facilities. The resultant report (ERAB, 1985a) was consequently rather more useful than the original NRC study and almost all its recommendations are being implemented.†

The review of the physics survey provides an interesting contrast. A panel of twelve (which included four non-ERAB members to augment the coverage of physics subfields) was set up to identify which of the Brinkman recommendations should be endorsed, modified or rejected by DOE (ERAB, 1987, p.1). However, the task was intrinsically more difficult than that facing the materials facilities panel. As the ERAB Chairman noted in his letter setting up the review:

> The NRC report does not assess priorities or make detailed funding recommendations. To conclude that all of the recommendations contained in the NRC report are so important that they must be supported without prioritizing would be of little use to the Department. [Schoettler, 1986]

After an initial briefing by DOE staff and other agency officials, six subpanels were given responsibility for preparing assessments of the main physics areas. These provided inputs to subsequent discussions of the full panel which attempted to arrive at overall priorities for the field. However, the exercise was little more successful than the Brinkman study. Although it managed to limit the number of proposed major new facilities to four (excluding the Superconducting Super Collider or SSC), the report merely listed these in order of suggested starting date rather than scientific importance (ERAB, 1987, p.iv). The estimated cost of implementing all the recommendations was again enormous, requiring a budgetary increase of 50 per cent in real terms (exclusive of the $5,000 million SSC) over a six-year period compared with the doubling over four years advocated by Brinkman. However, at ERAB's insistence panel members were forced to consider a zero-growth scenario. After what was described as a 'tortuous' series of meetings,* they reluctantly concluded that in such an eventuality the SSC should not be built and funds should be transferred from national laboratories to strengthen work in smaller university groups (ERAB, 1987, p.v). It is too early to judge the full impact of the panel's report, although the recommendation that the SSC should not proceed at the expense of other research has been accepted by both DOE and the Administration. However, there is some scepticism as to whether anything remotely approaching the funding increase proposed will be forthcoming.

(b) *Department of Commerce and National Institute of Standards and Technology*

The Department of Commerce (DOC) is among the smaller departments in terms of R&D, with a budget of just over $400 million in 1987. Besides being responsible for the National Institute of Standards and Technology and the National Oceanic and Atmospheric Administration, it is in principle the only federal agency with a sufficiently broad charter to address all aspects of economic competitiveness including the linkage between industrial policy and government support for innovation. Yet, according to certain critics, DOC has been 'the largest single defaulting agency in terms of a lack of research policy'.* During the Reagan administration, the dominant view within the agency was that a policy of targeting R&D support would be counter-productive (e.g. Merrifield, 1983; DOC, 1987b, p.3). The arguments against government priority-setting[53] include: (i) the failure of previous attempts (for example, DOE's costly programme on synthetic fuels during the 1970s initiated in response to exaggerated predictions for the price of oil); (ii) the fact that priorities inevitably become distorted by politics, with the consequence that money is wasted, disillusion sets in and cutbacks follow, the net effect being a series of 'stop-go' cycles; and (iii) the tendency among nations to target similar areas, resulting in an overconcentration on those sectors and a failure to obtain an adequate return on investment. Instead, emphasis has been placed on removing barriers to competitiveness, especially the anti-trust laws discouraging collaborative R&D (e.g. DOC, 1987a; n.d.).

Despite this opposition to R&D planning, DOC has recently become more interested in longer-term issues[54] and has been exploring the merits of requiring its agencies to produce an annual list of issues and objectives (for the next five years or so) relating to identified goals (DOC, 1984). The first two attempts at this were not very successful, with little integration being achieved between the various goals. The third, however, represented a marked improvement and reportedly proved useful to agencies in framing their budget proposals.*

A central problem confronting DOC in defining an overall strategy is the diversity of organizations for which it is responsible. Faced with the difficulty of deciding among competing claims for resources, the agency's Deputy Secretary set up several coordinating committees in 1986, including one which was asked to identify promising emerging technologies and then to rank them in priority order. The first step was for the staff officer responsible for the exercise (the head of planning at the National Bureau of Standards, NBS) to compile a preliminary list of technologies for consideration by the committee. This was deliberately kept short and at a high level of aggregation (for example, ceramics). The criterion for inclusion was likely impact on GNP by the year 2000, and this was also employed in ranking the technologies. The latter was undertaken by means of a Delphi exercise in which committee members agreed on an initial ranking, discussed it with agency colleagues and other experts, and after further debate re-ranked the technologies at the next meeting. The process was then repeated, yielding an ordered list of seventeen technologies. However, since some rankings were very close, it was decided finally to group technologies into three categories. Those judged likely to have the highest impact included advanced materials, genetic engineering and optoelectronics (NBS, 1987, p.8).

From early in the study, it was clear that the United States often lags behind other competitors, especially the Japanese, in exploiting R&D. The committee therefore examined the barriers impeding commercialization of new technologies, identifying and ranking ten factors such as the high cost of capital and tax regulations. The results were well received by the Deputy Secretary, who requested further information. After a short explanation of each barrier had been prepared, he next sought suggestions for overcoming them from a high-level DOC policy advisory group. Three brainstorming sessions were held, and a brief summary of the resultant recommendations was appended to the Emerging Technologies report.

The report was widely referred to in speeches by senior officials (e.g. Brown, 1987), generating much publicity and numerous requests for copies.[55] It also had a fairly immediate impact, with ceramics and automated manufacturing, for example, benefiting from increased support. Yet the report's importance should not be overstated. The effort involved (a committee of around ten experts meeting on eight occasions) was modest, the approach relatively unsophisticated[56] and the results unsurprising. In ranking technologies, no use was made of quantitative data on likely impact (for example, on size of markets), committee members instead relying on their collective expertise. Furthermore, soon after they had agreed on a list of important emerging technologies, it became clear that they had failed to appreciate the significance of high-temperature superconductivity.[57]

The other main research foresight undertaken within DOC has concerned the National Institute of Standards and Technology (NIST).[58]Divided into four laboratories (for example, the Institute for Computer Sciences and Technology) and with a budget of over $100 million, NIST's mission includes meeting the R&D needs of industry in relation to standards and production technology methods. This means that NIST has to initiate research in new areas perhaps six to eight years before firms seek assistance, and must therefore engage in long-range planning.

In 1979, Congress instructed NBS (as it then was) to set up a Planning Office which might take a longer-term policy focus and give greater emphasis to economic factors. Three plans were produced, each providing an assessment (based on the views of experts) of the likely supply and demand for R&D over the next twenty years. The second of these, for example, identified five priority areas such as advanced materials (NBS, 1981, p.v). The results provided a strategic context for budget-setting both within NBS and externally in DOC, OMB and Congress. The plans also helped foster communication between the NBS director and research groups. However, they were abandoned in 1984 after the third had fuelled OMB's concern about agencies using such documents to lobby for funds prior to completion of the budget approval process.[59*]

The Planning Office organized various more specific foresight exercises in the early 1980s, including an internal review of prospects for optoelectronics (Tassey, 1985). Studies were also commissioned from Ecosometrics (1984) and J. F. Coates Incorporated. The latter compiled an inventory of 200 scientific developments likely to affect NBS, categorized according to whether they involved: (i) research already being pursued; (ii) an incremental extrapolation from current NBS work; or (iii) a radically new departure (Coates, 1980). The approach was based on scanning the literature and discussions with experts, particularly editors of technical journals who are well placed to comment upon the interface between basic and applications-oriented research. Among the promising areas suggested for consideration were a number in biotechnology which were later subject to more detailed review (Coates, 1981). However, the findings reportedly had little impact on NBS programmes in this area.† Since the abandonment of the long-term plans, NIST has engaged in less foresight, reflecting DOC's lack of interest and OMB's opposition. Instead, it has reverted to a responsive mode, relying primarily upon researchers to propose new programme initiatives.

(c) Department of Defense

The Department of Defense (DOD) has by far the largest R&D budget of any federal agency ($36,000 million in 1987), although 90 per cent is devoted to development programmes. Basic research accounts for only $900 million, half going to seventy intramural laboratories and half to support work in universities and industry. Extramural programmes are managed by several essentially independent organizations — the Army Research Office, Air Force Office of Scientific Research, Office of Naval Research, and Defense Advanced Research Projects Agency (DARPA), with the Strategic Defense Initiative Organization now assuming an increasing role. The primary mechanism for distributing extramural funds involves grants and contracts awarded to researchers in response to unsolicited proposals, although general fields of interest are made known in advance. In 1986, the three services and DARPA launched the University Research Initiative which has considerably expanded DOD's academic programmes, especially in multidisciplinary areas (Teich, 1986a, pp.14–16).

DOD was responsible for much of the earliest systematic work on technological forecasting, financing several comprehensive exercises. For example, in 1963 the US Navy conducted a vast study which involved virtually every industrial contractor to the military, produced 13,000 pages of reports and cost $3.5 million. This examined what potential technologies might be expected from basic science, and how their subsequent incorporation in products would enable new military systems to be built. The study was updated every two years until the start of the 1970s, and reportedly had a significant impact on applied research funding, although less so on basic activities.* We shall, however, focus here on more recent foresight initiatives within DARPA and the Air Force Systems Command.

As the central R&D agency for DOD, DARPA performs a function similar to that of the corporate research division in many companies. Its task is to pursue innovative longer-term research promising an order of magnitude improvement over existing technologies, and for this it has a budget of around $1,000 million a year. Programmes fall into two categories — basic technology and demonstration projects. The former involves long-term research organized into nine areas (for example, applied and computational mathematics). The large numbers of unsolicited proposals are judged on the basis of technical merit, the proposer's track record and potential impact of the work (DARPA, 1988).

Over recent years, prospective analysis has centred around the Long Range Research and Development Plan, which used to be drawn up by the Technical Assessment and Long-Range Planning Office before it was disbanded. Responsibility is now shared by other DARPA Office Directors. Among the inputs to the plan are analyses of projected threats, US tactical and doctrinal evolution, anticipated state-of-the-art advances in technology and the longer-term impact of current R&D (ibid., p.4). Although the agency was previously a leading proponent of research foresight and conducted several pioneering studies, the view of senior officials today is that they do not produce sufficiently novel findings to be of continued interest to DARPA. Instead, the current philosophy is to rely on the knowledge of experts, for instance using brainstorming sessions with academic and industrial researchers to generate innovative ideas. Although monitoring external research to identify gaps in agency expertise (a recent example was electrochemistry), DARPA now makes little use of systematic foresight.

The US Air Force's enthusiasm for foresight, in contrast, remains undimmed. Following major studies in 1944 and 1964 (the latter entitled 'Project Forecast'),[60] the USAF in 1985 embarked upon Project Forecast II, a large-scale exercise aiming to identify radically new 'high-leverage' technologies with the promise of revolutionizing fighting capabilities over the next two decades (AFSC, 1987a, Annex A). As such, it complemented the annual planning process conducted by Systems Command relating the Air Force Science and Technology Program to user needs, a process which tends to overemphasize evolutionary and short-term technological developments (AFSC, 1987b, p.1).

Planning and definition of the study took place in the first half of 1985, the approach being modelled on that employed twenty years earlier. One task involved selecting participants. Experience from Project Forecast had shown that the involvement of industrial experts from rival contractors tended to result in ideas being suppressed because of concern about commercial secrecy. Participation was therefore limited to USAF personnel,[61] although inputs were made by industrialists and leading academics through a preliminary survey seeking ideas on promising technologies.

The main part of Forecast II centred around an intensive brainstorming and analysis effort lasting several months. An off-site location was chosen so that those involved would not be distracted by their normal duties. The 175 participants were divided into eighteen panels. Ten focussed on research fields (for example, life sciences), each consisting of around eight senior staff from USAF laboratories who would be in a position to commit the resources required to implement the project's eventual recommendations. There were also five mission panels (on theatre-war, for example) drawn from various USAF commands, providing a user-pull dimension to counterbalance the technology-push of the technical panels. Finally, there were panels responsible for systems analysis, assessment of resource implications, and a 'red team' to evaluate the threats posed by potential enemies. The exercise was overseen by two advisory groups, one comprising senior USAF officers and the other former military personnel (AFSC, 1987b, p.1).

The starting point for panels was a list of 2,000 possible new science and technology 'ideas' generated from several sources. Besides the initial survey, these included: (i) briefings by DOD officials; (ii) discussions with companies (held with a few at a time in order to minimize confidentiality problems); and (iii) prior consultation with laboratory staff and others within Systems Command. For each idea, panels prepared a short summary of potential advantages over existing technology and likely costs. These were evaluated by the technical panels, which combined some and rejected others. They were also screened by senior study managers to ensure that the ideas were novel and had significant potential — those offering only short-term or incremental advances were rejected. The military capabilities associated with each idea were then assessed by the mission panels. After the first few weeks of effort, the number of ideas had been reduced to 400.

A formal matrix-based methodology was next used to rank the contribution of each technology idea to a given system, and the contribution of each system to a specific military capability (employing a ten-point scale in both cases). In addition, technologies (and systems) were assessed by senior management in terms of whether they were: (i) 'pervasive' — that is, likely to contribute to several systems (or capabilities); or (ii) 'enabling' — i.e. critical to the development of a particular system (or capability). This resulted in another 200 ideas being dropped.

The remainder were subject to a further review which revealed that many were related and could therefore be combined. The existing summaries of comparative advantages were considerably expanded before repeating the matrix-evaluation procedure. This time, any system failing to score highly in respect of at least one military capability was filtered out. This left a final list of seventy new technologies and systems, with pervasive technologies generally ranked towards the top. Among the most promising areas were photonics, non-linear optics, artificial intelligence and high-energy chemical propellants (AFSC, 1987b; 1987c).

Since the study was completed in early 1986, emphasis has been given to implementing the findings. Several hundred briefings were first held within the Air Force and outside to develop awareness of the conclusions and build consensus around them. An implementation plan to reorient USAF laboratories was also drawn up which included a commitment that by 1988 10 per cent of the Science and Technology budget would be devoted to Forecast II initiatives, rising to 40 per cent over four years. However, by late 1987 Systems Command laboratories had already 'reached almost triple the first-year goals' (AFSC, 1987c, p.2), a reflection of the success in involving — and hence gaining the commitment of — those responsible for distributing R&D resources.

After the study, a thorough review was carried out. The lessons identified included the need to draw up a well-defined study plan at the outset, the advantages of a distraction-free location with plenty of space for brainstorming, and the benefits of using standardized software for handling the large volume of data involved in the matrix-evaluation procedure. One problem encountered during the first set of rankings arose when the results were disclosed to participants who immediately wanted to alter them: 'The result was that "motherhood" ideas won out ... When the top twenty technologies came out, we got garbage. We lost around two weeks of momentum because of this'.* However, such pressures to reformulate results running counter to conventional wisdom were minimized when the ranking exercise was repeated by restricting circulation of the overall findings to programme managers. Another lesson was that the criteria for evaluating ideas must be adequately defined and tested beforehand. Finally, whereas panels in the earlier Project Forecast had prepared independent reports, this was not done in Forecast II, thereby avoiding problems stemming from participants coming to identify too strongly with *their* ideas.

Several weaknesses were identified, however, one being that outsiders invited to attend the preliminary discussions were given insufficient time to prepare. Another was that the exercise should not have had two programme managers. Finally, presentations for the mid-term review to brief the senior advisory groups on progress took two months to prepare and caused all other work to come to a halt (AFSC, 1987a, Annex I).

Overall, Forecast II represents an extremely comprehensive research foresight exercise which has had a major impact on the distribution of R&D resources. Less clear is how far any lessons can be generalized in that it is difficult to conceive of circumstances outside the armed forces where so many experts could be 'locked up' for several months of brainstorming, or where a similar level of resources ($3 million not including experts' salaries) could be made available for a single study. Most importantly, there is a higher degree of central coordination and planning in the Air Force, coupled with less Congressional involvement in decision-making on *research* (as distinct from development) than in civil agencies. Nonetheless, the exercise does highlight the benefits from employing formal foresight techniques in research planning.

6. RESEARCH FORESIGHT BY CONSULTANCIES

Although no US manufacturing firms were covered in the study, the fact that organizations offering futures-oriented services are flourishing is one sign that science-based companies are continuing to make use of research foresight in strategic planning. We concentrate here on the foresight efforts of three consultancies — Computer Horizons Incorporated (CHI Research), SRI International and Forecasting International[62] — but also look briefly at a major exercise recently launched by the American Society of Mechanical Engineers.

Over the last ten years, CHI Research has developed a unique data-base on patenting in the United States. In addition to details on innovating firms and the characteristics of inventions, information is also available on the references made by patent examiners to previous relevant patents, and to the research literature which provides the scientific basis for the invention. Such data have several foresight-related uses: for example, heavily cited patents can be clustered to construct 'maps' of technological domains, enabling the relative strengths of companies or countries to be ascertained (e.g. Narin and Olivastro, 1988a). Moreover, the citations to basic research may reveal which fields are making important impacts on patenting in a given sector. By monitoring changes over time, one can thus identify the scientific areas contributing to new or fast developing technologies (Narin and Olivastro, 1988b).[63]

More recently, CHI Research has extended the analysis to examine the 'science-intensity' of patents broken down by country, industrial sector or company. Science-intensity is based on the relative number of citations made to the research literature, and may thus indicate how closely the technology embodied in patents is dependent upon basic science. Preliminary results suggest that, among leading industrial nations, US-invented patents are the most science-intensive, and the level has been increasing rapidly during the 1980s (Narin *et al.*, 1989). Such analyses of the links between science and technology are often complemented by bibliometric data[64] compiled from specialized abstracting services or drawn from the Science Literature Indicators Data-Base, a secondary analysis of the *Science Citation Index* carried out biennially by CHI Research under an NSF contract. Although the company has conducted studies for several governments, its main clients are companies seeking research intelligence for use in strategic planning.

Building on its long-established general interest in futures-oriented work, SRI International has established a small Science and Technology Policy Program, one element of which is to develop a capacity to identify exploitable areas of science. A technique often employed is co-citation analysis, details of which can be found elsewhere (e.g. Franklin and Johnston, 1988). Co-citation data were used, for example, in an assessment of West German science which revealed that, although generally strong in chemistry and biomedicine, the country's position in newer and fast-growing specialties within these disciplines was less secure (Coward, 1987b, pp.4–5). A similar approach was adopted in a study of Indian science for NSF aiming to identify areas of strength which might form the basis for collaborative efforts. The results showed India to be carrying out frontier research in a number of fast-growing physics and materials science specialties (Coward, 1987a, p.8). SRI has also now begun to combine literature-based indicators with other techniques. Among several projects under way is one focussing on Latin America, in which bibliometric data are being coupled with surveys of experts to assist US agencies in selecting promising areas for cooperative research.*

An important contribution to foresight has been made by Forecasting International. Founded by Marvin Cetron, who played a key role in early forecasting efforts by the US Navy during the 1960s, the company has since carried out numerous studies for government agencies and companies which have involved research foresight to varying degrees. Normally, a combination of methods is employed: (i) trend extrapolation (exploring a range of optimistic and pessimistic assumptions covering most foreseeable contingencies); (ii) trend correlation — i.e. comparing the results from two or more extrapolations; (iii) growth-analogy techniques (especially S-shaped curve analysis); and (iv) modified Delphi techniques.

Finally, mention should be made of a programme initiated in 1986 by the American Society of Mechanical Engineers (ASME) to develop models for the quantitative assessment of technological choice (Garfield, 1987). ASME commissioned Wassily Leontief (the pioneer of input–output analysis) and Faye Duchin, director of New York University's Institute for Economic Analysis, to develop a techno-economic model for examining the effects of R&D and innovation on individual industrial sectors and on the economy overall. It was felt that such a model might assist the federal government and larger companies in evaluating new technologies and hence determining the most effective allocation of research resources (Gilman, 1986). In order to compile the required data inputs, teams of economists and engineers have interviewed large numbers of industrialists, ASME having placed the expertise of its members at the disposal of the researchers. Initial work has concentrated on four sectors (biotechnology, microelectronics, materials and energy) but the coverage may subsequently be extended (Garfield, 1987). No results have yet been reported so it is too soon to judge how fruitful this approach will prove in relation to research foresight.

7. CONCLUSIONS

We have seen how considerable effort is devoted in the United States to identifying research opportunities, but in general this appears not to be closely integrated with policy-making or budget-setting. A central problem is that OSTP, the body supposedly overseeing federal science and technology policy, has largely abdicated responsibility for this task (Knezo, 1988, p.20). Instead, agencies must respond to three interest groups — the scientific community, Congress and OMB. Of these, OMB is primarily interested in constraining federal expenditure, while Congress places power in the hands of numerous committees, which gives rise to a 'fractionated' view of research. The political process has therefore failed to provide any clear sense of strategy or coordination of priorities.[65] Research policy, to the extent that it exists at all, is to be found mainly at the level of agencies, and even there it relies heavily upon the entrepreneurship of scientists in generating new ideas. Such 'policies' are consequently often lacking in coherence and only loosely linked to national needs (Bloch, 1988).

In this environment, the incentives to engage in research foresight are small and the disincentives great. Indeed, since our previous study, there appears to have been a swing away from more formal foresight in US government agencies. Recent attempts to look into the future have mostly relied upon expert panels or the BOGSAT approach referred to in Chapter 2—systematic data-driven approaches have been eschewed (except by the Air Force). It is necessary, therefore, to consider what factors have contributed to this situation and the general lessons which can be drawn:

(a) Perhaps the underlying reason for the limited policy impact of research foresight has been the absence of prior consensus on the need to set priorities at a national level. Since 1945, the United States has tried to pursue a comprehensive approach of supporting research in all fields. Only today is it being confronted with the reality that this can no longer be afforded. As certain influential officials (e.g. Bloch, 1988; Press, 1988; Rosenzweig, 1988) have pointed out, explicit choices must be made — for example, whether to build the Superconducting Super Collider or to sequence the human genome. Those arguing for selectivity have previously been very much in a minority, and foresight exercises have generally been used more to lobby for extra funds than to set priorities (Bloch, 1988, p.9). Whether the situation can continue now that the powerful Congressional budget committees have begun to express concern[66] remains to be seen.

(b) Related to this has been the lack of a specific 'client' with well-defined needs for foresight, especially in regard to formulating inter-field priorities. This stems from the philosophy of pluralism, the ineffectiveness of OSTP, and the fragmented way in which OMB and Congress treat research. Decisions on research funding clearly have to be made in the Executive and Legislative Branches, but many foresight exercises have failed to provide an *appropriate* product. Reports should not be too long and academic (contrast, for example, the first two Outlooks and the Brinkman physics survey with the Research Briefings and the DOC report on Emerging Technologies). Nor should they be too specialized, as later Research Briefings became. Most importantly, they should not be politically unrealistic in their financial demands, as NRC field surveys have tended to be.

(c) One reason for the limited impact of the early Five-Year Outlooks was that they were contracted out by OSTP to NSF, which in turn subcontracted them to NRC and others, giving rise to a tenuous relationship between those preparing the reports and policy-makers. Conversely, the success of Forecast II is partly attributable to the fact that senior R&D managers were integrally involved. The institutional location of those responsible for foresight is thus crucial. While retaining some autonomy from day-to-day decision-making, they must nevertheless be sufficiently close to ensure the results can be integrated into the policy and budget-setting process.

(d) In the absence of any federal science and technology policy, there is heavy reliance on the annual budget-setting cycle. This has given rise to an emphasis within agencies, the Executive Office and Congress on the immediate future at the expense of the longer term. In addition, OMB has sometimes viewed long-term planning as advance lobbying, although agencies are now required to submit five-year budgetary projections. Such factors have greatly reduced the scope for foresight.

(e) Aside from Forecast II and the OTA study on new materials, few exercises to identify long-range priorities for strategic research have systematically taken account of demand-pull factors. Conversely, science-push considerations have too often dominated. In the view of some, the field surveys have been 'hijacked' by the research community. One reason for this is the dual allegiance of NRC to the federal government and to the national academies with their professional constituencies. Where the desire of agencies for priorities related to national needs has run counter to the wishes of researchers, the latter have generally prevailed.

(f) Perhaps the main problem with the NRC surveys has been their failure to appreciate the significance of emerging interdisciplinary areas. The field-based approach employed has tended to yield conservative conclusions that reflect initial assumptions about the existing disciplinary structure of research. For example, the Brinkman study resulted in separate reports by the seven panels on areas like nuclear physics, linked rather unconvincingly by an 'overview' containing little in the way of a discussion on priorities. Field surveys such as those for chemistry and engineering, which deliberately refrained from organizing panels along disciplinary lines, were on the whole more successful.

(g) The reluctance of the research community to employ foresight in priority-setting has often only been overcome by the promise of additional funds. For example, NSF's interest in longer-term planning was revived by the promise to double its budget. Even then, researchers may still attempt to sidestep the issue of priorities. ERAB's experience shows that a strong chairperson can be invaluable in coercing participants in foresight exercises to go through the necessary 'tooth and claw' process before eventually arriving at an ordered set of suggested priorities.

(h) As the futurist, Walter Hahn (1986, p.24), has pointed out, 'the initiation, pursuit and success of futures activities seems to be unusually dependent on the presence, personality and persistence of specific individuals.' Hahn himself provided these qualities at the Congressional Research Service during the brief life of the Futures Research Group, while Kenneth Gordon (formerly head of the NBS Planning Office) was central in the DOC Emerging Technologies exercise as well as the NBS long-term plans.

(i) A powerful external sponsor has also often been necessary in overcoming the scepticism and even hostility of agencies and the scientific community. This is illustrated by the decline in impact of the Research Briefings after 1985 when no one was willing to take on Keyworth's earlier role of sponsor. Likewise, the demise of the Five-Year Outlook stemmed partly from the fact that it never had a strong product champion.

(j) More generally, US experience shows the importance of professional competence in the use of forecasting and survey techniques, as well as familiarity with the principles of research policy. Although good graduate programmes in science and technology policy exist, there is widely perceived to have been some stagnation of academic work in the field following financial cutbacks.* This may have contributed to an evident failure to experiment with the new approaches to research foresight being developed overseas.

(k) Political ideology has exercised a dominant influence in limiting the introduction of foresight techniques, with centralized planning by government rejected in favour of leaving matters to the 'market'. This has been particularly true under the Reagan administration. In the case of fundamental science, the *laisser-faire* philosophy has been interpreted as meaning that federal agencies (especially NSF) should operate in a responsive mode, while for mission-oriented research a system involving multiple centres of decision-making is presumed to equate with a 'free-market' solution. Yet some are now coming to recognize that foresight and priority-setting need not be synonymous with centralized top-down planning, and that the private sector (especially given current financial uncertainties) cannot be expected to take a sufficiently long-term view to assume responsibility for promising strategic research.

(l) Finally, the adversarial basis of policy-making in the United States sets an intrinsic limit on the impact likely to be achieved by foresight. With the exception of Forecast II (and the armed forces are clearly a special case), there is little evidence that foresight studies 'drive' the research agenda of agencies.[67] Instead, policy is shaped by a wide array of factors (not least the powerful lobbies to which Congress is subject), with foresight representing just one among many inputs. Given this political context, the scientific community has recognized the benefits of using foresight studies as a platform for advocacy — for expounding the case for their area to be judged before the 'court' of Congress and the Executive Branch. A field survey or Research Briefing has therefore been viewed more as an opportunity for researchers to advance the best possible case for their discipline rather than as a means of systematically appraising priorities for the future taking into account both scientific promise and national strategic needs.

NOTES

* Refers to material collected during interviews in 1987 and early 1988, while † denotes information drawn from subsequent correspondence.

1. A significant fraction of basic research in the United States is conducted by industry — especially major high-technology firms such as IBM and Du Pont.
2. The term 'National Academy complex' is used as shorthand for the National Academy of Sciences, National Academy of Engineering and Institute of Medicine, together with the National Research Council.
3. Limitations of time prevented any manufacturing companies from being included in this study. A description of the research foresight efforts of three firms during the early 1980s can, however, be found in Irvine and Martin (1984, pp.88–93).
4. Obligations in 1987 were $56,500 million compared with outlays of $53,500 million (NSF, 1987a, pp.20-1). Budgetary figures given elsewhere are for obligations.
5. Some informal 'tracking' of allocations for *basic* research is, however, undertaken as the President's budget is formulated. This helps determine whether NSF — the 'flywheel' — might take up any slack or fill any gaps.†
6. A widely circulated report by the Council on the health of universities recommended substantially increased federal support (Packard and Bromley, 1986).
7. The Board consists of twenty-four members appointed by the President, together with the Director of NSF.
8. An excellent description of the Congressional budget-making process can be found in Teich (1986a; 1986b).
9. One exception was in 1984, while more recently the House and Senate Budget Committees have again raised the need to consider all R&D together.
10. For example, in the late 1970s the Joint Economic Committee of Congress undertook a two-and-a-half year study on Research and Innovation: Developing a Dynamic Nation. This reviewed over forty innovation studies and all existing foresight or outlook reports. In addition, three alternative scenarios were put forward, together with an analysis of major world issues (Joint Economic Committee, 1980). Like the larger Special Study on Economic Change of which it formed a part, it was almost completely ignored by Congress as well as by the Executive Branch and industry.†
11. The report was concerned more generally with the health of US research and the adequacy of existing mechanisms for formulating national science policy and determining priorities (GAO, 1987, p.1).
12. During the mid-1970s, CRS established a Futures Research Group to look into the longer-term, provide forecasting expertise and develop a Futures Information Retrieval System (FIRST). The group produced various reports, including one of the earliest analyses of foresight in the private sector and its potential use by government (Foresight Task Force, 1983). However, the Director of CRS disbanded the group in the early 1980s (Hahn, 1986, p.20).

13. Professional societies also have an influence on science policy. The American Physical Society, for example, convened a study group in 1988 to 'identify, assess and recommend ... an effective method for strategic planning of physics research, namely the identification of needs and the allocation of resources over the next decade' (*Physics Today*, 1988, p.111).
14. They were also expensive — the first cost $470,000 and the second $705,000 (House Committee on Science and Technology, 1984). This compares with $120,000 for the third Outlook.
15. The 1976 Act also mandated OSTP to produce an Annual Science and Technology Report. Like the Outlook, this was initially subcontracted to NSF (1978; 1980b; 1981b). OSTP then reassumed responsibility with the intention of using it to publicize the Executive Branch's research achievements, and produced two somewhat more 'glossy' reports (OSTP, 1982; 1983). When the next issue failed to appear on time (in breach of the Congressional statute), its title was changed from 'Annual' to 'Biennial'. However, even then the 1983–4 volume was published a year late (OSTP, 1986), and at the end of 1988 that for 1985–6 was still awaited. Since reports have concentrated on the present rather than the longer-term future, they are not considered here. Exceptionally, however, the expert panels convened by NSF to help OSTP in preparing the 1981 and 1982 volumes did attempt to examine a number of emerging issues (e.g. NSF, 1982).
16. The Executive Director of COSEPUP, having previously worked on Capitol Hill, knew that reports had to be kept short and simple.*
17. Although preparation of a fifth Outlook was reported to be under way in 1987 (COSEPUP, 1987, p.3), this was apparently never completed.
18. Other criticisms can be found in Blanpied and Leshner (1981) and Chubin and Rossini (1981). The Outlooks were also subject to review in evaluations commissioned by NSF from science-policy graduate students. These concluded that, while the Outlooks provided adequate summaries for the less informed, they failed to produce anything very new or useful in relation to resource allocation by agencies; they also tended to summarize conventional wisdom while avoiding criticism; and, most importantly, they did not attempt any proper foresight in respect of emerging socio-economic needs.*
19. This led one Committee member to remark that the question at issue was not their utility to the Executive but to Congress, which had enacted the legislation requesting they be prepared (House Committee on Science and Technology, 1984, p.1).
20. One example was the 1983 Briefing on computers in design and manufacturing which had to be rewritten twice. However, from the discussions which ensued came the idea of Engineering Research Centers—widely seen as one of Keyworth's successes.
21. In later years, an effort was also made to brief other interested parties, for example in Congress and industry.
22. It is perhaps significant that OSTP staff were unable to locate copies of recent Briefings when the Office was visited in 1987.
23. The 1987 Research Briefings are likely to be the last. The continued existence of COSEPUP (at least in its present form) is also under review by a National Academy panel.
24. For an analysis of lessons from the approach adopted by OTA in its studies, see Procter (1987). Among the conclusions are that policy options should be identified early on, that each study must have at least one clear client (and preferably several with differing points of view), that adequate resources need to be devoted to developing a study plan, and that workshops are often a cost-effective alternative to commissioned studies.
25. COSPUP's origins are described in Irvine and Martin (1984, pp.68–9).

26. Summaries of the earlier surveys can be found in Lowrance (1977), together with an analysis of their impact and shortcomings. See also Irvine and Martin (1984, pp.70–6).
27. Each report is generally referred to by the name of the person chairing the committee, a convention employed here, even though that name does not normally appear on the front cover.
28. Besides surveys of broad fields, NRC has conducted many future-oriented studies of more specific areas such as physics at the laser-atomic frontier (CAMS, 1986) and theoretical atomic, molecular and optical sciences (CAMS, 1987). The approach in the former involved organizing a workshop and commissioning experts to make presentations. The latter was prepared by a panel of eight academics, but also made use of a survey of theorists in fifty leading university departments.
29. This was the explanation offered by NRC officials for the fact that only five members came from industry while seven were university professors from California.
30. After a five-day workshop, an interim report was produced as one of the 1983 Research Briefings. Although the group responsible did not completely overlap with the survey committee, most of the recommendations were absorbed in the Pimentel report.
31. Pimentel also co-authored a popular version of the report (Pimentel and Coonrod, 1987), 20,000 free copies of which were distributed to schools and colleges.
32. All twenty-three members of the survey committee were academics, which would in any case have made it difficult to assess future demands for behavioural and social science research.
33. See the criticism in Irvine and Martin (1984, pp.72–5). Physicists have the dubious distinction of having produced the lengthiest surveys, the first volume of the Bromley report alone extending to more than 1,000 pages, while the Brinkman findings ran to eight volumes. As one official commented, this meant that little energy was left for distilling the 2,000 pages into a good overview.*
34. Somewhat surprisingly, this argument proved quite successful, mathematics soon being rewarded with a large increase in support. However, interviewees doubted whether such an approach would work in other studies.
35. That is, *intra*-field priorities (the surveys are clearly inappropriate for generating inter-field priorities).
36. In an area like the Strategic Defense Initiative, it may be completely impossible to reconcile the conflicting interests. This might explain why since 1984 the National Academies have not even attempted to provide advice to government on SDI (in contrast, for example, to OTA).
37. Similarly, the failure to engage in systematic foresight may reflect the opposition of scientists on survey committees rather than reluctance on the part of NRC staff, some of whom would prefer reports to be more soundly based on analysis and data.*
38. NSF officials argued that the payback is often greatest from the least expected areas, citing examples such as the work on high-temperature superconductors at the University of Houston. This was supported for many years because of its ‘good, solid craftsmanship’ rather than any promise of future strategic benefits. See also NSB (1987, p.5).
39. For example, the Division of Policy Research and Analysis is not engaged in foresight, instead concentrating on constructing an integrated data-base on US science and technology to help in budget-planning (see e.g. PRA, 1987).
40. Another indication of the revived interest in seeking more systematic information for use in long-range planning is the series of JTECH reports prepared for NSF on trends in Japanese basic technology.

41. Although a key theme was the establishment of new Science and Technology Centers, the plan explicitly states that it 'is not NSF's intention to choose specific topics for ... Centers' (NSB, 1987, p.14).
42. During the early 1970s, there had been an earlier attempt by NSF to stimulate Research Applied to National Needs (Lomask, 1976, chapter 14). According to Ryer (1985), the main reason for the demise of the RANN programme was strong opposition from the basic science community.
43. In 1985, NSF asked the National Research Council to identify priority topics for the third programme announcement. The resulting list (NRC, 1985) was not, however, used on the grounds that more exciting ideas emerged from the proposals received from researchers.*
44. A two-stage review procedure is employed in judging ERC proposals, the first part involving technical panels and the second a 'blue-ribbon' panel, industrialists being well represented in both (NRC, 1986, pp.44-50). Centres are then subject to periodic evaluation, as a result of which funding for two of the first six is to cease after five years.
45. For details of the fairly elaborate annual budget-setting process (which in fact takes three years to complete), see NIH (n.d.) and Moskowitz (1988).
46. As one official put it, 'the [NIH] Director's job is to act as a *moderator* of what institutes are trying to do and as a *broker* between NIH and the outside world, rather than as an *architect*'.*
47. Another major foresight effort was that involved in formulating the National Cancer Programme. Forty working sessions involving 250 experts produced over 3,000 pages of material, on the basis of which a five-year strategic plan was drawn up together with an operational plan (DHEW, 1974). These were subsequently revised annually.
48. The third plan was later updated for a further two years (NAEC, 1987).
49. Long-range plans produced by certain other institutes are, however, less well regarded than those of NHLBI and NEI; some cover just one part of the institute's responsibilities or involve only a limited section of the scientific and professional community, while others suffer from poor 'packaging'.*
50. After considerable debate as to which institute should be given responsibility for the initiative, it was eventually decided that a new genome office should be located in the Office of the Director (Roberts, 1988, p.878). Whether this presages a shift in the balance of power away from individual institutes remains to be seen.
51. Participants failed to distinguish what is important from what is likely to be affected by R&D, with the result that the voting process began to produce nonsensical results.*
52. The report admits that 'no attempt ... [has been] made to rank these [R&D] issues in terms of their central importance', adding somewhat feebly that this is because they '*all* require attention' (ERAB, 1985b, p.7, emphasis added).
53. 'This, however, does not mean that the government can't support selected areas of technology, *provided* that they are (1) initiated, (2) managed and directed, and (3) primarily funded by the private sector.'†
54. Various foresight studies had previously been carried out by the Office of Technology Assessment and Forecast — for example, to explore the use of patents in identifying 'active technologies'(DOC, 1981).
55. The conclusions were written up in an internal DOC paper, which was deliberately kept short. Many of those requesting copies of 'the full report' were surprised that the twenty-page document was all that was available, erroneously assuming that it was drawn from a longer publication containing detailed quantitative analysis.†
56. DOC has since undertaken a series of more sophisticated 'competitive assessments' of rapidly developing technologies. Conferences are generally held to

discuss the results and catalyse the formation of R&D consortia among companies working in these fields.†

57. After the discovery had been announced, committee members debated whether to add superconductivity to the list. They eventually decided against this, but included a note referring to superconductors as a 'potential emerging technology' (NBS, 1987, p.2).
58. In 1988, Congress changed the name from the National Bureau of Standards (NBS), reflecting the increased emphasis on technology transfer and industrial support, especially in relation to emerging generic areas.
59. As a result, the third plan was never formally released.
60. In 1970, Project Recast retrospectively assessed Project Forecast. Analysis of the predictions revealed that, although the main directions had been correct (for example, in the area of composite materials), many had been too conservative.*
61. At about the same time, however, the Aeronautical Systems Division organized a brain-storming workshop where professional futurists, academics and several well-known science-fiction writers joined USAF personnel in speculating on possible, very long-term developments in technology and society during the coming century — for example, laser-powered engines to enable fighters to fly at up to Mach six. (This Futurist II workshop followed a similar exercise in 1982 where aerospace industry representatives discussed trends in areas like artificial intelligence and transatmospheric vehicles.) Participants judged the Futurist II exercise to have been a success.† However, the resultant report was so highly classified that even those who contributed were denied access to the conclusions,† which makes it difficult to establish whether the exercise had much impact on Forecast II.
62. Also visited was J. F. Coates Incorporated, one of whose studies (for the National Bureau of Standards) is described earlier.
63. See the description of the study to identify West German technological strengths in Chapter 4.
64. CHI Research has also been a pioneer of scientometric indicators for evaluating past performance in basic research, undertaking an important series of studies for the National Institutes of Health (e.g. NIH, 1986).
65. As one interviewee remarked, 'there is no strategy in US science, only tactics'.
66. In 1988, the Committees commissioned a report from the National Academy complex on possible mechanisms for setting research priorities (NAS/NAE/IOM, 1988). This was published after our study had been completed.
67. While many of the agency officials interviewed described the field survey reports as generally useful *to others*, none could give any instance where they had made a direct impact on their own views or decisions.

REFERENCES

Adams, R. McC., Smelser, N. J. and Treiman, D. J. (eds) (1982), *Behavioral and Social Science Research: A National Resource*, Washington, DC, National Academy Press.

AFSC (1987a), *Project Forecast II*, Andrews Air Force Base, Air Force Systems Command (Classified).

—— (1987b), *Project Forecast II: Executive Summary*, Andrews Air Force Base, Air Force Systems Command.

—— (1987c), *Air Force Science and Technology: The Legacy of Project Forecast II*, Andrews Air Force Base, Air Force Systems Command.

Amundson, N. R. (1988), *Frontiers in Chemical Engineering: Research Needs and Opportunities*, Washington, DC, National Academy Press.

Baldwin, E. and Hill, C. T. (1988), 'The Budget Process and Large-Scale Science Funding', *CRS Review*, vol.9, no. 2, pp.13–16.

Blanpied, W. A. and Leshner, A. I. (1981), 'Origins and Aims of the Five-Year Outlook', *Technological Forecasting and Social Change*, vol. 20, pp.145-54.
Bloch, E. (1988), 'People and Responsibilities: Science Policy in the 1990s' (mimeo), Henri Sack Memorial Lecture, Cornell University, April.
Brinkman, W. F. (1986), *Physics Through the 1990s*, Washington, DC, National Academy Press.
Bromley, D. A. (1972), *Physics in Perspective*, Washington, DC, National Academy of Sciences.
Brown, C. J. (1987), 'News Briefing on Emerging Technologies' (mimeo), June, Washington, DC, Department of Commerce.
CAMS (1986), *Physics at the Laser-Atomic Frontier*, report by the Committee on Atomic and Molecular Science, Washington, DC, National Academy Press.
—— (1987), *The State of Theoretical Atomic, Molecular, and Optical Sciences in the United States*, report by the Committee on Atomic and Molecular Science, Washington, DC, National Academy Press.
CBO (1988), *Risks and Benefits of Building the Superconducting Super Collider*, Washington, DC, Congressional Budget Office.
Chubin, D. E. and Rossini, F. A. (eds) (1981), *Special Issue of 'Technological Forecasting and Social Change': The Five-Year Outlook on Science and Technology in the United States*, New York, Elsevier.
Coates, J. F. (1980), *Developments in Science and Technology Likely to Affect the National Bureau of Standards*, Gaithersburg, National Bureau of Standards.
—— (1981), *Implications of Biotechnology for the National Bureau of Standards*, Gaithersburg, National Bureau of Standards.
Cort, H. R. and Cort, E. C. (1984), 'An Evaluation of the Five-Year Outlooks from the Users' Point of View', testimony in House Committee on Science and Technology (1984).
COSEPUP (1983), *Frontiers in Science and Technology: A Selected Outlook*, New York, Freeman.
—— (1984), *Research Briefings 1984*, Washington, DC, National Academy Press.
—— (1985a), *The Outlook for Science and Technology 1985*, Washington, DC, National Academy Press.
—— (1985b), *New Pathways in Science and Technology: Collected Research Briefings 1982–84*, New York, Vintage Books.
—— (1985c), *Research Briefings 1985*, Washington, DC, National Academy Press.
—— (1986), *Research Briefings 1986*, Washington, DC, National Academy Press.
—— (1987), 'Chairman's Report to the Membership of the National Academy of Science' (mimeo), April, Washington, DC, Committee on Science, Engineering and Public Policy.
—— (1988), *Research Briefings 1987*, Washington, DC, National Academy Press.
Coward, H. R. (1987a), ' A Bibliometric Assessment of Indian Science', in *Indian Scientific Strengths: Selected Opportunities for Indo-US Cooperation*, Washington, DC, National Science Foundation.
—— (1987b), *Assessment of FRG Strengths in a 1984 World Model of Science*, Arlington, SRI International.
DARPA (1988), *DARPA 1988*, Arlington, Defense Advanced Research Projects Agency.
David, E. E. (1984), *Renewing US Mathematics: Critical Resource for the Future*, Washington, DC, National Academy Press.
DHEW (1974), *National Cancer Program: The Strategic Plan*, Washington, DC, Department of Health, Education and Welfare.
DOC (1981), *Technology Assessment and Forecast: Tenth Report*, Washington, DC, Department of Commerce.
—— (1984), *Agenda for Long-Range Productivity Planning*, Washington, DC, Department of Commerce.

—— (1987a), *Cooperative R&D in Industrial Competitiveness*, Washington, DC, Department of Commerce.
—— (1987b), *Technology, the Engine of Global Competitiveness*, Washington, DC, Department of Commerce.
—— (n.d.), *Antitrust and Other Barriers to Computer Integrated Flexible Manufacturing Systems*, Washington, DC, Department of Commerce.
Ecosometrics (1984), *Technologies in the Service Sector*, Gaithersburg, National Bureau of Standards.
ERAB (1981), *Federal Energy R&D Priorities*, Washington, DC, Energy Research Advisory Board.
—— (1985a), *Review of the National Research Council Report 'Major Facilities for Materials Research and Related Disciplines'*, Washington, DC, Energy Research Advisory Board.
—— (1985b), *Guidelines for DOE Long Term Civilian Research and Development*, Washington, DC, Energy Research Advisory Board.
—— (1987), *Review of the National Research Council Report 'Physics Through the 1990s'*, Washington, DC, Energy Research Advisory Board.
ESRC/NSF (1985), *Rise and Fall of a Priority Field*, Strasbourg, European Science Foundation.
Field, G. B. (1982), *Astronomy and Astrophysics for the 1980s*, Washington, DC, National Academy Press.
Foresight Task Force (1983), *Foresight in the Private Sector: How Can Government Use It?*, House of Representatives, 98th Congress, 1st Session, Washington, DC, US Government Printing Office.
Franklin, J. J. and Johnston, R. (1988), 'Co-Citation Bibliometric Modelling as a Tool for S&T Policy and R&D Management: Issues, Applications and Developments', in van Raan (1988).
GAO (1981), *Major Science and Technology Issues*, Washington, DC, General Accounting Office.
—— (1987), *US Science and Engineering Base: A Synthesis of Concerns About Budget and Policy Development*, Washington, DC, General Accounting Office.
Garfield, E. (1987), 'Needed: Information on Technology's Impact', *The Scientist*, vol.1, no.13 p.9.
Gerstein, D. R., Luce, R. D., Smelser, N. J. and Sperlich, S. (eds) (1988), *The Behavioral and Social Sciences: Achievements and Opportunities*, Washington, DC, National Academy Press.
Gilman, H. (1986), 'Coming, A Model to Measure Effects of Scientific Research', *Wall Street Journal*, 3 October, p.31.
Goodwin, I. (1988), 'Erich Bloch: On Changing Times and Angry Scientists at NSF', *Physics Today*, vol.41, no.8, pp.47–52.
Hahn, W. A. (1986), 'Futures in Politics and the Politics of Futures', School of Government and Business Administration, George Washington University.
Healy, M. J. (1987), *US Planning for Basic Research: Identifying Areas of High Scientific Opportunity*, Washington, DC, Embassy of Australia.
House Committee on Science and Technology (1984), *Report on Hearing: Five-Year Outlook on Science and Technology, January 24, 1984*, Washington, DC, US House of Representatives.
Institute of Medicine (1984), *Responding to Health Needs and Scientific Opportunity: The Organizational Structure of the National Institute of Health*, Washington, DC, National Academy Press.
Irvine, J. and Martin, B. R. (1984), *Foresight in Science: Picking the Winners*, London, Pinter Publishers.
Joint Economic Committee (1980), *Research and Innovation: Developing a Dynamic Nation*, Washington, DC, US Government Printing Office.

Keyworth, G. A. (1983), 'Federal R&D and Industrial Policy', *Science*, vol.220, pp.1122–25.

Knezo, G. J. (1988), *Analysis of Science and Technology Policy*, Washington, DC, Congressional Research Service.

Ling, J. G. (1984), testimony in House Committee on Science and Technology (1984).

Lomask, M. (1976), *A Minor Miracle: An Informal History of the National Science Foundation*, Washington, DC, National Science Foundation.

Lowrance, W. W. (1977), 'The NAS Surveys of Fundamental Research 1962–74 in Retrospect', *Science*, vol.197, pp.1254–60.

Merrifield, D. B. (1983), 'Forces of Change Affecting High Technology Industries', *National Journal*, 29 January, pp.253–56.

Moskowitz, J. (1981), *Diabetes and Atherosclerosis*, New York, Juvenile Diabetes Foundation.

—— (1988), *Activities of the Office of Science Policy and Legislation*, Washington, DC, National Institutes of Health.

NAE (1983), *Guidelines for Engineering Research Centers*, Washington, DC, National Academy of Engineering.

NAEC (1976), *Vision Research Program Planning*, report of the National Advisory Eye Council, Bethesda, National Institutes of Health.

—— (1977), *Vision Research — A National Plan: 1978–1982*, report of the National Advisory Eye Council, Bethesda, National Institutes of Health.

—— (1983), *Vision Research — A National Plan: 1983–1987*, report of the National Advisory Eye Council, Bethesda, National Institutes of Health.

—— (1987), *Vision Research: A National Plan — 1987 Evaluation and Update*, Bethesda, National Institutes of Health.

Narin, F. and Olivastro, D. (1988a), *Identifying Areas of Leading Edge Japanese Science and Technology*, Haddon Heights, NJ, CHI Research.

—— and —— (1988b), 'Technology Indicators Based on Patents and Patent Citations', in van Raan (1988).

—— Rosen, M. and Olivastro, D. (1989), 'Patent Citation Analysis: New Validation Studies and Linkage Statistics', to be published in proceedings of the Leiden Workshop on Science Indicators: Their Use in Science Policy and Role in Science Studies.

NAS (1987), *Science and Technology Centers: Principles and Guidelines*, Washington, DC, National Academy of Sciences.

NAS/NAE/IOM (1988), *Federal Science and Technology Priorities*, Washington, DC, National Academy Press.

NBS (1981), *Long Range Plan of the National Bureau of Standards*, Gaithersburg, National Bureau of Standards.

—— (1984), *National Bureau of Standards*, Gaithersburg, National Bureau of Standards.

—— (1987), *The Status of Emerging Technologies: An Economic/Technological Assessment to the Year 2000*, Gaithersburg, National Bureau of Standards.

NHLBI (1984), *Tenth Report of the Director, National Heart, Lung and Blood Institute*, Bethesda, National Institutes of Health.

NHLI (1973), *National Heart, Blood Vessel, Lung and Blood Program*, report of the National Heart and Lung Institute, Washington, DC, Department of Health, Education and Welfare.

NIH (1986), *An Analysis of Research Publications Supported by NIH*, Bethesda, National Institutes of Health.

—— (1987), *Biennial Report of the Director, National Institutes of Health*, Bethesda, National Institutes of Health.

—— (n.d.), *Description of the Planning Function in the Office of the Director, NIH*, Washington, DC, National Institutes of Health.

NRC (1982), *Outlook for Science and Technology: The Next Five Years*, San Francisco, Freeman.
—— (1985), *Subject Areas for the Third Program Announcement for the Engineering Research Centers*, Washington, DC, National Academy Press.
—— (1986), *The New Engineering Research Centers: Purposes, Goals and Expectations*, Washington, DC, National Academy Press.
NSB (1979), *Status of Science Reviews 1979*, Washington, DC, National Science Board.
NSB (1983), *Status of Science Reviews 1984*, Washington, DC, National Science Board.
—— (1987), *National Science Foundation Long Range Plan, FY 1988-1992*, Washington, DC, National Science Board.
NSF (1978), *Science and Technology: Annual Report to the Congress*, Washington, DC, National Science Foundation.
—— (1980a), *The Five-Year Outlook: Problems, Opportunities and Constraints in Science and Technology*, vols 1 and 2, Washington, DC, National Science Foundation.
—— (1980b), *Science and Technology: Annual Report to the Congress*, Washington, DC, National Science Foundation.
—— (1981a), *The Five-Year Outlook on Science and Technology, 1981*, Washington, DC, National Science Foundation.
—— (1981b), *Annual Science and Technology Report to the Congress, 1980*, Washington, DC, National Science Foundation.
—— (1982), *Emerging Issues in Science and Technology, 1981*, Washington, DC, National Science Foundation.
—— (1984), *Program Announcement: Engineering Research Centers, Fiscal Year 1985*, Washington, DC, National Science Foundation.
—— (1987a), *Federal Funds for Research and Development: Fiscal Years 1986, 1987 and 1988*, vol. XXXVI, Washington, DC, National Science Foundation.
—— (1987b), *Annual Report 1986*, Washington, DC, National Science Foundation.
OSTP (1982), *Annual Science and Technology Report to the Congress: 1981*, Washington, DC, Office of Science and Technology Policy.
—— (1983), *Annual Science and Technology Report to the Congress: 1982*, Washington, DC, Office of Science and Technology Policy.
—— (1986), *Biennial Science and Technology Report to the Congress: 1983-84*, Washington, DC, Office of Science and Technology Policy.
OTA (1986a), *New Structural Materials Technologies: Opportunities for the Use of Advanced Ceramics and Composites*, Washington, DC, Office of Technology Assessment.
—— (1986b), *Research Funding as an Investment: Can We Measure the Returns?*, Washington, DC, Office of Technology Assessment.
—— (1987), *What OTA Is, How OTA Works*, Washington, DC, Office of Technology Assessment.
Packard, D. and Bromley, D. A. (1986), *Report of the White House Science Council Panel on the Health of US Colleges and Universities*, Washington, DC, Office of Science and Technology Policy.
Pake, G. E. (1966), *Physics: Survey and Outlook*, Washington, DC, National Academy of Sciences.
Physics Today (1988), 'Group Will Investigate Strategic Planning of Physics Research', vol. 41, no.6, p.111.
Pimentel, G. C. (1985), *Opportunities in Chemistry*, Washington, DC, National Academy Press.
Pimentel, G. C. and Coonrod, J. A. (1987), *Opportunities in Chemistry Today and Tomorrow*, Washington, DC, National Academy Press.

PRA (1987), *Strategic Academic Research Planning System*, Washington, DC, National Science Foundation.

Press, F. (1988), 'The Dilemma of the Golden Age' (mimeo), Presidential Address to the National Academy of Sciences, April.

Procter, M. (1987), 'Probing the Core of Controversy: Issue Management in the Office of Technology Assessment', paper presented at the European Congress on Technology Assessment, Amsterdam, February.

Puckett, A. E. (1987), *Directions in Engineering Research: An Assessment of Opportunities and Needs*, Washington, DC, National Academy Press.

Rettig, R. A. (1984), testimony in House Committee on Science and Technology (1984).

Roberts, L. (1988), 'Watson May Head Genome Office', *Science*, vol.240, pp.878–79.

Rosenzweig, R. M. (1988), 'Thinking about Less', *Science, Technology and Human Values*, vol.13, pp.219–23.

Ryer, D. (1985), 'NSF's RANN Program', in ESRC/NSF (1985), pp.53–63.

Schoettler, J. H. (1986), letter to Dr B. Ancker-Johnson, 29 May, included in ERAB (1987).

Seitz, F. and Eastman, D. E. (1984), *Major Facilities for Materials Research and Related Disciplines*, Washington, DC, National Academy Press.

Smelser, N. J. and Gerstein, D. R. (eds) (1986), *Behavioral and Social Science: Fifty Years of Discovery*, Washington, DC, National Academy Press.

Staats, E. B. (1980), *Statement of the Comptroller General of the United States before the Subcommittee on Science, Research and Technology on Long-Term Planning for National Science Policy*, Washington, DC, General Accounting Office.

Tassey, G. (1985), *Technology and Economic Assessment of Optoelectronics*, Gaithersburg, National Bureau of Standards.

Teich, A. H. (1986a), *Support for Basic Research: A Review of the Federal Role*, Washington, DC, National Science Foundation.

Teich, A. H. (1986b), 'Coordination of United States Research Programs: Executive and Congressional Roles', *Science and Technology Studies*, vol.4 no.2, pp.29-36.

Van Raan, A. F. J. (ed.) (1988), *Handbook of Quantitative Studies of Science and Technology*, Amsterdam, Elsevier.

Westheimer, F. H. (1965), *Chemistry: Opportunities and Needs*, Washington, DC, National Academy of Sciences.

6 Research foresight in Japan

1. INTRODUCTION

Among the eight countries studied, Japan has the most developed approach to research foresight, government and industry sharing a common concern with longer-term planning. This can partly be explained culturally in terms of the emphasis on consensus in Japanese society, and partly historically through the policy since the Meiji restoration of monitoring overseas R&D in order to acquire the technological capacity to preserve economic independence from the West. In an increasingly competitive global technological climate, it is accepted that those failing to engage in foresight will suffer the fate of the surfer on the beach — 'unless you are ready for the next wave, you're dead'.[1]

The mid-1980s have seen a reorientation of science policy as the nation has recognized it is no longer a 'technological follower' (CST, 1984a).[2] Indeed, a recent survey of researchers showed many regard Japan as being ahead of Europe in various fields (although not in life sciences and chemistry), and leading the United States in a growing number (STA, 1987b, pp.11–12). Stress is now being placed upon the indigenous development of basic technology and the underlying fundamental research (CST, 1984a), especially given protectionist threats from the United States and EEC and complaints about Japan 'free-riding' on their R&D efforts. Priority has also been given to strengthening the links between universities, government laboratories and industry. Furthermore, to overcome the insularity of Japanese basic research and improve creativity, attempts have been made to open up the R&D system to foreign scientists and stimulate international cooperation (STA, 1987b). The targeting of life sciences for international collaboration (notably through the Human Frontier Science Programme) has, however, brought further criticism from countries who see this field as one of their remaining comparative advantages over Japan.

In line with the belief that R&D is central to future industrial competitiveness, overall expenditure rose by 75 per cent over the period 1980–6 to reach 9,200 billion yen (2.75 per cent of GDP). The private sector share was nearly 80 per cent, extremely high in international terms, with an increase in basic programmes contributing to a near doubling of spending during the 1980s (Statistics Bureau, 1988; STA, 1987b). Government support, although rising more slowly, has also shifted towards basic research, with the Ministry of Education, Science and Culture (Monbusho) being favoured in recent budgets relative to the two agencies primarily responsible for more strategic and applied research — the Science and Technology Agency and the Ministry of International Trade and Industry (MITI). Expenditure on energy and other

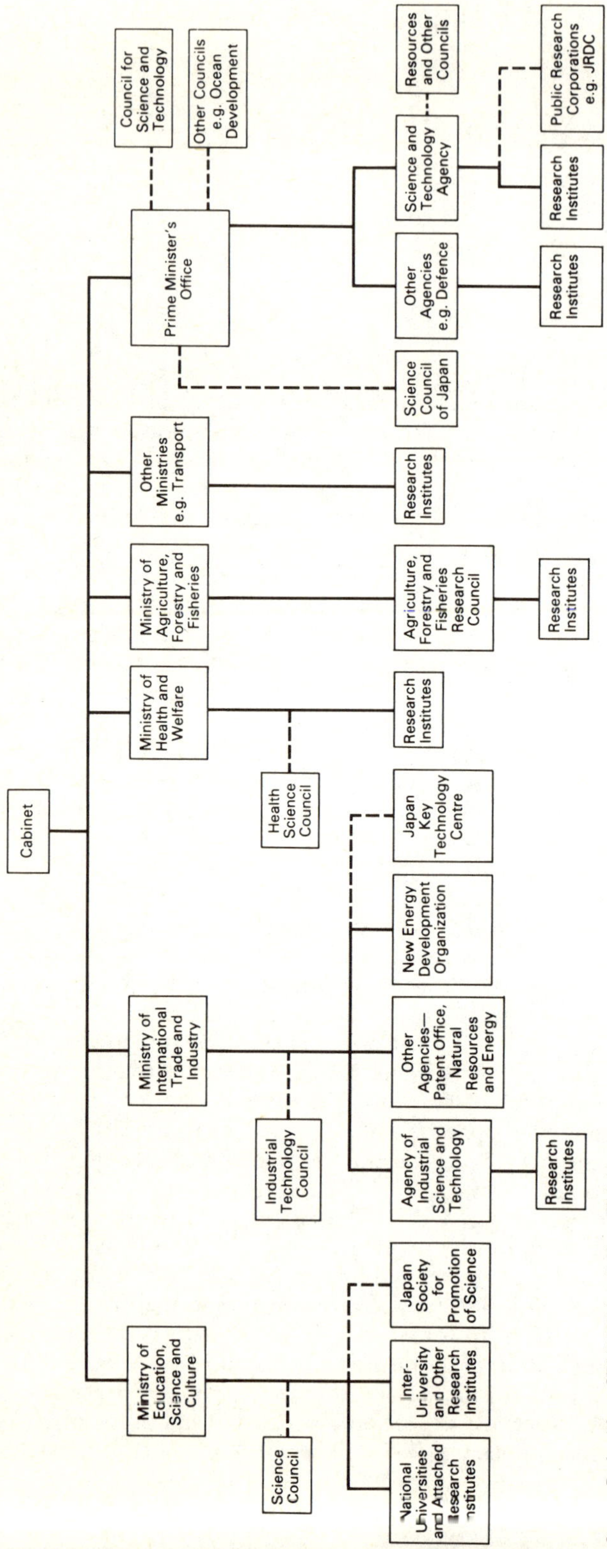

Source: Derived from STA (1987a) and other official reports.

Figure 6.1 Outline of government R&D system in Japan

large-scale development projects has been reduced, while support for long-term basic technology programmes has grown (Ministry of Finance, 1987). An awareness that the most promising opportunities often emerge from the confluence and fusion of previously separate research fields (STA, 1986; Kodama, 1986) has stimulated efforts to coordinate R&D programmes across scientific disciplines, technological and industrial fields, government agencies, and the private and public sectors. This has led to wider use of systematic foresight in establishing priorities for the interdisciplinary programmes which 'must be pursued increasingly from a long-range viewpoint if this nation ... is to continue to grow' (Ministry of Finance, 1987, p.2). There has also been greater stress on planning and evaluation to improve the efficiency of R&D programmes (ibid., p.2).

An outline of the Japanese R&D system is shown in Figure 6.1, the centrality of consensus-generating mechanisms being reflected in the reliance upon high-level advisory councils to frame research policy both nationally and within agencies. In what follows, we first discuss the use of foresight in attempts by the Council for Science and Technology to establish national guidelines for R&D. The activities of the Science and Technology Agency (STA) are also considered, in particular its series of thirty-year technology forecasts. We then examine the small but growing role prospective analysis plays in setting policy for basic academic research within the Science Council of Japan and Monbusho. Next, foresight in strategic research and basic technology is reviewed, coverage including MITI's Agency of Industrial Science and Technology, the Research Development Corporation of Japan, and a number of public and private-sector think-tanks. Finally, after discussing the research foresight undertaken in science-based companies we summarize the lessons which can be drawn from Japan.

2. FORESIGHT IN NATIONAL PRIORITY-SETTING

The process of framing national research policy has three main characteristics: first, an integration of discussions on broad R&D priorities and likely budgetary requirements. Second, a stress on consensus-seeking advisory councils with the breadth and seniority of membership to ensure agreed policies are implemented. Third, a clearly structured hierarchical process in which: (i) overall guidelines and priorities are established by the Council for Science and Technology (CST); (ii) policies for designated fields are set by other councils (such as the Space Activities Commission), with CST being responsible for research in life sciences, materials, energy and disaster prevention; and (iii) within the overall policy framework agreed by CST, individual ministries and agencies determine their own R&D priorities under the guidance of sectoral advisory councils. Overall coordination is provided by the Science and Technology Agency for all areas of government R&D apart from academic research (STA, 1987c; Tanaka, 1987). Given that the Japanese feel 'the utilization of councils in the process of policy planning is very effective' (STA, 1987c, p.27), it is necessary to focus particular attention on their role in foresight.

(a) Council for Science and Technology

As 'the supreme deliberative council on science and technology policy', CST is responsible for generating national consensus on 'long-term and comprehensive goals for R&D', and ensuring that funding priorities and the research infrastructure are appropriate for these needs (STA, 1987c). Reflecting the importance accorded to R&D, its membership includes the Prime Minister (who chairs key meetings) and the Ministers of Finance[3] and Economic Planning. Although formally the Council only advises the Prime Minister, its recommendations are invariably later enshrined in government policy (Hirasawa, 1988, p.4).

The long-term horizon for policy decisions means that prospective analysis is essential. The way foresight is integrated into decision-making in Japan is well illustrated by the process for carrying out CST's most important task — producing its 'overall guidelines on science and technology policy'. The main features are as follows:

(1) The CST Council first ensures broad agreement exists among government departments, R&D agencies and industry on the remit and orientation of the guidelines exercise. (The Council also later acts as the 'authorizing committee' for the final report.)
(2) The influential CST Committee on Policy Matters (which consists of six council members and eight leading researchers) then establishes a working group to manage the exercise and supervise drafting of the report. This has the linkages vertically (with government) and horizontally (with agencies, companies and the research community) necessary for negotiating the compromises required to secure consensus on the guidelines.
(3) Responsibility for most of the concrete tasks is devolved to two subcommittees on 'general planning' and 'research activities'. These draw in further external members, including experts with forecasting and science-policy skills. The subcommittees are serviced by officials from the STA General Policy Bureau[4] who play a coordinating role in the exercise. Inputs are provided by: the annual CST 'hearings' with ministries on their budgetary plans and R&D priorities; consultations with agencies, companies and scientific interest groups, carried out both informally and through meetings addressing specific issues of concern; discussions with invited technical experts from government, industry and the universities (often taking the form of seminars on the prospects for particular fields); analyses of findings from previous foresight studies (especially those undertaken by STA to obtain comprehensive overviews of future trends); and studies commissioned from external think-tanks.

The most recent CST guidelines were published in 1984 and recommended a shift towards creative longer-term research, increased internationalization of the scientific system, and ensuring R&D was in harmony with future societal needs. Other tasks were to improve the basic research infrastructure, strengthen cooperation between public and private sector R&D, and introduce more systematic procedures for priority-setting and evaluation in government agencies (CST, 1984a). Among the foresight studies providing background information for the report was one commissioned from the

Institute for Future Technology concerning 'the systematic formulation of R&D themes in science and technology from the viewpoint of social needs' (IFTECH, 1982). This employed a two-stage procedure in which government, academic and industrial officials were first surveyed on socio-economic needs over the forthcoming decade and beyond. In a second survey, experts on social and economic issues ranked the importance of each of the identified needs on a four-point scale, while scientists and R&D managers proposed research programmes to meet those needs. By means of a matrix relating social and economic needs to future R&D requirements (ibid., pp.25–6), around 150 research themes (including twenty with a basic orientation) were identified as having potential for major socio-economic impact. The findings helped determine the 101 priority fields judged by the CST report as likely to provide the 'basic seeds' for technological innovation in the period up to the twenty-first century. Many were in interdisciplinary materials, information technology, electronics and life-sciences areas (such as bioelectronics and fundamental brain mechanisms).

According to the Chairman of the Committee on Policy Matters, the guidelines exercise required appreciable resource and personnel inputs.* The wide-ranging consultations also meant that it took a long time to negotiate agreement among ministries and agencies. However, such exercises are carried out only every seven years and, without a high level of commitment, the guidelines would have little impact. Following further discussion, the CST recommendations were accepted by the Cabinet in early 1986, thus becoming the basis for national science policy in the period to 1991 (Cabinet Council, 1986). It should be stressed that the 'guidelines' are just that — they are not intended to impose rigid planning directives on agencies. Generally, however, the themes are later taken up in the annual STA white papers on science and technology — for example, the issue addressed in 1987 was 'the internationalization of Japan's science and technology' (STA, 1987b).

Systematic foresight techniques are also routinely used in more specific studies initiated by CST to address urgent policy issues or to frame basic priorities in the fields for which it is responsible. These are normally carried out by *ad hoc* working groups which may commission studies from consultants or think-tanks, as with a recent report establishing guide-lines for research evaluation. This set out criteria for the ex-ante assessment of projects (for example, 'likely direct and indirect scientific or technological impacts', or 'possible future social constraints on implementing anticipated results'), and thus explicitly recommended the incorporation of prospective analysis in research management (CST, 1986; Irvine, 1988; Tanaka, 1988). Other influential studies have covered opportunities and needs (over the next twenty years or so) in fields such as life sciences[5] and cancer research, but some have addressed social priorities like 'the ageing society' (see CST, 1988a). Such studies typically involve marshalling large amounts of quantitative and qualitative information, often including cross-national comparisons of R&D trends.

Foresight inputs to policy formulation are also made by expert seminars. Since 1982, CST has organized an annual forum to assess research prospects in emerging strategic areas. Its 1987 forum on 'future materials' was attended by over sixty Japanese and foreign researchers drawn from materials science and 'technologically converging fields' (e.g. life sciences and electronics), together with members of the Committee on Policy Matters and STA officials. The three-day discussion focussed on six promising areas, one of which was artificial cell functions. Here, the task was 'to search for self-organizing industrial materials for the 21st century, including medical and electronics materials, by utilizing the self-organizing ability of ... artificial cells' (STJ, 1987a, p.45). The forum provides a valuable opportunity for interaction between government officials, industrialists and the scientific community in discussing long-term research options.

Finally, CST has at its disposal limited resources within the Special Coordination Funds for Promoting Science and Technology, which it has increasingly used to support innovative cross-disciplinary programmes in materials and life-science fields aiming to encourage collaboration among universities, government laboratories and industry (STJ, 1987b). In this way, CST has been able to initiate exploratory work in areas cross-cutting the responsibilities of agencies, as well as to conduct feasibility studies to assess research opportunities and needs. One important recent case concerns the Human Frontier Science Programme (discussed below on pp.168-70), where disagreement between MITI and STA would otherwise have made it impossible to initiate the international feasibility study required by government before agreeing to commit funds. This example highlights two limitations in the CST approach to priority-setting: first, the traditional autonomy of Japanese ministries means that it is sometimes difficult to resolve policy issues involving fundamental conflicts of interest. Second, the emphasis on consensus tends to result in lengthier decision-making than in Western countries, while the scope for ranking priorities and for proposing radical policy alternatives is often more limited (Tanaka, 1987). Given the current desire to foster scientific creativity, the conservative limitations of consensus processes are increasingly of concern, although few would wish to sacrifice their advantages in relation to policy implementation. The government has signalled its desire (through the Ad Hoc Council on Administrative Reform) for stronger coordination by CST (NSF, 1985a), especially in relation to academic and other basic research, and this may eventually result in more proactive policy depending less on an agreed synthesis of the plans of departments and agencies (Kono, 1986).

(b) Science and Technology Agency

(i) Coordination of national research policy

Established in 1956, the Science and Technology Agency has two principal responsibilities: (i) supporting R&D in generic fields cross-cutting the interests of different ministries (for example, materials) and in capital-intensive areas of 'big' science and technology (such as nuclear and aerospace research); and (ii) overall planning and coordination of national scientific and technological activities (STA, 1987a).[6] This dual mission has in the past led to friction with other ministries over departmental responsibility for R&D initiatives, and it was therefore decided in 1985 to separate the planning and research support functions.[7] A General Policy Bureau was set up (replacing the existing Planning Bureau) with a remit limited to national coordination and planning of research, including servicing CST and administering the Special Coordination Funds Programme.

A major part of STA's support for CST involves research foresight. The General Policy Bureau is not only responsible for administering *ad hoc* studies commissioned by CST, but also for providing information and policy advice on R&D trends and emerging issues. Considerable effort is devoted to regular surveys of 'seeds' and 'needs' across the entire spectrum of scientific and technological fields. One influential recent study was a review of 'future directions of basic research to induce innovative technologies' undertaken in 1985–6 by the National Institute of Resources (a think-tank attached to the General Policy Bureau). The main objective was:

> to find and plant the seeds of potential new technologies so they might grow and flower in the next 10 to 15 years. Industry in Japan is either catching up or has overtaken that of the other technologically advanced countries and led to problems, such as trade friction or high-technology friction. Japan has no choice but to move towards a policy that places more emphasis on basic research. [STJ, 1985, p.34]

The study was organized in a similar way to the CST guidelines exercise. Overall responsibility was vested in a committee of eight eminent researchers drawn from universities and government laboratories, thus ensuring that the final report had the requisite scientific authority. A further eighteen distinguished scientists (including several from industry) acted as 'senior investigators', forming a working group together with staff from the National Institute of Resources (NIR). They provided further scientific legitimacy as well as ensuring inputs from both science-push and demand-pull perspectives. This group was assisted by over fifty 'associate investigators' representing active younger scientists (of assistant professor or equivalent status) familiar with state-of-the-art developments and opportunities in key fields. Expertise on foresight techniques was provided by NIR staff, while several private-sector think-tanks (including Technova and Nomura Research Institute) undertook reviews of overseas developments and studies of promising opportunities in specific areas of industrial technology.

The first task was to identify research themes likely to provide the sites for important future developments. This was approached by dividing 'high technology' into the physical, chemical and biological sciences, and interviewing large numbers of researchers to obtain their personal views on 'themes with a promising future'. This marked rather a change from the procedure employed previously in that it was decided not to select experts from a representative sample of organizations, who would then report prediscussed views agreed with colleagues and senior management. While the latter procedure had proved effective in identifying lines of future technological development (where the direction and dynamics of emerging trends were already apparent), STA had concluded it was too conservative an approach for more basic research. As a result, the interviews with researchers (who were selected primarily on the basis of their scientific track-record) were organized to preclude prior consultation. This yielded fifty-two candidate research areas (for example, artificial metal enzymes).

In-depth studies were next conducted on the current state of research in each area and the prospects for breakthroughs with socio-economic potential. Assessments were obtained from leading science-based companies as to market and other factors likely to influence the commercialization of potential developments. This was followed by an appraisal of measures to promote initiatives in the more promising lines of research and to determine the general priorities that should be recommended for future government funding.

One feature of the survey was that 'oxide superconductors of high critical temperature' was among the fifty-two research themes identified in the initial phase. This was well before the announcement of the IBM discovery of high-temperature superconductivity in 1986, and provides evidence of how foresight can focus attention on active research areas showing promise for yielding breakthroughs.[8] Indeed, Japanese companies (such as Hitachi) and researchers (notably at Tokyo University) were already working in the field, and, while impressed with IBM's scientific success, saw this as a discovery waiting to be made rather than an essentially unpredictable breakthrough.*

According to STA officials, the study had an impact on CST discussions of new research initiatives, influencing selection of topics for the Special Coordination Funds Programme. It also helped convince the government that industry supported an expansion of the innovative JRDC Exploratory Research for Advanced Technology Programme and establishment of the Frontier Research Programme (which includes work on the 'artificial brain'), as well as demonstrating consensus within the research community on priorities.

More generally, it has been recognized that the increasing need to focus foresight on science-push factors calls for new approaches to setting research policy, and a broadening of the analytical capability available to CST within the General Policy Bureau. As a result, the National Institute of Science and Technology Policy (NISTEP) was established in 1988 (replacing the National Institute of Resources). The aim is to strengthen STA's science-policy effort, especially in such areas as international cooperation, R&D indicators and the growth dynamics of scientific and technological fields (STA, 1988). NISTEP will also undertake research foresight and may assume responsibility for initiatives like the STA long-term forecast considered next.

(ii) Thirty-year technology forecast

Unlike most other STA prospective studies, the thirty-year technology forecast is targeted neither at a particular audience nor a specific policy process. Rather, the aim is to make available a comprehensive overview of trends in technological innovation 'so as to contribute to the planning and formulation of the government policies on science and technology and to provide private enterprises with guidelines for activities in science and technology' (STJ, 1987c, p.34). The first in the series of quinquennial forecasts was completed in 1971 (STA, 1972), and this stimulated much of the subsequent growth in Japanese foresight activities.[9] The origins of the survey lay in recommendations made in 1968 by the Future Research Operational Group (established under the auspices of the Japan Techno-Economics Society), which 'studied technological forecasting and the development of creative power as a means of shifting Japanese technological development from reliance on imported technology to a self-reliant approach' (JATES, 1986, p.8). Discussions with US forecasting experts convinced this committee of senior industrialists that there was much to be gained from government initiating a comprehensive Delphi exercise on the nation's technological future.[10] The importance now accorded to research foresight in Japan therefore stems as much from accumulated experience as from any intrinsic cultural predisposition towards studying the future.*

The forecasting approach recommended to STA was based on four principles which have continued to guide even the most recent survey undertaken in 1986–7 (IFTECH, 1988a). First, the forecast must consider future economic and social needs as well as developments in science and technology. Second, it should cover holistically all areas of research rather than focussing on individual fields. Third, the relative importance of different R&D tasks needs to be evaluated and priorities established. Finally, forecasts of specific research developments should be seen as having two aspects — an exploratory or predictive element relating to experts' expectations of change and a normative element which involves setting objectives and time-scales within which they are to be achieved (STA, 1972, p.7).

The structure and organization of the forecast also largely remain unchanged, although tasks like survey design, data-processing and analysis of results were entrusted to the Institute for Future Technology (IFTECH) in the most recent exercise. Responsibility for managing the survey and 'authorizing' the results remained with a Steering Committee (consisting of fifteen forecasters, academics, R&D managers and industrialists) reporting to the Director-General of STA.[11] Its members in turn headed eleven working groups dealing with specific R&D areas (e.g. life sciences). The sectional groups each comprised between six and nine experts drawn from the public and private sectors (in order to balance science-push and market-pull perspectives), with overall coordination undertaken by IFTECH. STA officials joined the working groups as observers to ensure close communication with IFTECH during the exercise.[12]

A two-stage Delphi procedure was employed involving the following tasks:

(1) Each working group identified R&D developments in their area likely to contribute to meeting societal needs over the next three decades.[13] They began by examining whether the 800 survey topics included in the 1982 forecast were still relevant and what modifications were required in view of subsequent progress, any developments likely to be commercialized within five years being excluded from further consideration.[14] Drawing upon the collective knowledge of its members, each group identified new topics for inclusion, a time-consuming task taking up all four meetings in the first year. This yielded 1,071 topics — 57 per cent new, 25 per cent modified, and the remainder unchanged.[15]

(2) A list of experts to be consulted in the Delphi exercise was compiled by IFTECH. The selection criteria employed attempted to ensure: (i) representative coverage across all main R&D fields and socio-economic sectors; (ii) a balance between industrial respondents and those from government and academia; (iii) avoidance of conservatism by involving a considerable number of younger researchers and excluding anyone no longer active in their field;[16] and (iv) inclusion of non-conventional perspectives from futurologists, journalists and even science-fiction writers. Inputs to the selection process were made by the working groups, government agencies, academic societies and industrial research associations.

(3) Invitations to participate were then sent by STA to 6,000 identified experts. Because only 2,650 responded positively (fewer than expected),[17] a further 500 names had to be added without first sending out invitations in order to avoid possible sampling errors.

(4) Questionnaires were circulated to the 3,150 experts requesting a range of judgements on the 1,071 R&D topics. These covered: (i) their potential societal importance (ranked on a three-point scale); (ii) foreseeable constraints on realization (technical difficulties, socio-cultural acceptability, economic/market problems, or other factors); (iii) the most appropriate mechanism for promoting R&D (domestic programme, technology transfer from abroad or international collaboration); and (iv) the sector which should primarily be responsible (national or prefectural government, joint public/private sector or private industry). Respondents also ranked their ability to assess future developments in respect of each R&D topic (on a four-point scale) and where appropriate added comments to clarify the reasoning behind their judgements. The survey results were analysed by IFTECH and summaries of the aggregate responses to each topic prepared (including statistical profiles of estimated realization dates).

(5) In the second stage of the Delphi procedure, respondents were invited to reassess their views, bearing in mind the general overall judgement. If continuing to hold markedly divergent opinions, they were asked to give reasons. To provide guidance on future policy, respondents were also invited to propose appropriate government initiatives (research or training programmes, improved coordination or 'other measures') to help realize each development in the predicted period. Table 6.1 contains illustrative forecasts for the health and medical care sector: for example, it can be seen that a cure for AIDS was predicted for around 1997 (with 50 per cent of experts specifying dates within the range 1993–2001). The main factor likely to limit progress was judged to be shortage of funds, while the preferred R&D support mechanism was multinational collaboration across both public and private sectors. Such a forecast is clearly more than a simple prediction, and should rather be seen as a statement of informed opinion on what might realistically be achieved given appropriate action to mobilize resources and coordinate R&D efforts.

As regards the lessons for research foresight, three specific features of the 1986–7 exercise merit discussion. First, increased emphasis was placed in the published report on expert commentary on projected trends in key fields and industrial sectors. This was prepared by working groups using statistical analyses of survey responses, and reportedly added to the credibility of the predictions as well as making it easier to interpret the significance of particular developments. Media interest focussed especially on a review of 'hot areas of research' likely to impact on a broad range of industries and socio-economic sectors.*

Second, the involvement of a specialized policy research institute like IFTECH enabled greater attention to be paid to conceptual and theoretical questions than previously.[18] A more analytical approach could therefore be

Division	Topic	Round	Number of respondents	Degree of expertise (%) High	Medium	Low	None	Degree of importance (%) High	Medium	Low	None	realization or reasons for realization (%) Technical	Social	Economic	Other	of promoting R&D (%) Domestic development of technology	Introduction of technology from abroad	International joint development	promoting R&D (%) Public sector	Public and private sector	Private sector	government (%) Funding	Staffing	Organization	Other	Unnecessary
Health and medical care / Improvements of therapeutic methods	Performance of kidney, heart, lung, and other organ transplantation in Japan with a regularity on the order of that in the United States today.	1	140	22	32	34	13	55	36	7	1	3	86	1	1	31	26	35	34	57	4					
		2	108	20	31	35	15	64	29	6	0	4	93	0	0	29	21	44	28	67	2	6	3	56	7	1
		X	25					92	8	0	0	0	100	0	0	32	16	48	16	80	4	0	8	68	4	4
	Development of agents to suppress rejective reactions which outperform ciclosporin.	1	110	21	28	19	31	62	33	5	0	70	6	4	4	27	15	52	9	68	19					
		2	88	20	29	20	31	66	32	2	0	86	5	3	0	22	15	60	3	77	18	41	14	11	2	6
		X	26					85	15	0	0	73	4	12	0	19	8	69	4	73	23	38	12	8	4	19
	Widespread use of hetero-organ transplantation as a temporary means of treatment.	1	117	19	26	28	27	21	38	25	15	38	38	1	3	15	15	47	21	51	5					
		2	93	17	27	29	27	19	41	24	14	40	45	0	1	13	10	61	20	59	5	20	5	27	8	3
		X	22					32	23	23	23	36	45	0	0	23	0	55	9	59	9	9	9	36	5	5
	Perfection of systems to lengthen organ preservation, enabling worldwide supply of all kinds of organs for transplantation.	1	127	18	31	31	21	44	38	13	4	26	55	7	2	5	3	80	30	57	2					
		2	100	17	28	34	21	44	37	12	5	28	59	6	1	7	1	84	27	85	2	16	1	46	6	2
		X	21					57	24	10	10	57	43	0	0	10	0	81	10	81	5	33	5	43	5	0
	Possibility of autotransplantation of tissue damaged by injury or disease, using tissues extracted and cultivated in advance.	1	118	15	29	30	26	34	48	14	3	48	14	22	4	30	8	53	24	60	8					
		2	98	15	28	34	23	32	52	15	0	62	11	20	2	32	5	59	24	66	6	32	5	32	4	0
		X	19					58	32	11	0	79	5	11	0	32	5	58	26	63	11	37	11	26	5	0
	Development of effective anti-cytomegalovirus agents.	1	84	13	21	19	48	42	45	13	0	76	4	2	2	29	11	57	10	73	15					
		2	69	10	24	20	46	42	46	12	0	88	1	3	0	26	7	61	9	68	20	46	13	4	3	6
		X	13					54	31	15	0	92	0	0	0	31	15	54	0	62	38	38	15	0	0	23
	Widespread use of pharmaceuticals highly effective against hepatitis and other viral diseases.	1	124	14	34	29	23	73	23	2	0	77	2	4	2	27	9	60	8	77	9					
		2	96	13	36	26	24	76	20	2	0	90	1	3	0	23	4	68	4	81	10	47	10	9	1	2
		X	17					82	18	0	0	88	0	0	0	35	6	59	0	76	24	35	18	24	0	6
	Development of antifungal agents of assured effect and low toxicity.	1	113	13	31	28	29	47	47	4	0	77	2	4	2	30	11	53	6	73	16					
		2	87	11	35	23	31	47	48	3	0	91	1	2	0	34	5	55	3	72	21	47	11	5	1	3
		X	14					71	29	0	0	86	0	0	0	21	7	64	0	64	36	50	21	0	0	14
	Mass in-vitro production and employment of all types of therapeutically effective antibodies based on gene manipulation	1	127	15	34	30	21	60	34	3	1	68	4	11	2	28	7	58	7	77	9					
		2	99	16	33	29	22	66	30	2	0	88	3	3	1	24	3	69	3	83	10	47	6	9	3	4
		X	20					90	10	0	0	85	0	5	0	40	0	60	0	80	15	55	10	10	5	10
	Establishment of cure for AIDS	1	124	11	31	36	23	45	47	6	1	73	6	3	2	10	23	58	21	69	3					
		2	99	13	33	32	22	49	44	4	0	91	2	2	0	8	15	72	15	78	3	45	8	17	0	0
		X	16					75	19	6	0	81	6	0	0	13	19	63	31	56	13	44	6	19	0	0

Time of realization (Year)

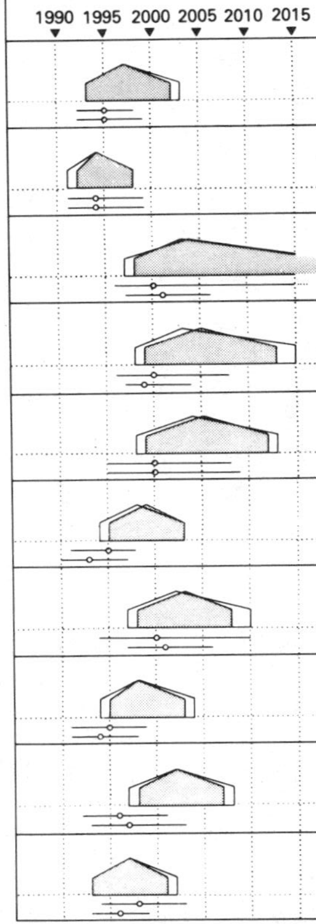

Table 6.1 Illustrative results from 1987 STA thirty-year forecast in the health and medical care sector

Note: (i) The non-shaded areas within charts represent the distribution of views on estimated realization time in the first round of the Delphi survey, while shaded areas refer to modified judgements made in the second round. The two lines under charts refer to the distribution of judgements made in each round (with the upper line relating to the initial survey) by those describing themselves as having 'high expertise' on the topic in question (see IFTECH, 1988a p. 16). Responses are limited to the two middle quartiles, with the apexes of charts (and emphasized dots on bar lines) depicting the median points in each set of judgements.

Source: IFTECH (1988a, pp. 164–5). Reproduced with the kind permission of the Institute for Future Technology.

adopted to identify common factors underlying apparently diverse technological developments — for example, the forecasts for 116 topics in microelectronics, optoelectronics, bioelectronics and information processing were structured in terms of six core trends (such as large-scale integration and miniaturization). A wider range of data analysis was also undertaken, yielding a variety of policy-relevant findings concerning: (i) the characteristics of high priority areas (fifty-one topics in ten fields related to basic work on 'elucidation of natural principles and phenomena'); (ii) the limitations on forecast targets being achieved (technical factors were cited in 63 per cent of cases, economic in 20 per cent and social in 4 per cent); and (iii) measures to promote R&D (domestic initiatives were judged most effective for more than 70 per cent of topics, while in not a single case was it felt necessary to rely on imported technology).[19]

Third, the fact that fifteen years had elapsed since the initial survey meant it was possible to conduct an evaluation of the accuracy of previous forecasts. Wide variations across fields were found among the 400 topics for which assessments could be made. Success in terms of projected targets being achieved (irrespective of timing) was highest in health and medical care (94 per cent), life sciences (88 per cent) and mineral and water resources (85 per cent). In contrast, the record for topics in energy (36 per cent) and transportation (36 per cent) was especially poor. A different picture emerged, however, if the date predictions were realized was also taken into account (using a criterion of ±5 years). This showed forecasts to have been most accurate for telecommunications (77 per cent), agriculture, forestry and fisheries (76 per cent) and information, electronics and software (67 per cent), and least successful in transportation (14 per cent), life sciences (25 per cent) and energy (27 per cent). The sharply lower success rate for life sciences was because many developments had been achieved much earlier than anticipated following rapid scientific progress in the field as a whole. According to the STA official responsible for the survey, this highlights a general rule that 'the more basic topics with a very steep realization time have tended to be predicted most accurately'. As a result, he was confident, for example, that 'finding a cure for AIDS by the date predicted [1997] has a high probability of realization'.* In contrast, forecasts in fields characterized by lengthy time-horizons for development programmes or infrastructural investment have been least successful, with those for energy particularly influenced by unforeseen fluctuations in the price of oil.

Probably the most significant finding from the assessment of forecast accuracy concerns the success rates achieved by specialists on particular topics compared with respondents whose expertise is in other areas. According to IFTECH officials, there was little difference in the predictive performance of the two groups. However, experts were slightly less accurate in their forecasts than the overall set of respondents for each R & D area, tending to be over-optimistic in estimating realization times. While non-experts had less in-depth knowledge of a particular topic, they were often well-informed on developments in related fields (perhaps being more aware of opportunities for cross-disciplinary interaction) or in a better position to identify barriers to implementation (experts in clinical medicine specialties, for example, tended to be more conservative than their colleagues in the basic life sciences).[20] The clear lesson is that foresight panels should be constituted to draw on a wide range of disciplinary expertise even when concerned with very specific research topics.*

More generally, the 1987 forecast opened up a debate on the strengths and weaknesses of consensus approaches to foresight, especially their utility in framing policy for more basic science. Worries had previously been expressed that Delphi approaches may yield overly conservative predictions, and this became a matter of concern in late 1986 following the IBM breakthrough in superconductivity. At the time, IFTECH was analysing the second round of the Delphi exercise, including responses to two topics on potential developments in the field: 'the time to realize superconductivity with a transition temperature of more than 20°K, and 'the time of commercialization of ceramic superconductors with critical temperatures above liquid nitrogen which will be used for industrial electric machines'. While the prediction for achieving superconductivity was 2002 (and 2011 for commercialization), one respondent aware of the still-unpublished results indicated that such a breakthrough had already occurred and he could not therefore accept the majority view (IFTECH, 1988b). This caused some embarrassment among officials responsible for the survey, who quickly decided to undertake revised forecasts for two topics on the commercialization of superconductivity research. These suggested much shorter time-horizons, while recognizing that fundamental technological problems remained to be solved (for example, in durability of performance and manufacturing).

Views of officials on the policy implications of the superconductivity episode differ. On the one hand, IFTECH staff argue that:

> The Delphi technique is essentially a way to sum up the accumulated experience of large numbers of experts. It is not the way to identify the especially novel ideas held by only a few creative individuals since the approach is based on consensus formation. You cannot, therefore, use Delphi to forecast fundamental scientific discoveries, although it can be used to predict the timing of important breakthroughs where pre-established trends are already apparent. The strength of the STA survey is that it satisfies the need to identify basic technological trends in a comprehensive manner, and it is therefore of greatest use to industry.*

Other interviewees, especially from companies, agreed that the experience provides a note of caution on placing too much confidence in intrinsically uncertain predictions for basic science, as well as highlighting the conservative tendencies of consensus mechanisms for establishing longer-term policy. However, in their view the STA survey had once again proved its value as a tool in research *foresight,* even if the superconductivity prediction was somewhat inaccurate. The discovery was not regarded as totally unexpected since 'pre-established trends' had been apparent (e.g. in French and Danish research undertaken about 1980)* and even the 1981 survey had recognized this by including topics on superconductivity (STA, 1982). Along with other STA forecasts, it had therefore promoted a wider appreciation of the opportunities opening up in the field. Most importantly, the 1987 survey picked up information on a breakthrough known to only a few people, and helped transmit it to the Japanese R&D community which mobilized rapidly to explore the commercial possibilities. As a senior official from Sumitomo commented:

> To tell you the truth, the accuracy of the predictions in the STA forecast is not always high so we do not necessarily believe the precise details and dates. However, this is not a problem to us. The value of the forecast is that it provides a comprehensive overview of the technologies likely to become important, as well as giving a rough idea of when. We in our company have a very narrow eye-view of specific fields and the forecast helps draw our attention to relevant developments in other areas. Along with other firms, we had already identified superconductivity as a priority area in the early 1980s so the discovery was by no means surprising.[21*]

The thirty-year forecast is thus seen as primarily playing an *indicative* role, focussing attention on emerging R&D opportunities and needs and the policy initiatives required. Its principal strengths are, first, that the scientific community is periodically forced to think seriously about emerging issues, and about how its work should relate to future socio-economic priorities. Without this, more immediate problems would tend to crowd out consideration of longer-term strategy, the forecast providing an opportunity for scientists to make a case to government and the public at large as to the importance of basic research.

Second, the involvement of a broad constituency of experts from science-based companies and other research users enables government to assess in general terms the demands likely to be placed on the national R&D infrastructure. This is valuable in ensuring that research-support mechanisms and programmes are consistent with projected industrial and societal needs (for example, requirements for skilled scientists in specific fields). The forecast consequently provides essential background information on long-term trends to CST and other advisory bodies responsible for framing R&D policy.

Third, the STA forecast embraces developments across the full spectrum of research fields and industrial sectors. Such a holistic approach enabled the Japanese in the early 1970s to be among the first to appreciate the significance of emerging interdisciplinary areas like information technology[22] and mechatronics, and hence to derive many of the subsequent benefits. The most recent survey highlights the importance of 'chematronics' (the fusion of IT and chemistry/materials research), biotronics (the convergence of IT and biology) and various life-science fields (which accounted for the largest number of R&D topics ranked 'highest priority'). Such findings have proved useful both to companies (primarily in planning strategic research) and to agencies like MITI involved in launching interdisciplinary initiatives such as the Basic Technology for the 21st Century Programme (which typically require support from industry and the scientific community to be demonstrated before the Ministry of Finance accepts the case for funding).

Finally, the ability of the STA survey to generate consensus among Japan's leading academic, government and industrial researchers enables it to play an integrative role in shaping an overall national vision of how R&D can contribute to creating a desirable future. This is extremely important in informing the public about problems and possibilities, as well as focussing attention on the need to make social choices (for example, in respect of the ethics of transplantation) if certain types of research are to be continued. One result has been to generate a greater and more informed societal commitment to science and technology than exists in other industrial nations.[23*]

3. FORESIGHT IN BASIC ACADEMIC RESEARCH

The Japanese university system was originally modelled along German lines and therefore traditionally enjoyed considerable academic freedom and institutional autonomy. Enforced integration into the military-industrial complex during the Second World War strengthened hostility towards externally imposed research priorities, and reinforced an anti-industrial ideology within academia which subsequently hindered the development of links with companies. The postwar administrative structure accorded Monbusho special status among government agencies, enabling universities to remain outside the growing coordination of national policies which took place after STA was established in 1956. The Science Council of Japan (SCJ) has been central in defending academic freedom, using its position as 'the collective voice of the university research community' to resist external planning of university science during the 1960s and 1970s. The SCJ policy on research planning was set out in its Draft Basic Act for Science Research submitted to government in 1962:

> For sound development of science research, it is indispensable to guarantee the freedom of researchers. Thus, it is one of the key responsibilities of the state to respect to the utmost the freedom of researchers ... The basic principles for scientific research shall be worked out from the national standpoint and in a far sighted perspective by an authoritative body representing all the scientists in the country. [SCJ, 1975, p.108]

The importance of basic research for future competitiveness has, however, resulted in renewed demands during the 1980s for relevance on the part of the universities. This has been coupled with proposals for restructuring the higher education sector to promote greater creativity in both research and postgraduate training. These have been recurrent themes in government science-policy reports (e.g. CST, 1984a), with the Educational Reform Commission seeing the task of securing agreement from universities on the need for reform as a top priority (NSF, 1985b). As its former Chairman stressed, 'it is time to change the attitudes in universities and education from uniformity to individuality since industry is now at a world level and creative people are required'.* Although resistance to change has continued, budgetary pressure seems to be having an effect: the willingness of government (especially the Ministry of Finance) to increase expenditure on academic research has rapidly followed a shift in the attitude of SCJ towards longer-term planning and priority-setting, and the acceptance by Monbusho that socio-economic criteria have at least some part to play in resource-allocation decisions.

(a) Science Council of Japan

A major task of SCJ is to advise government on basic research policy and priorities across the entire range of natural and social science fields, its Chairman being an ex-officio member of the Council for Science and Technology. Traditionally functioning as a representative body of the scientific community, its 739 affiliated societies have a combined membership of nearly half a million. The governing SCJ Council was formerly directly elected by scientists and this provided a focus for reaction against plans to restructure the university system. As a result, the Council was controversially dismissed in 1983 and the organization placed under closer government jurisdiction, with the Prime Minister's Office now approving nominations for members put forward by academic societies.

Previous hostility towards research planning means that the only substantive foresight undertaken by SCJ has concerned the latest 'general plan of activities' (SCJ, 1988; Kondo, 1988). Such plans are prepared triennially (in line with the term served by Council members) to advise government on problems facing science and future investment needs. The reorganization of SCJ enabled a more systematic approach to consultation to be adopted in the 1988 exercise, and the resultant plan represents its most comprehensive assessment to date of scientific opportunities and priorities. The first step was to survey the views of 840 academic societies (including non-affiliated organizations) and 2,000 SCJ members on: (i) the relative international strength of Japanese science; (ii) fields showing particular scientific promise; (iii) potential bottlenecks to progress (e.g. instrumentation); (iv) research areas likely to impact on industry and society; and (v) the implications for funding and other policy decisions (e.g. international collaboration). SCJ's 180 discipline-based 'research liaison committees' then analysed the questionnaires, the results being used to draft summaries of trends in seventy-one fields. After review by the SCJ Council, these were published as an appendix to the overall plan (SCJ, 1988).

According to the SCJ President, Jiro Kondo, the plan represents a landmark in that the scientific community was for the first time able to agree an overall strategy and priorities for the future. While this 'unofficial white paper' reports optimism among researchers that Japan will continue to strengthen its position in fundamental science, certain problems remain to be addressed: (i) increasing the level of separately budgeted finance for research and training programmes in priority fields; (ii) reducing the current emphasis on 'big science'; and (iii) introducing improved mechanisms for planning and evaluation in the academic system.[24] While certain of the more radical recommendations are opposed by sections of the academic community, they reinforce parallel policy developments in Monbusho which have been favourably received by government. In particular, Kondo feels that a useful start has been made in tackling the difficult question of priority-setting in basic science, even though considerable progress still remains to be achieved. The underlying problem is that the diverse disciplinary membership of SCJ means that only fairly general priorities can be identified — for example, the fields given highest priority within engineering included superconductivity, biosensors, molecular devices and optical computing. The academic societies remain opposed to the establishment of overall priorities across fields. Indeed, a number were from the outset critical of any attempt to identify promising research areas, although few persisted in refusing to collaborate.*

Aside from the general plan, SCJ normally takes a long-term perspective in its *ad hoc* reviews of specific fields or emerging policy issues. These rely primarily on committee discussions but also make use of surveys and other quantitative data. Especially influential studies include assessments of instrumentation needs in big science (one of which led, for example, to the construction of the Tokyo University radio-telescope), while a 'committee on the ageing society' has been examining the research needs of the health system in the twenty-first century (SCJ, 1986).

(b) Ministry of Education, Science and Culture (Monbusho)

Monbusho is responsible for both infrastructural and project-based funding of academic research. Several support mechanisms are employed: (i) institutional finance for national universities primarily distributed on a per capita basis in the form of 'general funds' to departmental research groups (the *koza* system) but also as 'specific funds' for items like equipment; (ii) grants-in-aid for public universities (operated by prefectural governments) and private universities; (iii) support for various inter-university research institutes providing central facilities (e.g. the National Institute of Polar Research) and a large number of laboratories attached to individual national universities; (iv) grants for specific projects and programmes; (v) graduate and postgraduate fellowship schemes; and (vi) subsidies to the Japan Society for the Promotion of Science (JSPS).

Monbusho has so far made little use of research foresight. This stems both from its long-established policy of distributing most funds through the *koza* system (thus allowing professors autonomy to define their own lines of research) and an ideological preference for a responsive mode of operation relying upon academics and their professional associations to generate initiatives. As a result, the general view within the Ministry has been that:

> In an agency whose main mission is to support fundamental research, one cannot establish policy and priorities in the traditional Japanese way of AIST and STA. Outside groups like think-tanks lack the necessary expertise and so have little role to play. We must resort to the opinions of researchers — they are the think-tank.*

Overall policy is determined by Monbusho on the basis of recommendations from its Science Council, and from fifteen science advisers who provide guidance on general issues as well as on 'research trends in specific fields of science' (Monbusho, 1986, p.3).[25] Science Council membership is drawn largely from academia and this led industry and government to doubt its ability to recommend measures for tackling the structural problems facing university research: the lack of flexibility in channelling resources to more productive institutions; difficulties in targeting support on fields with high scientific or strategic promise;[26] an education system ill-equipped to train creative researchers; inadequate mobility and recruitment of 'new blood' (leading to institutional ossification and low scientific output); and limited cooperation with industry and government laboratories.[27]* Major changes have nevertheless occurred since 1984 following a seminal Science Council report on 'improving the scientific research system' (MSC, 1984). While still according 'respect for the researcher initiative', the report argued that changes in the nature of scientific activity (such as the increased pace of progress, specialization and instrumental sophistication, and the emergence of 'composite' and interdisciplinary fields), together with greater societal expectations (in respect of the potential socio-economic impact of strategic areas), call for more targeted promotion of research within a coordinated overall strategy. This advice has since resulted in a number of Monbusho initiatives which include:

(1) A major 'priority areas' programme introduced in 1987 to promote research on topics 'in which there are strong academic and social needs' (Monbusho, 1987, p.4).[28] Areas are chosen by the Science Council on the basis of proposals from researchers, academic societies and its own subcommittees, the forty-one selected in 1987–8 spanning the natural and social sciences.

(2) More stress on developing academic–industrial collaboration through a variety of support mechanisms: a 'joint research programme' (introduced in 1983 prior to the formal release of the Science Council report); 'university–industry joint research centres' (the first three of which were established in 1987 in strategic fields like mechatronics — NSF, 1987b); and increased funding for 'developmental grants' (oriented to practical application of academic research) and for several JSPS programmes aimed at stimulating university–industry cooperation.

(3) Greater emphasis on research cooperation, primarily in areas requiring large teams or costly facilities like world climatic change or nuclear physics (JSPS, 1986).

Most importantly, the Science Council report argued that any move by Monbusho towards targeted programmes would benefit from the use of research foresight:

> To meet new research and social needs, it is necessary to promote scientific research on a priority basis from a long-term and integrated viewpoint in all areas where promising seeds ... can be expected ... To identify important areas ... it is necessary to accurately grasp trends in scientific research ... and do sufficient preliminary research and preparatory study work ... It is necessary to make projections on the number of researchers needed from a long-range viewpoint. [MSC, 1984]

In preparing its recommendations, the Science Council undertook exploratory studies of trends in electrical engineering, linguistics, biology, solid-state physics, chemistry and economics (Monbusho, 1983; 1984). Carried out by expert panels, these examined whether existing support mechanisms could meet the future needs of each discipline. Although promising areas were identified, no attempt was made to generate priorities or to produce a synthesis report. Nor were systematic foresight techniques used. The significance of the exercise was rather that it represented the first attempt by Monbusho to undertake disciplinary reviews, and that the results provided evidence on the value of prospective analysis. This conclusion was reinforced by another study commissioned from consultants which made projections on the supply and demand for researchers to the turn of the century, taking into account such factors as economic growth, expenditure on training and academic research conditions (MSC, 1984).

Interviewees reported considerable resentment among academic scientists at the recent growth in targeted initiatives, and this may explain Monbusho's continuing lack of enthusiasm for research foresight. One exception, however, is the expansion in activities of the JSPS Advisory Committee on University–Industry Cooperative Research, whose main function is to identify and promote promising new fields. Two research development committees have been exploring the prospects for protein technology and 'the interface between biological functions and electronics', while around forty university–industry cooperative research committees are involved in discussions on the basic science needed for progress in strategic technologies like nanometer structure electronics (JSPS, 1986). A second exception is the advocacy of foresight by a more recent Science Council report outlining priorities for biosciences (MSC, 1986). Besides repeating the call for restructuring of the university system, it implemented its recommendation that such reviews should identify in detail the lines of research meriting future support. Experience with strategic planning in fast developing fields cutting across disciplinary and generic boundaries highlighted the requirement that:

> In promoting bioscientific research, it is necessary to identify and indicate key themes to be studied ... [and] to accurately grasp research trends and related circumstances. For this purpose it is advisable to have professional research groups survey and investigate research trends at home and abroad. [MSC, 1986, pp.3 and 6]

This report has been described as 'one of the most comprehensive and detailed prescriptions for [university research] reforms to appear', and one likely to encounter resistance from 'tradition and organizational inertia' (NSF, 1986a, p.1). Whether its recommendation that professional think-tanks be employed to analyse research trends will be implemented remains uncertain, given the widely held view that only academic scientists can legitimately perform this task. Whatever the outcome, the fact that a significant proportion of priority programmes are in the biosciences partly reflects the impact that can be achieved by well-documented analyses of prospective research opportunities, while the political realism shown by the Science Council has undoubtedly contributed to securing the recent increase in Monbusho's budget.

4. FORESIGHT IN STRATEGIC RESEARCH AND BASIC TECHNOLOGY

In contrast to Monbusho, mission-oriented agencies routinely employ foresight as an integral part of the process of formulating and planning new programme initiatives. A comprehensive treatment of Japanese foresight in strategic research and basic technology would therefore need to cover: (i) the fifteen ministries and agencies with sectoral R&D responsibilities; (ii) the technologically-oriented research financed by MITI's Agency for Industrial Science and Technology (AIST) through its network of national institutes, the Large-Scale R&D Programme and Basic Technology for Future Industries Programme (AIST, 1987a); and (iii) the support provided by the Science and Technology Agency for advanced technology (e.g. nuclear R&D) and interdisciplinary fields (e.g. life sciences) through its intramural laboratories, attached public research corporations and the two major research programmes it administers (Exploratory Research for Advanced Technology and Special Coordination Funds) (STA, 1987a). Given such a range of activities, some selectivity is inevitable and coverage of foresight has been restricted to MITI and STA. These were chosen, first, because their 1987 R&D budgets of \$1,700 million and \$3,200 million, respectively, dwarfed those of other mission-oriented agencies (which combined amounted to \$1,900 million). Second, they have both been at the forefront of moves to stress longer-term strategic research at the expense of more applied R&D, moves which have resulted in interesting foresight lessons. We first focus on MITI, in particular its Basic Technology for Future Industries Programme and involvement in establishing the Human Frontier Science Programme. This is followed by a discussion of the STA Exploratory Research for Advanced Technology Programme. Finally, the contributions of public-sector and private-sector think-tanks to research foresight are considered.

(a) Ministry of International Trade and Industry

Research planning in MITI is formally the responsibility of a hierarchical set of advisory councils and official committees. This often gives outsiders the impression that strong central government coordination of industrial R&D exists in Japan. However, intertwined with the official system is an extensive network of informal 'light touch committees' which ensure that the bottom-up element in planning is dominant. These unofficial committees bring together key industrial interests and generate the consensus required on new R&D initiatives if government funding is to be secured.

At the top of the formal planning hierarchy is the Industrial Structure Council (ISC) in which companies are well represented. It advises the Minister for International Trade and Industry on overall policy, and is responsible for the long-term 'visions of MITI policies' produced at the beginning of each decade. The last of these, published in 1980, recommended a major policy shift to 'forward engineering' in order to cope with the increasingly knowledge-intensive needs of Japanese industry. This entailed redirecting resources towards basic research on 'next generation' technologies, and restructuring the R&D system so that more emphasis is placed on radical rather than incremental technical change (MITI, 1980). Work began in early 1988 on the 'vision for the 1990s', the central themes of which are likely to be 'creativity in research', 'new institutional structures for advanced R&D and technological education', 'the role of technology in improving the quality of life', and 'internationalism in R&D'.[29*]

The role of the vision is 'to produce a consensus overview on future goals and priorities for MITI' and this means all relevant interest groups must be consulted.* The process of preparing the '1980 vision' began by 'collecting together the matters to be discussed'. This made use of results from a questionnaire survey of industrialists, government officials, scientists, economists, journalists and other opinion-formers, which sought views on key issues for the future. Questionnaires were also sent to foreign experts and to MITI officials located overseas. Working groups were next established under ISC subcommittees to deal with specific topics like 'future generation technology' and 'international issues'. These engaged in extensive consultation, especially with industry, and drew on inputs from informal committees set up to address particular problems. Ministry officials were responsible not only for coordinating the exercise but for providing background material, such as data on R&D trends, social and economic indicators, and analyses of results from previous foresight exercises, as well as for managing several studies commissioned from think-tanks. They also helped prepare the 'draft vision' which was then subject to further discussion within and outside ISC.

Similar consultative procedures are employed by ISC in producing more regular policy guidelines — for example, a report on objectives for Japanese industry during the period to the end of the century (MITI, 1986). This translated into agency policy the 1986 government R&D guidelines, providing a context for the 'annual visions' submitted to the MITI Minister. These are linked to the budget-setting process which begins in March when internal discussions are held within each MITI division (and in relevant advisory committees) to review existing policies and consider proposals in the pipeline for new initiatives. Negotiations to decide overall priorities then take place (with external interest groups being mobilized to support particular proposals) so that a draft budget is agreed by August. ISC next incorporates this into its 'annual vision', which in 1988 took the theme 'Harmonization of International Economy and Society' (MITI, 1988a).

The Industrial Technology Council (ITC) is the second level in the hierarchy of policy advice, providing guidance on 'important matters related to scientific technology ... in response to inquiries from the Minister for International Trade and Industry' (AIST, 1987a, p.27). This includes advice on R&D priorities for AIST, which means that ITC is central in the budgetary discussions. Each of its ten committees oversees an area of AIST activities (e.g. the National Programme Development Committee is responsible for Large-Scale Projects), while most of its forty or so subcommittees deal with specific projects (e.g. the Large-Scale Project on Interoperable Database Systems). The subcommittees in particular help guide and plan future research, drawing on large numbers of industrial, government and academic experts as members. A major aim is to establish consensus between AIST institutes and industry on goals and technical targets for new research, with MITI officials playing a central role in bringing together relevant interest groups to identify policy needs and priorities.

ITC makes greatest use of longer-term research foresight in: (i) *ad hoc* policy reviews responding to requests from the MITI Minister (for example, a report on the 'Future Course for Policy on the Development of Industrial Technology' produced jointly with the Industrial Structure Council — MITI, 1984); and (ii) studies carried out by committees established to explore the prospects for specific areas (such as electro-optics). Inputs are made here by the AIST Technology Research and Information Division, which undertakes extensive monitoring of global R&D trends (e.g. AIST, 1986a) as well as more detailed surveys of developments in emerging fields.

One interesting recent survey is a 'Study on Technological Trends Using Databases' commissioned from the Japanese Systems Development Research Institute (JSDRI, 1987). This was undertaken in connection with preparing the first MITI white paper on 'Trends and Future Tasks in Industrial Technology', a major theme of which was the importance of technological fusion[30], with chematronics identified as a fundamental technology for the twenty-first century (MITI 1988b). The aim of the JSDRI study was to explore the use of bibliometric and patent data-bases to identify sites where fusion is occurring among scientific fields linked to strategic industrial technologies. The approach involved using the field classifications contained within publication abstracts like *Compendex* and *Chemical Abstracts* as the vertical axes of matrices for each technology area, while sets of keywords representing both classical scientific and generic fields provided the horizontal axes. On-line searches were undertaken (using such keywords as AI or VLSI, for example) to establish how many 'hits' were made on papers included in each abstract category (e.g. 'chemical engineering and processing'). Trends in cross-impacts of keywords and index classification terms were then analysed to identify convergent research areas. A similar methodology was also employed by JSRDI to study joint patenting among companies in different sectors, enabling patterns of technological fusion to be identified in fields like fibre-optics where the wide range of technical skills required often encourages industrial collaboration. MITI officials regard the JSRDI work as methodologically promising, but stressed that further research is necessary to establish the meaning and validity of the empirical relationships identified using literature-based and patent techniques.*

To understand the role played by foresight in MITI policy-making, as well as the pervasive influence of informal industrial committees, let us focus on two initiatives. The Basic Technology for Future Industries Programme (BTFI) provides an example of Japanese 'best practice' in integrating foresight into the research planning process, while the Human Frontier Science Programme (HFSP) illustrates the political problems which can arise in cross-sectoral programmes requiring agreement among several agencies, and, in this case, negotiations with foreign governments.

(i) Basic Technology for Future Industries Programme

The BTFI programme was established in 1981 to develop 'revolutionary basic technology essential to the establishment of new industries which are expected to flourish in the 1990s' (AIST, 1987b, p.1), and was the principal mechanism for implementing the MITI policy of 'forward engineering' outlined in the '1980 vision'. Fourteen projects were being supported in 1988 in the fields of materials, biotechnology and novel electronic devices. Each is in an area which has 'theoretically or experimentally shown potential for application in new industrial technologies' (ibid., p.1), but which would be too risky or costly for individual firms to undertake alone. R&D is conducted from the basic stage to the point it is beginning to yield technologies or materials ready for practical application. Projects have a planned duration of ten years and are executed by specially constituted 'research associations' bringing together consortia of companies and institutes (Irvine, 1988). The ITC Future Technology Development Committee and its six subcommittees are responsible for strategic planning. Total expenditure on BTFI up to its completion in 1991 is likely to be around $850 million (110 billion yen).

Industrial 'light touch committees' were crucial in framing the programme, as was a Study Group for the Formulation of a Long-Term Industrial Technology Development Programme (AIST, 1981). Since the extensive foresight activities carried out in connection with the establishment of BTFI are described elsewhere (Dore, 1983; Irvine and Martin, 1984), we shall concentrate on the lessons that can be drawn from activities begun in 1986 to create a follow-on Basic Technology for the Twenty-First Century Programme (BTTCP).

As with BTFI, the planning process initially concentrated on three tasks: (i) generating consensus within industry on broad areas where potential breakthroughs might have extensive cross-sectoral impacts; (ii) identifying project themes which match commitments to collective research by industrial consortia with the technical capabilities of AIST institutes; and (iii) demonstrating to government that a national need exists for the programme. Over the last decade, MITI's position has perhaps begun to weaken as industry has attained a leading-edge position in many basic technologies, making it harder to enlist the support of companies in arguing for new programmes (Sigurdson, 1986, pp.135–6). The agency has therefore begun to focus more on niche activities, particularly in fusion fields where coordinated programmes of strategic research by consortia from several sectors can lead to major benefits even for larger firms. This reorientation is reflected in recent MITI industrial policy which includes encouraging the globally dominant electronics corporations to 'pull through' medium-sized chemical and biotechnology firms to the international forefront. Chematronics has thus been targeted as the core theme of BTTCP, just as previous MITI R&D initiatives have formed part of an integral vision of the future. Such visions are invariably shared by industry:

> Three kinds of technology, microelectronics, biotechnology and advanced materials, are coming into our sight today. Though each was developed in different fields and is further advancing respectively, there exists a common undercurrent, chemical technology. Therefore the key technology for the creation of the industry of the twenty-first century is chemistry. [Suzuki, 1987, p.45]

Signals began to be picked up from industry around 1983–4 that chematronics was likely to become strategically important. This became apparent, first, from the converging technological trends emerging from several foresight studies (e.g. NIRA, 1985a; STA, 1982; STJ, 1985), which were reinforced by the results of a survey commissioned from the Industrial Development Research Institute (IDRI, 1985) as an input to selecting projects for the latter phase of the BTFI programme. The survey was managed by a high-level steering committee, to which four working groups on electronics, mechatronics, new materials and biotechnology reported. Drawing on both quantitative information and extensive consultation, each group distinguished the factors structuring the growth trajectories of technologies in their area (e.g. greater speed, higher precision and increased capacity/density, in the case of electronics hardware). This framework was then employed to agree a total of 181 priority topics for future technology development (e.g. novel superconducting and superlattice devices). The underlying research needed for progress in each development topic was next identified (for superconducting devices, this included work on thin-film formation and materials reliability), together with examples of projected applications (such as supercomputers and magnetic sensors). The fact that seventy-two of the priority topics were in new materials, and many others relied upon interdisciplinary research in chemistry, electronics and the life sciences, provided clear evidence for the probable future importance of chematronics.

A second signal received by MITI came directly from industry through the emergence of collaborative research between electronics and chemicals companies (FTMCI, 1985a; Hirooka, 1984), and from increased interest in the new interdisciplinary field shown by industrial associations (e.g. FTMCI, 1985b). The agency subsequently lent its support to the establishment in 1986 of the Association for the Progress of New Chemistry (ASPRONC) which has been responsible for much of the informal planning for BTTCP.[31] Since then, four 'frontier technology study committees' have been discussing possible collaborative projects in advanced chemistry, new materials, electronics and information, and life sciences. The main chemical and materials firms have all participated, as have several electronics companies, with evening meetings held every two to three weeks. The task was first to explore possible precompetitive research of common interest, and then reach consensus on the broad outlines of projects and an appropriate organizational framework for collaboration (with AIST institutes generally taking on a coordinating function). The attendance of MITI officials as observers has facilitated communication between these informal groups, the agency and AIST laboratories.

The activities of these embryonic 'light touch committees' have been invisible in the *formal* planning process which began early in 1987 with the establishment of a Round-Table on Basic Technologies Towards the Twenty-First Century. Chaired by an eminent engineer, its eleven members included directors of major chemicals, materials and electronics corporations. Inputs were made by a working group of sixteen technical experts from AIST laboratories, universities and companies, including prominent younger researchers. There was extensive cross-representation among firms involved in the round-table and working group and those already participating in the ASPRONC study committees. In this way, the pre-discussed views of industry were incorporated within the MITI vision for the programme, which appeared during late 1987 in the form of a widely distributed report identifying forty-six technologies (e.g. molecular arrangement control technology) required for Japan to maintain its position as a leading nation in the impending 'third industrial revolution' (AIST, 1987c). To those familiar with the process for producing the vision, it is perhaps unsurprising that 'in every one of the 46 research domains singled out by the report, chemistry-based new materials development is cited as the central item on the agenda' (Maruta, 1987, p.156).

It is also instructive to focus in greater detail on the organization of the round-table and its political context. Responsibility for the exercise lay with the AIST Bureau for Next Generation Industrial Technology, which also serviced the working group. Although certain planning tasks were undertaken by three sub-groups on new materials, electronics and information and biosciences, it was recognized that overall conclusions could only be drawn by the working group as a whole 'since electronics people are, for example, interested in bioscience and may have different but relevant ideas. Discipline-based reports would have been no use'.* An interdisciplinary perspective was also maintained through joint meetings of the round-table and working group. Prior to initiating the round-table, AIST officials had coordinated six months of discussions within the agency's laboratories to identify possible projects for the planned programme, ensuring they took account of the activities underway in ASPRONC. A summary of the results from these discussions was provided to the round-table in the form of a 'preliminary proposal' at the beginning of its deliberations, together with background material like the 1985 IDRI survey of emerging technologies. According to an industrial participant, however, the proposal was 'rather underdeveloped and primitive. It was necessary to add many new ideas, significantly develop existing ones and undertake general setting of priorities'.*

Interviewees from both MITI and industry stressed that the significance of the round-table was its role in demonstrating to government the existence of consensus between industry and the research community that chematronics is a fundamental technology for the future.[32] Its task was also 'to *indicate* the basic themes of future work and their potential significance. It *did not decide* on the work. These details we will discuss later' (emphasis added).* Foresight exercises oriented towards producing overall visions thus have a strong advocacy function.[33] As one official remarked: 'the round-table was an integral part of the process of convincing the Ministry of Finance that important national opportunities exist and there is consensus that the fields in the new AIST programme are those needing to be excavated by industry'.[34]*

In securing government support for an initiative like BTTCP, it is also necessary to demonstrate consensus within MITI as to its priority relative to other proposed programmes. This involves a lengthy process of internal competition mediated by the ITC. Even after the political go-ahead has been obtained, competition continues in relation to decisions on project selection and budgetary support, with officials mobilizing 'their' consortia of institutes and companies to exert whatever influence is possible.

Provisional consortia were in 1988 engaged in negotiations over conditions for participation in projects (a difficult task when two direct industrial competitors wish to be involved) and in finalizing details of their proposals so that work can begin promptly once budgetary authorization has been given. As in the BTFI programme, proposals must outline the main goals and technical targets to be achieved over the ten-year duration of the project, these being progressively refined as work moves from the initial basic phase (where a pluralist policy is adopted of allowing rival groups to explore alternative research approaches) to increasingly focussed work on the most promising technological options in the second and third stages (Irvine, 1988).

In short, research foresight has been central in planning and launching the BTTCP programme, making important contributions in: (i) providing a holistic overview of promising interdisciplinary research opportunities; (ii) giving a structured framework for evaluating their relative technological merits and potential socio-economic impacts; (iii) generating consensus among AIST laboratories and companies with differing scientific and technological interests on the benefits of collaboration; (iv) demonstrating the existence of that consensus to government; and finally (v) helping frame individual projects and set agreed technical targets and goals. While the more visible elements of the process (especially the round-table exercise) might easily lead foreign commentators to the mistaken impression of top-down priority-setting, the informal planning mechanisms exercise a controlling influence and ensure a strong bottom-up orientation. Indeed, the loss of MITI's previous technological dominance over industry (which is now ahead of most universities and government laboratories even in more basic research) and the reduction in its ability to provide effective R&D subsidies have together resulted in even stronger reliance on bottom-up processes in order to seek out niche areas like fusion research fields. Companies have demanded greater control of intellectual property from BTFI and BTTCP, and MITI decided in 1988 to devolve its extramural R&D support activities to a semi-public 'special corporation'. This will overcome restrictions on public ownership of patents and give greater flexibility in allocating the increased resources likely to be required by BTTCP than is currently allowed under the strict limits on government expenditure.*

(ii) Human Frontier Science Programme

The planning process involved in the Human Frontier Science Programme (HFSP) has, in contrast, been far more problematic, but it nevertheless enables useful lessons to be drawn about the conditions under which research foresight can prove effective. The origins of the programme stem from a combination of political, bureaucratic and technological imperatives existing in the mid-1980s: (i) pressure on Japan, especially from the United States, to increase spending on basic science; (ii) the need for MITI to initiate new programmes to replace the reduced support government had indicated would be available for applied R&D; (iii) the recognition by industry that interdisciplinary research on biotronics, homeostasis and chematronics was crucial for the future development of information, communications and control technologies; and (iv) the desire of Nakasone to establish a prestige project matching the European Eureka and US SDI initiatives. These imperatives were rather more complex than those shaping earlier MITI programmes and this is one reason major political problems have arisen.

According to senior officials, plans for HFSP were initiated following criticism of Japan for 'free-riding' on US basic research at an informal intergovernmental meeting in mid-1985. The US delegation made it clear that the issues of trade conflict and an internationally 'parasitic' research policy were linked. Following political consultations, Nakasone invited the Vice-Minister of MITI to establish a Committee on Technological Development and International Cooperation Towards the 21st Century to develop an appropriate Japanese response. This was headed by the previous AIST Director-General, Seiichi Ishizaka, and consisted of nine eminent academics and industrialists (AIST, 1986b). The guidelines agreed were that any proposal should: (i) be extremely large-scale (at least $5 billion) so that it would be seen as recycling a significant part of Japan's trade surplus; (ii) have the potential to make major impacts on science and technology in a wide range of non-defence areas; and (iii) focus primarily on basic rather than applied research.*

The report which finally emerged was, however, rather more technologically oriented than originally envisaged:

> We propose here to formulate an international research and development programme to create new areas of industry under our country's initiative in order to unify and put together mankind's knowledge and wisdom to create a new industrial society ... Research and development themes worthy of a challenge to the frontier ... are all tasks in the interdisciplinary areas of advanced application technologies simulating the function of living organisms, and are expected to serve as a major key to a new system of industrial technology for the 21st century. [AIST, 1986b, p.1]

Among the recommendations outlined in the report was that a Human Frontier Research Institute be set up with significant involvement by Japanese industry — an idea hardly in accord with internationalist thinking. One reason for the change in orientation was Nakasone's view of the initiative as a strategic response to Western threats, despite pleas from scientists that it would be politically inopportune to propose anything other than a strictly basic programme. As one remarked:

> When I met Nakasone in early 1986, he told me that Japan was located in a deep valley overlooked by the two big mountains of Eureka and SDI. Japan therefore needed its

own strong project to propose to the Tokyo Summit — the Human Frontier Programme. By then, his idea was fixed and I could do nothing about it.*

A second factor was the role played by MITI officials in orchestrating industrial and political pressure for the potential technological benefits from the programme to be stressed.[35] The hope was that this would deflect criticism from STA, Monbusho and other ministries that HFSP should not be its responsibility, and hence make possible the long-resisted move into the life-sciences area. In arguing their case, MITI officials drew on the foresight study undertaken by IDRI as an input to planning projects for the BTFI programme. Some forty of the 181 development topics judged likely to yield industrial benefits called for interdisciplinary life-sciences research where breakthroughs in biology were the key determinant of progress. Several were related to information technology — for example, 'construction of a hybrid computer capable of imitating the structure and function of the brain' was seen as likely to lead to the development of intelligent supercomputers, biosensors, actuators and biomolecular electronic devices (IDRI, 1985).

Despite extensive industrial and political support for HFSP, and demonstrable evidence as to its future importance, MITI encountered strong resistance from several sources after circulating its initial plans calling on government to finance an international feasibility study. The academic community, fearing a targeted AIST programme would threaten discipline-based work in the life sciences, argued that Monbusho should be responsible for any new initiative.[36] STA proposed its own Human and Environmental Sciences Programme stressing earth sciences as well as biology to fit in with existing international programmes. Each of the protagonists lobbied government heavily, especially the Ministry of Finance, with Monbusho and STA playing on the concern of foreign agencies like the US National Science Foundation that the programme hardly gave the appearance of a benevolent attempt by Japan to make a fairer contribution to world science.[37] This was recognized by a MITI official: 'From the outside, the Human Frontier Science Programme appears like a 'black-hole' to grab foreign basic knowledge because Japanese industry knows that control of the human mind and life-science technology will be crucial in the 21st century'.*The failure to achieve consensus meant that announcement of the planned programme by Nakasone at the 1986 Tokyo Summit had to be postponed to the following year in Venice. By then, CST had assumed overall responsibility for HFSP, financing the work of a planning group within MITI (which had been placed under the joint supervision of STA). However, the damage had been done and foreign scientists attending the international planning meetings (CST, 1987; 1988b) were suspicious of what they saw as the self-interested Japanese motives underlying the programme. As a result, the initiative soon lost domestic political support, especially after the retirement of Nakasone (who had a strong personal investment in its success), while its increasingly 'pure' orientation reduced the scope for direct industrial involvement. The likely budget progressively declined to a level an order of magnitude less than initially planned, so that it became scarcely worth the effort for overseas organizations like the US National Institutes of Health to participate.[38] In late 1988, it was still intended to launch a much reduced HFSP, with an international panel administering a European-based fund supporting collaborative research between Japanese and foreign scientists. The lower political priority now accorded to the initiative means that any increase in the size of the fund will depend largely on success in persuading foreign governments and firms to contribute (CST, 1988b, p.6).

The experience with HFSP has highlighted the problems that can exist in research planning when consensus among several government departments is necessary, as well as revealing the difficulties likely to be faced by foreign organizations in relating to such decision-making processes. However, it also shows just how pluralistic the Japanese R&D system has become in that all three rival agencies have since 1986 set in place their own interdisciplinary life-science programmes. Monbusho has funded an initiative on brain science (e.g. mapping cerebral-nerve circuitry), while STA has established a Frontier Research Programme in which foreign scientists have been invited to participate in projects on homeostasis (to replicate basic life-science processes) and on the 'artificial brain' (RIKEN, 1986). Most importantly, MITI is supporting work on biodevices (as part of the BTFI programme) to develop the underlying technologies required for future biocomputers, with synthetic brain and nervous systems similar to those of living organisms being seen as the key to high-level performance in pattern recognition and learning (MITI, 1987a). The end result is that the foresight activities undertaken by MITI have undoubtedly achieved policy impact, even if this has been more diffuse than would have been the case had its strategic vision of HFSP ultimately proved acceptable.

(b) Exploratory Research for Advanced Technology

Examination of the role of foresight in the ERATO programme is instructive in that it focusses attention on the problems of conservatism that can arise when using consensus mechanisms in research planning. ERATO was launched in 1981 with the aim of 'fostering the creation of advanced technologies and advancing future interdisciplinary scientific activities while searching for a better system of scientific research' (JRDC, 1988, p.2). This was in response to the recognition that radical institutional reorganization of research was necessary if Japan was to develop an indigenous capacity to create the innovative 'seeds' required for the science-based industries of tomorrow. According to Shogo Kurachi, former President of the Research Development Corporation of Japan (JRDC) which administers the programme, the problems to be overcome were fundamental:

> The fact that there has not been a single major Japanese contribution to basic scientific inventions has ... exacerbated trade friction. It is a puzzle and a source of embarrassment to the Japanese why the Japanese people, who are capable of producing many high quality goods at low cost, have not been able to create radically new technology. It is an inevitable fact that the Japanese economy must be supported by advanced science and technology, and, therefore, it is imperative that the Japanese people must strive to create and nurture original and basic scientific and technological innovations. [Kurachi, 1986, p.1]
>
> The various factors which contributed to the realization of efficient mass production technology are, at the same time, hindering the development of radically

new technologies ... People in Japan follow its leader. Group orientedness is an invisible but gigantic pillar which supports Japanese industries. On the other hand, in such a society independent action by a member of a group is looked at with disfavour. Therefore, if a person comes up with a creative idea and wishes to realize it, he is not likely to be rewarded for his action. He may even be driven out of the group as an alien element. In such a society, creative ideas are hard to be recognized and realized. Thus if radically new technologies are to be developed in Japan, it is necessary to organize a new group dedicated to the creation of new technologies outside of the group. Members of the new group are encouraged to engage in research as freely as possible. [Kurachi, n.d., pp.1-2]

As a result, the structure of the ERATO programme differs from the traditional model of government-funded research in several respects: (i) emphasis is placed on selecting talented research directors who are then granted considerable autonomy in running five-year projects once the initial plan has been agreed; (ii) each director recruits gifted scientists (preferably aged under 35 years) from industry, universities and government laboratories to conduct research under a system aiming to exploit individual creativity to the full (restrictions on employing foreigners have also been lifted); (iii) projects are managed flexibly, with objectives modified in line with progress achieved; and (iv) research is undertaken principally in rented private-sector laboratories which are closed on completion of projects (STA, 1987a). The underlying theme of most projects is the physical, chemical and biological properties of new materials. Projects typically have an annual budget of $3 million which supports a team of up to twenty researchers, around 60 per cent of whom are drawn from companies.

The novelty of ERATO means that it has attracted a certain mystique in Japan as much as overseas, especially concerning its apparent reliance on a Western-type model of undirected research where foresight plays little or no role. Such an impression is easily gained from Genya Chiba, the programme's director:

We have negated the consensus system. Science is art, and, like the theatre, we have producers, directors, and performers. Only young people can participate. We appoint the directors and they choose their teams. There are no committees. We give them themes to work on, but we make the definitions as vague as possible. The directors are given five years to see what they can do! We don't know where they are going to end up. It's a bit like saying, sail from Yokahama and when you get there tell us where it is. [Chiba, cited in Veitch, 1986]

However, Chiba might also have mentioned the influences exercised by the playwright, the potential audience, and those managing and financing the play. These have ensured that foresight has played much the same role as in the BTFI programme, especially at the inception stage of ERATO. Here, various surveys of R&D trends and opportunities by STA helped both define the organizational innovation required to promote creativity in the scientific system, and establish the need for a coordinated long-term initiative in interdisciplinary materials research. They consequently provided 'a broad overview and background to the decision to set up ERATO, a market survey of expectations', as well as 'a mechanism to gain the public, industrial and political support needed to set up the programme'.* Similar surveys (especially by the National Institute of Resources — STJ, 1985) were instrumental in making the case for an expansion of the programme during its second phase between 1986 and 1991 (with funding increasing by 40 per cent over two years to reach $30 million in 1988).

Formal foresight studies initiated by JRDC have also contributed to project selection within ERATO. In particular, the Japan Techno-Economics Society (JATES) has undertaken a series of surveys to identify promising research topics. In one of these, questionnaires were sent to 800 experts in the private sector, government and academia, seeking views on fields which: (i) might provide the 'seeds' for radically new industrial technologies; (ii) already have a cadre of active researchers who could be brought together in a coordinated project; and (iii) are sufficiently intellectually interesting to attract the best academic scientists (including foreigners) yet promise the long-term technological benefits needed to persuade leading companies to contribute creative researchers. Such surveys have moreover been valuable in securing industrial commitment to projects and establishing contact with potential academic participants.

A formal approach has also been adopted in the all-important task of selecting project directors. JATES has, for example, undertaken studies to identify potential candidates with a track-record in pioneering novel scientific or technological developments, especially where this has been achieved in the face of criticism from an entrenched, conservative school of thought. Another necessary qualification is that the work of directors should be 'endorsed by the younger generation', and citation analysis has been employed to assess the nature and extent of their scientific influence on the research community.* However, such information alone is insufficient, and needs to be supplemented by extensive consultation with academic and industrial experts undertaken by JRDC programme officers. This enables detailed checks to be carried out as to whether shortlisted candidates have the requisite combination of intuition, creativity and management skills.

Where ERATO differs from other strategic programmes is that little attempt is made to steer research once projects have been initiated, and no formal technical targets are set as in the parallel BTFI initiative. This is not to imply that economic and technological benefits are not anticipated — the Teraherz project (on molecular electronics) is, for example, 'intended to create a new field ... expected to make a remarkable contribution to faster large-capacity communications' (STJ, 1987d, p.73). Without such expectations, it would neither have been possible to obtain government funding nor the high level of industrial support which now exists for the programme. Indeed, evaluation of the initial projects completed by 1987 showed they had yielded both high-quality basic research and patented technology (Irvine, 1988, p.66). Furthermore, companies see ERATO as providing valuable 'postdoctoral training' for their researchers not available within the academic system.

Foreign commentators may well explain the achievements of the ERATO programme in terms of 'how to disappear into the great blue yonder and still end up, probably, a winner' (Veitch, 1986). Yet the important point is that project directors 'sailing from Yokahama', to use Chiba's analogy, have been supplied with the navigational equipment and crew to ensure their voyage of discovery starts and continues to progress in a promising direction. Research foresight thus plays a central role in what should be seen as a novel extension of existing Japanese R&D management practices to cope with the uncertainties of longer-term strategic research. ERATO is not an attempt to transplant Western-style scientific autonomy where the research community largely dictates the direction and priorities.

(c) *Role of think-tanks*

Japan is unique in its extensive network of around 300 public and private-sector think-tanks (NIRA, 1987a), many of which have developed levels of expertise in monitoring and forecasting R&D unmatched even by the large US consultancies. As we have seen, they contribute considerably to the research foresight undertaken by public-sector agencies. Five main roles can be distinguished, the first of which is to ensure that industry's views are reliably reflected in government foresight studies. Widespread support for new initiatives must be shown to exist during the process of advocacy and competition for funds, with preference given to proposals supported by evidence on the long-term need for R&D. There is much to be gained, therefore, from employing the services of respected private-sector think-tanks able to draw upon their knowledge of technological and market trends developed through work for corporate clients. Such organizations often appoint retired senior industrialists to advisory positions, further enhancing the respect with which their work is held within government.

A second role is in studies of more specialized areas of industrial technology and the underlying basic research. A typical example is a recent Mitsubishi Research Institute (MRI) project for STA on emerging developments in composite materials relevant to next-generation rocket propulsion systems. This was largely based on interviews with thirty materials scientists, but the US patent system was also scanned to gain an overview of technological trends, while information on progress in relevant basic science areas (like thermo-mechanical research) was obtained through keyword searches of several literature abstracts. Scenarios were then constructed of alternative technological possibilities for reaching target materials performance goals. These were based on analyses of likely development trajectories of related technologies, taking into account potential impacts and cross-linkages between fields.[39*]

The third contribution made by industrial think-tanks comes through their extensive monitoring of foreign R&D. Organizations like MRI and Nomura Research Institute are large enough to undertake routine tracking of global developments in a wide range of strategic research fields. Overseas offices in particular enable them to follow the activities of innovative laboratories and companies, using what the less charitable might term low-level (but perfectly legal) commercial espionage, but which the Japanese describe as 'holistic, observational scanning of state-of-the-art industrial information' (Irvine and Martin, 1984, p.130). This includes in-depth assessments by expert study groups of novel technological developments incorporated in prototypes and other advanced products exhibited at international trade-fairs like Hanover (to which one think-tank regularly sends a delegation of fifty). It also involves attending specialist conferences in emerging fields where breakthroughs are often first reported publicly (and where intelligence on work-in-progress can be gleaned through informal discussions with those attending). Such information is subsequently fed into the research foresight activities of agencies like MITI and STA through various mechanisms, complementing the comprehensive monitoring of overseas publications, reports and patents undertaken by the Japan Information Center of Science and Technology.[40]

The development and diffusion of improved forecasting techniques is a fourth area where think-tanks have had an impact on government foresight. Although a number of public and private sector organizations have been responsible for methodological innovation (often deploying techniques first developed within industry), the Japan Techno-Economics Society (JATES) has been pre-eminent. Established in 1966 in response to the growing use of technology assessment and long-term planning techniques in US companies, JATES is funded by corporate subscriptions (from 200 members) and is therefore primarily concerned with meeting industrial needs. The early work of the society (especially by its Future Research Operational Group, FROG) was instrumental in promoting the adoption by Japanese companies of strategic planning and forecasting methods, which, as Yamauchi (1983) shows, took place as late as the early 1970s. We have already seen, however, that one result of the FROG initiative was the STA thirty-year forecast. The influence on long-term planning by public-sector R&D agencies continues to be considerable, with JATES sponsoring regular conferences and workshops on forecasting methodology as well as study groups on topical issues (JATES, 1986). Also significant was the 'spin-off' from JATES of the Institute for Future Technology (IFTECH), which assumed responsibility for the growing number of forecasting projects commissioned by government.

Finally, think-tanks have largely been responsible for providing the broader analytical perspectives often required in government foresight exercises, especially in determining the social and economic factors likely to structure developing trends in science and technology. JATES has again made important contributions (for instance, organizing symposia on issues like 'science and technology for the new age'), as has IFTECH (through projects on such topics as 'future technology for preservation of the environment and prevention of pollution' — IFTECH, 1979). A leading role has also been played by the National Institute for Research Advancement (NIRA), a joint public-private sector organization established by Parliament in 1974 to analyse the 'rapid change in the development of a new industrial society, new sciences and technologies, new styles of local communities and new lifestyles', and to propose policies to meet the 'urgent need ... for various systemic reforms to deal with such changes and cope with the new economy and society' (NIRA, 1987b, pp.2–3). Funded by interest from a 30,000 million yen endowment fund (33 per cent of which was contributed by industry), NIRA is currently working on a number of future-oriented programmes, including one on 'new developments in science and technology and policy measures to cope with them' (NIRA, 1987b). Much of the institute's work on research foresight has, however, been carried out under contract by other think-tanks. Among the more influential of these studies have been projects on:

(1) 'Progress in science and technology — its impact on social and economic systems' (which developed scenarios for Japanese socio-economic

development in the twenty-first century, and the technological, industrial and other policies required to achieve a desirable future — MRI, 1977);

(2) 'Comprehensive policy guidelines for Japan in the changing world — national priorities for the 21st century' (which reviewed alternative approaches for coping with emerging global issues in such spheres as trade, security and technology — NRI, 1977);

(3) 'The long-term international impact of Japan's high-technology industries' (a quantitative assessment of the economic effects of potential advances in electronics and information technology, bioengineering, materials and new energy sources — FAIS, 1986);

(4) 'A comprehensive study of microelectronics' (a series of related studies on the 'seeds', 'needs' and socio-economic impact of electronics technologies — NIRA, 1985a).

The last of these projects illustrates the sophisticated conceptual and methodological approach to research foresight typically employed by the larger think-tanks. Several separate lines of investigation were pursued to arrive at a multifaceted picture integrating the perspectives of historians of technology, sociologists, economists, scientists, and industrialists from the manufacturing and service sectors. These included:

(1) An analysis of the historical development of electronics over the last century, identifying the nature and origins of fundamental breakthroughs[41] and the 'environmental factors' facilitating successful creative research (IFTECH, 1983).

(2) An assessment of the likely social impacts of electronics, drawing on various inputs:
  - a questionnaire survey of domestic and foreign experts on negative and positive effects of emerging electronics technologies (NIRA, 1985b);
  - a review of the potential impact on employment, social welfare, economic growth and decision-making structures of projected advances in electronics technology up to the year 2000 (NRI, 1983);
  - more specific studies of the probable effects on employment of progress in microelectronics (JATES, 1982), mechatronics (KPC, 1983) and information and communications technology (MKI, 1984);
  - an extensive programme of lectures by corporate and other experts on socio-cultural and political issues likely to arise with the adoption of new developments in electronics (including the legal and ethical problems associated with information storage and transmission), and on the impact of emerging technologies in the home and within the main industrial and socio-economic sectors.

(3) A major study on 'long-term development strategies for electronics' carried out by Hitachi Research Institute. A panel of ten experts, mostly industrialists, first identified several hundred development targets for the electronics industry that would meet 'social needs' in the period to the end of the century. This was undertaken with the aid of literature and patent analysis (which enabled a map to be constructed, for example, of

the linkages between patents cited more than five times by subsequent patents). The results of a questionnaire survey of 500 Hitachi research staff were then used to evaluate possible new technologies in terms of: (a) their *application rate* (i.e. the percentage of identified development targets to which each technology might contribute); and (b) their *indispensability* (an indicator of the likely competitiveness of each technology for realizing the various development targets). The study is regarded as significant in Japan because it represents a pioneering attempt to use quantitative approaches in the demand side of technological forecasting. Among the areas identified as being in greatest need of further research were electronic devices, optoelectronics, bioelectronics, AI and new materials (HRI, 1983).

(4) A follow-up study by Hitachi Research Institute entitled 'in search of future technology in electronics'. This used the demand-side projections as a basis for forecasting the 'seeds' of new technologies, and for assessing the opportunities arising from 'the fusion of electronics with border technologies such as optics and biotechnology' (NIRA, 1985a, p.52). A survey was carried out of 280 electronics companies and 128 public-sector laboratories in which researchers were asked: (a) to rank specified development needs across various fields like advanced materials (for example, 'to achieve high-density recording'); (b) to assess the indispensability (or degree of necessity) and application rate of 'current', 'new' and 'potential breakthrough' technologies;[42] and (c) to estimate the impact of future electronics developments in terms of improved functions and performance. Examples of the forecasts produced in Figure 6.2 show that breakthroughs in materials and biotechnology were judged likely to impact widely on electronics over the next two decades. Overall, the study concluded that 'materials will act as the key technology for achieving future breakthroughs in electronics' (NIRA, 1985a, p.56), with superconductive materials identified as important for both information processing and instrumentation/control (HRI, 1984). Although only made in 1984, this prediction has already proved remarkably prescient.[43]

(5) A study by Mitsubishi Research Institute of 'the international evolution of Japan's technology policy' considering the extent to which future strategy for electronics R&D should involve international cooperation. Inputs included reviews of: (a) existing cross-national programmes in high technology fields; (b) the technology policies of major industrialized nations; and (c) the comparative scientific and technological strengths of these countries relative to Japan, the aim being to identify potential fields for collaboration (MRI, 1984).

NIRA then distilled the results of the various studies into a widely disseminated report on the opportunities and problems likely to arise as the microelectronics revolution continues over the next two decades (NIRA, 1985a). Besides providing a systematic, quantitative basis for identifying the research fields underpinning emerging electronics technologies, the report

(a) Biotechnology

Method	Directions of technological development	Technological themes
Use of biological substances	Improvements in sensing capability	Increased ability to discriminate between specific substances
	High-density recording	Molecular memory (1 billion times that of silicon devices)
	Affinity with human tissue	Embedded ultra-miniature computers for restoration of impaired functions
Imitation of biological functions	Large volume data processing	Parallel data processing
	Selection of information	Ability to amplify and emphasize important information

(b) New Materials

Technological need		Breakthrough technology	Materials
Communications	Large-volume transmission	Milliwaves, laser beams	Compound semi-conductors (InP, GaAlAs, etc.)
	Long-distance transmission	Low-loss optical fibre	Non-silicic material
Information processing	High-speed operations	Compound semi-conductor ICs	Compound semi-conductors (GaAs, InP, etc.)
		Josephson-junction devices	Superconductive materials (alloys, compounds)
	High-density recording	Perpendicular magnetic recording	Perpendicular magnetized film
		Magneto-optic recording	Magneto-optic recording materials
		Molecular memory	High polymers, biological substances (protein)
Instrumentation and control	Improvement in sensing performance	Josephson-junction devices	Superconductive materials (alloys, compounds, etc.)
		Biosensors	Biological substances (micro-organisms, enzymes, etc.)
	Improved resistance to environmental conditions	Devices more resistant to environmental conditions	Compound semi-conductors (SiC, GaAs, etc.)

Source: HRI (1984) and NIRA (1985a, pp. 55–6).

Figure 6.2 Expected electronics breakthroughs in biotechnology and new materials identified by Hitachi Research Institute in 1984

focussed attention on the policies required to shape a desirable future for the nation. These included greater emphasis on organizational innovation in the R&D system, and the promotion of creativity among researchers, as well as the need to take 'prudent corrective measures' to ensure public commitment to scientific progress is not threatened by any adverse effects of new technology on legal, social and democratic rights (NIRA, 1985a). Hence, one function of this and similar NIRA studies is to contribute a technology assessment perspective to the foresight undertaken by government and industry. This complements the holistic and macro-level surveys undertaken by STA and consequently provides useful background inputs to the work of organizations like CST.

5. RESEARCH FORESIGHT IN INDUSTRY

The four firms visited — Hitachi, NEC Corporation, Sumitomo Chemical Company and Toray — primarily operate within the electronics, telecommunications and chemicals sectors.[44] As with most Japanese companies, research foresight is an integral part of the planning process and has become increasingly formalized as basic science has gained in importance during the 1980s. This integration arises from the organic linkage between planning of strategic research and that of company-specific core technologies and industrial capacity (depicted by Giget, 1988, in terms of a 'tree structure' the roots of which are constituted by generic technologies interrelated with scientific disciplines). Foresight activities are therefore typically carried out at several levels (or sites) within companies: (i) the corporate strategic planning office (whose head normally has access to the board); (ii) product development or new ventures departments; (iii) research planning or administration offices (within R&D divisions and individual laboratories); and (iv) research sections and groups within laboratories (where it forms part of the process of establishing budgetary priorities).

Over recent years, longer-term foresight in all four companies has concentrated on ensuring that the requisite competence is developed in the generic technologies and underlying strategic science likely to be required over the next ten to twenty years (the 'day after tomorrow' in Japanese terminology). The most formalized approach is employed by NEC which has identified thirty 'core technology' areas (e.g. organic electronic materials and pattern recognition) and predicts for both 'tomorrow' (the next five to ten years) and the 'day after tomorrow' how development of the underlying 'basic technologies' (voice, character, figure, image and object recognition in the case of pattern recognition) will provide future manufacturing opportunities in specific fields like information processing and transmission. Strategic research projects, which must promise a high degree of originality, generally fit within the core-technology framework and are carried out by the more creative staff. NEC has given such work increasing priority so that what is termed fundamental research accounted for 10 per cent of its 1988 R&D budget of $3,000 million.

NEC's first attempt to develop a longer-term R&D strategy was completed in 1975, and this was updated during 1983–5. The exercise involved two years of effort by eight staff from the R&D Group Planning Office, aided by thirty specialist researchers and 'high-flying managerial candidates' drawn from throughout the company (who were seconded on a half-time basis). The working group thus brought together the likely next-generation leadership of NEC, as well as balancing technology-push and market-pull perspectives. It also provided a basis for initiating the extensive consultations with operating groups and divisions necessary to generate corporate consensus. The working group met weekly (usually for an entire day) to coordinate the exercise. Not surprisingly, therefore, it cost several million dollars.

The approach employed was to focus first on trends in R&D demand within the company, and then extend the analysis to external markets for 'current', 'next' and 'future' generation technologies. A major task here was to undertake a portfolio analysis for each of the 300 strategic business units (SBUs) in the eighty NEC divisions, assessing the extent to which their activities would be strengthened over the medium and long term by the development of new and improved technologies. Inputs included in-depth interviews with staff in each SBU to identify relevant 'key technologies'.[45] This resulted in an unwieldy matrix listing up to twenty such technologies against each of the 300 SBUs. The large number of key technologies was then grouped into thirty core technologies and linkages identified with underlying basic technologies. The process was complex and called for considerable analytical expertise (a problem companies attempting to emulate the NEC approach have found difficult to overcome). However, this bottom-up synthesis of the judgements of knowledgeable staff in each SBU was seen as essential in order to produce meaningful results (since they are in the best position, for example, to judge how customer needs are likely to structure next-generation product development), as well as ensuring the corporate-wide commitment required to implement the R&D strategy.

The second step was to carry out a similar supply-side analysis for science and technology, taking into account the typical time-horizon of fifteen years for incorporating radically new electronics and communications technologies into products, and the fact that fundamental scientific breakthroughs may take as long as three decades to achieve significant commercial impact. Several tasks were involved:

(1) Examining the development dynamics of various sets of basic technologies and related scientific disciplines over the medium and longer term. This was carried out by: (i) consulting experts within and outside the company; (ii) tracing previous discoveries and developments in each area, together with emerging trends and cross-disciplinary impacts; (iii) reviewing papers presented at relevant international conferences; (iv) investigating trends in publishing and patenting; and (v) analysing results from the company's routine monitoring of global R&D, as well as foresight studies undertaken by think-tanks and government agencies.
(2) Evaluating the strategic implications of emerging research developments for meeting the company's longer-term basic technology needs.
(3) Assessing the extent to which NEC was already able to meet the medium- and long-term research requirements of SBUs, and deciding how best to develop necessary expertise currently unavailable internally or at too low a level relative to competitors.
(4) Establishing priorities in specific strategic fields (e.g. bioelectronics) to meet the company's future basic technology needs.

According to the architect of the long-term R&D strategy, Dr Uenohara, the 1985 exercise has proved 'eminently justified and likely to pay off' despite the costs incurred (Marsh, 1987, p.20). Besides providing guidelines for the annually updated medium-term R&D plan and influencing corporate planning more generally, it has had several other effects: (i) generating a consensus vision within NEC on the future importance of R&D; (ii) giving senior management a holistic overview of global technological trends; (iii) educating staff in operating divisions about commercial opportunities being opened up by R&D, and ensuring researchers are made aware of likely demands on their work; and (iv) breaking down interdisciplinary and divisional barriers within the firm. The latter has, according to Uenohara, been especially significant:

> The core technology analysis really broadened the scope of people's experience throughout the company. While before they fought to protect their own disciplinary castle, it opened their eyes to possibilities beyond their area. This has led to a recognition of the benefits of opening up collaboration and communication across organizational barriers. This has been a major gain, perhaps more important than the set of core technologies itself.*

Sumitomo Chemical Company and Toray have relatively similar planning philosophies to NEC, recognizing the need for systematic foresight to identify a core set of strategic technologies. Following a previous attempt to develop a corporate R&D plan five years earlier, Sumitomo was in 1988 in the middle of a more comprehensive exercise to produce a 'technological map for the 21st century'. This was being undertaken by a 'future action group', involving staff from the corporate planning office, research administration office and new product development department. The focus has been on identifying interdisciplinary bioscience and electronic materials areas where fusion with the company's established strengths in chemical process technology might create promising business opportunities over the coming two decades.*

Toray has a more established procedure for preparing its 'long-term research plan', with inputs made by the R&D planning department, corporate planning department, marketing development group and several consultative mechanisms like the annual laboratory managers' conference.[46] Since 1986, its corporate philosophy of 'creating new values through innovative concepts and technologies' has been associated with more focussed attempts to identify key technologies for all existing and planned product areas, as well as the basic research areas likely to underpin future activities. Like Sumitomo, it has concentrated on assessing the prospects for emerging fusion technologies, aware that its recent diversification has largely been based on integrating traditional disciplinary research strengths with new fields — for example, the fusion of film-fabrication and electrochemical technologies has yielded various novel electronics-related materials.*

Sumitomo and Toray were both aware, however, that their position as medium-sized firms necessarily limits the scope for longer-term planning compared to the major electronics corporations (which have a global lead in a broad range of technologies and enjoy R&D budgets around an order of magnitude larger). Following the 'yen-shocks' during the 1980s (which have made it uneconomic to continue bulk chemicals production), research foresight has focussed on identifying areas which build upon their existing technological advantages stemming from supplying advanced materials to the electronics industry (in order to avoid head-on conflict with stronger European and US competitors). Recognizing the importance of collaborative strategic research, both companies intend to play a central role in the chematronics projects within the Basic Technology for the 21st Century Programme. They also rely on sectoral foresight initiatives coordinated by industrial associations, especially the Federation of Technological Management of Chemical Industries which recently undertook studies on:

(1) 'the future development of the chemical industry' (which surveyed twenty-two Japanese companies on trends in R&D, likely sales in different market sectors, etc.);
(2) 'the chemical industry in the 21st century' (in which questionnaires were sent to the chief executives of a range of Japanese and foreign firms — FTMCI, 1985a); and
(3) 'technological management in the European and US chemical industries' (which compared eleven Japanese firms and ten foreign competitors in terms of R&D strategies, management and organization, and acquisition behaviour, with detailed projections made for sales distribution, markets, profits, assets, and level and nature of R&D — FTMCI, 1985b).

The results showed that the most promising prospects for the Japanese chemical industry lay in new functional materials (e.g. ultra-high purity gallium arsenide and superconducting ceramics), with important cross-linkages developing between materials and communications technologies. Foreign companies were judged not to have moved into these areas sufficiently rapidly, thus opening up significant opportunities for the domestic chemical industry (whose sales of plastics to the electronics sector were already proportionately double those of their European and US rivals). One result has been to reinforce the trend of chemical companies to invest in the interdisciplinary research areas underlying these technologies. Toray in particular increased funding of 'basic and exploratory research' from 13.1 per cent of its R&D budget in 1986 to 30.5 per cent a year later (Toray, 1987). Such a shift was only possible because consensus on the centrality of basic research reduced the risks involved, and because the chematronics initiative being launched by MITI guarantees long-term government support (as well as a conducive industrial policy). A consensus vision of the future can thus strongly influence the likelihood of that predicted future being realized.*

Toray and Sumitomo both stressed the value of information on the likely development of user-needs for longer-term foresight, as well as for their more extensive medium-term research planning efforts where detailed technical targets are often set for applied initiatives (Irvine, 1988). In this respect, the electronics industry has been a model customer in that it is exceedingly well informed about its future needs, while the systemic development dynamics of the underlying core technologies typically allow reasonable confidence in forecasts for the next five to ten years. The sophisticated understanding Japanese companies have of the dynamics of technological growth is clearly a major advantage they hold over most Western firms. So also is the way they are able to draw on systematic and informed inputs from corporate functions like marketing when carrying out research planning, the 'tree structure' ensuring that even longer-term foresight is firmly rooted to future commercial opportunities.

The two chemical companies also emphasized the importance of long-term perspectives on changing socio-economic structures as an input to research foresight. Sumitomo in particular has analysed the effects of the predicted Kondratiev long wave for the global chemicals industry, as well as the strategic implications for their future operations in Japan. Here, the foresight reports produced by Japan's unique network of long-term credit banks are often useful.[47] The Industrial Bank of Japan, for example, recently published a review of the technological and commercial prospects for new materials over the next decade (IBJ, 1986). The Long-Term Credit Bank of Japan has also undertaken several assessments of basic technologies like fine ceramics and ocean exploitation (LTCB, 1984b; 1986) in the context of its socio-economic projection that Japan will, at the end of the century, be characterized by: (i) a 'softened service-sector' economy; and (ii) much heavier reliance on high-technology industry (which is forecast to account for 30 per cent of GDP as opposed to 5 per cent in the 1970s and 15 per cent in the mid-1980s — LTCB, 1984a). Given the importance of macro-level socio-economic data for longer-term planning, Toray established its own think-tank (Toray Corporate Business Research) in 1987. Among other things, this publishes *World Techno-Trend*, a monthly report assessing the commercial implications of developments in strategic research (for example, recent articles have covered new functional materials, bioelectrocatalysis and biological response modifiers — Toray, 1987).

Hitachi has also been strengthening its longer-term research. Overall R&D expenditure was around $2,350 million in 1987 (having grown by over 10 per cent per annum during the decade), with projects having a time-horizon of over ten years now accounting for nearly 15 per cent of the total and planned to increase further. Where Hitachi differs from NEC and the two chemical companies is in placing less emphasis on formal foresight techniques to identify core technologies. An important role is nevertheless played by the corporate research and technology planning office in coordinating preparation of the annually-updated 'long-term R&D plan' (dealing with the next three to twelve years). This provides a focus for discussions within the company on strategic research opportunities and needs, as well as guidelines for laboratory department heads in producing their own long-term plans. A variety of forecasting techniques is used in assessing the future prospects for R&D, including the approach developed for NIRA by Hitachi Research Institute considered above. However, the company believes that relying primarily on consensus-seeking foresight exercises to establish priorities for basic research can result in conservatism, and, in any case, 'their problem is often that they have too much information on the future, not too little'.*

Hitachi has a longer tradition in basic research than most Japanese companies. From the time of its establishment in 1942, the central research laboratory (which accounts for around 25 per cent of R&D) was given the task 'to do research and plan 10 and 20 years ahead, and to make our technology truly creative and original' (Hitachi, 1987). Experience has shown that an essential element of foresight — the monitoring of emerging technological trends — is best undertaken in-house by researchers working at the international forefront. This contributed, for example, to the laboratory's decision to launch a programme on superconducting technology (which has yielded several innovative developments like the superconducting transistor) well before the IBM breakthrough in 1986. Granting scientists greater freedom than is normal in Japanese firms has proved successful largely because Hitachi at the same time established mechanisms to ensure research results are diffused effectively within the company. In particular, a key role is played by the '*tokken* system' which enables interdisciplinary teams from different corporate functions to work on priority projects.

The *tokken* system has also made it easier for Hitachi to mount a far-sighted experiment designed to overcome a tendency to focus on shorter-term projects detected within the central research laboratory. In April 1985, an 'advanced research laboratory' was established, bringing together 100 leading scientists from different fields (the number is set to double by 1995) to work on 'science and technology that will be the basis for new business of Hitachi Ltd in the 21st century' (Hitachi, 1988). Researchers have considerable freedom to propose projects in interdisciplinary fields like molecular electronics, advanced software science and various life-science areas (for example, cell fusion and genetic engineering), which are judged likely to form the basis of future-generation electronics, communications and computing technologies.[48] The laboratory will also act as a conduit for picking up information on emerging developments in relevant basic science disciplines. The Hitachi model quickly set a precedent, with thirty similar laboratories being established by other companies over the following three years.*

With more and more Japanese firms finding themselves in a position of world technological leadership, research planning is becoming increasingly difficult since the future is now less certain. The option of setting specific technical targets for projects secure in the knowledge that they could improve upon basic technologies already demonstrated overseas no longer exists. Considerable thought is therefore being given to questioning whether current planning and forecasting approaches are still appropriate. JATES, for example, has been engaged in a study to evaluate the research forecasting methods most commonly employed by firms. Around 100 Japanese companies from several industrial sectors have been surveyed about their use of eleven techniques, and the success achieved when applied to different R&D tasks (for example, new product innovation or development of novel technology — see JATES, 1987). Preliminary results show that: (i) companies are most familiar with Delphi methods, followed by trend analysis, relevance trees, scenario techniques and network analysis; (ii) the techniques most widely used to forecast developments in novel technologies are patent analysis, trend analysis and scenario-based approaches; (iii) by far the most effective technique overall is patent analysis, with trend analysis, technology portfolio appraisal and scenario methods also judged favourably; and (iv) greatest problems have been experienced in forecasting longer-term developments in emerging scientific and technological fields, especially where breakthroughs might potentially impact across a wide range of sectors. Here, Delphi techniques are judged most useful, despite all their difficulties, while trend analysis is preferred for research areas characterized by slower incremental progress.*

Besides strengthening in-house basic research, firms have also responded to technological uncertainty by expanding their monitoring of academic science, especially in the elite US and European institutions. Industry has benefited here from the investment by government in computerized research intelligence services at the Japan Information Center of Science and Technology. More emphasis is also being placed on attending specialist conferences, publication in prestige academic journals (to gain access to information transmitted through 'invisible colleges'), sponsoring leading scientists (to keep up with the latest research) and studying systematically the dynamics of emerging fields. Success in such tasks will depend, like corporate basic research itself, upon an increased supply of creative graduate scientists, and industry therefore strongly supports current measures to restructure the higher education system.

6. CONCLUSIONS

The huge investments by industry in basic research, coupled with the priority being given to upgrading the public R&D infrastructure, seem likely to result in Japan soon achieving the enviable position of being both a technological and scientific superpower (Narin and Olivastro, 1988; Martin *et al.*, 1987; Negishi, 1988). The competitiveness of other industrialized nations will suffer, especially if companies succeed in using foresight to target their longer-term research effectively, and if government is able to secure the changes needed in the academic system over the coming decade.

In drawing lessons from Japan, the uniquely conducive environment for undertaking research foresight must be stressed. It is widely accepted that longer-term research is essential for creating a desirable future, and that planning helps ensure scientific resources are channelled to meet important socio-economic needs. This is part of a more general cultural emphasis on looking to the future, enabling research foresight to be integrated with economic and social forecasting. Foresight by government is similarly linked with that undertaken in industry (especially by research associations like the Association for the Progress of New Chemistry). Also important is the priority placed on up-to-date information, and the consequent willingness to finance global monitoring systems to track emerging developments in strategic research. This provides a basis for supporting the large network of public and private-sector think-tanks so crucial in foresight.

Another feature is the emphasis on consensual decision-making, with the agreement of all relevant groups normally being a precondition for launching any research initiative. Moreover, to secure government funding, its future significance must be established and this places a premium on systematic foresight combining science-push and demand-pull perspectives. Unlike other countries, foresight is thus integral to the policy-making process for research.

Probably the main lessons from Japan concern the organization of foresight. First, although 'the policy-making process and the consensus-making process are one and the same', it is recognized that the organizational structures employed need to 'reflect the understanding that administrative authority and the necessary knowledge do not go hand in hand' (Hirasawa, 1988, pp.1–5). The Japanese have been especially successful in establishing a framework for reconciling the inherent tension of top-down versus bottom-up involvement in foresight, and in integrating the views of interested parties while still according importance both to systematic prospective analysis and specialist third-party expertise. This has been well-summarized by Hirasawa:[49]

> It is the norm in traditional organizations for a top-down process to be used to establish the basic principles and framework and for a bottom-up process to be used to define the precise measures and themes ... In Japan the top-down and bottom-up processes are better harmonized than in Europe and the United States [where] excessive influence is exerted by lobbyists ... In the bottom-up process, it is important, as a rational screening in the middle stages becomes necessary, to develop techniques for policy evaluation. In the top-down process, what is most important is to develop information channels to meet the necessity of maintaining routes for effective and detailed communication between the upper and lower levels in the hierarchy ... [The] process of evaluation and consensus is achieved in a smooth interactive way which does not exclude the individuals making the proposal, contrary to what happens when the proposals are submitted for evaluation to outside entities using conventional evaluation techniques. [Ibid., pp.1-5]

Second, it is clear that the integration of research foresight activities at different levels has brought significant advantages. Holistic exercises like the STA thirty-year forecast give a comprehensive overview of emerging trends and opportunities, informing the national policy guidelines set by the Council for Science and Technology as well as the macro-level foresight undertaken by individual agencies like MITI in producing periodic sectoral visions and other scene-setting exercises. The latter provide a context for more focussed meso-level foresight which is often oriented to identifying promising research opportunities in order to demonstrate the merits of a proposed initiative. Such activities are again not formally tied to specific resource-allocation decisions and therefore 'can be free, active and liberal'.* Foresight at a micro-level is more closely linked to such decisions and is often conducted within a framework for establishing consensus not only on priorities but also on the technical goals for research.

A third lesson concerns the highly effective division of labour generally employed in research foresight: (i) the exercise is overseen by a high-level committee with the *authority* to ensure results are implemented; (ii) a 'working group' is given executive responsibility, its members possessing the forecasting, scientific and industrial expertise required for the results to be accepted as *legitimate* by all concerned; (iii) detailed research and data compilation are carried out by think-tanks or external consultants, thus ensuring *technical credibility;* and (iv) the use of consensus-seeking mechanisms throughout the foresight process generates *commitment* to the resultant recommendations, as well as introducing a strong bottom-up orientation.

Fourth, the experience with holistic and macro-level foresight has demonstrated that the most promising research areas (like information technology during the late 1960s and chematronics today) arise from the fusion of several existing fields. Interdisciplinarity has therefore become central in selecting members for foresight committees, and both government agencies and industry are now studying quantitatively (for example, through bibliometric methods) and qualitatively (through historical case-studies of emerging fields) the cross-impacts and linkages between different sciences and technologies. The finding by the most recent thirty-year forecast that experts in a particular field have, if anything, been less accurate in their predictions than respondents with a broader range of expertise demonstrates the advantages of seeking judgements from all those with some understanding of developments in related areas. Equally, users of technology are often better able to assess the wider systemic factors likely to retard or accelerate the impact of research developments, making it even more crucial to combine demand-pull and science-push perspectives in foresight.

Japan's evident success with foresight raises the question of the extent to which other countries can transplant such procedures to their own research-policy environment. While socio-cultural and political differences certainly exist, these should not be overemphasized. On the one hand, some argue that:

> Owing to national personality traits, we can get a good level of cooperation in the official STA 30-year survey. This is a typical Japanese approach to forecasting in that scientists are willing to make judgements based on their feelings and intuition. European and US researchers are generally unwilling to engage in activities that are not strictly logical and not underpinned by scientific evidence. As a result, they might not cooperate so much or so easily.*

Other Japanese officials, in contrast, take the position that:

> What is possible for us in research foresight is intrinsically possible in other countries. On many issues, it is easy for us to achieve consensus, while in others we find it extremely difficult to arrive at agreement. Some say the Japanese are different from other people in that they do not argue or criticize publicly. But it does happen that no consensus is possible [on certain issues].*

Indeed, the South Korean government replicated the latest thirty-year survey in 1987, beginning after IFTECH but finishing earlier much to the chagrin of STA officials. In 1988, the South Africans were also contemplating launching a similar exercise. The key issue may thus be not so much one of intrinsic cultural specificity, but how to *adapt* Japanese approaches for use elsewhere. Here, the many possibilities available will probably only become apparent by going through a learning process with experimental studies.

Any attempt to explore the use of Japanese approaches should particularly take into account the problems which have become evident as the country has moved from a 'technological follower' to 'technological leader' role. As an experienced MITI official has noted, this transition has meant 'the Japanese [R&D] system has begun to work negatively. The merits have turned into demerits' (Tanaka, 1988, p.23). Initiatives like ERATO are an attempt to reorient the research system towards greater creativity, while concern about conservatism in longer-term foresight is being met by increasing the emphasis placed on the judgements of younger scientists and those with international reputations. A remaining institutional problem, however, is that 'the nature of R&D has changed towards a more interdisciplinary direction and has become more related to many government organizations, thus exacerbating the struggle amongst them' (ibid., p.23). This was especially evident in the planning process for the Human Frontier Science Programme where strong interdepartmental rivalries have matched those in Western countries over priority programmes in cross-disciplinary areas like biotechnology.

The sceptic might conclude that these emerging problems demonstrate that Japanese foresight approaches are fundamentally unsuited to 'leading' scientific nations. This would be to underestimate the probability that the

Japanese will solve the difficulties they currently face. Rather, what Western nations should perhaps be asking is whether the transition in their position to that of 'technological followers' means that the traditional Japanese approaches to foresight are in fact likely to become eminently suited to their needs.

NOTES

* Refers to material collected during interviews in Japan in 1988, while † denotes information drawn from subsequent correspondence.

1. Oral statement by Professor F. Kodama (Saitama University), cited in Dixon (1988).
2. Among more important English-language documents on recent Japanese science policy are: STA (1983; 1985; 1986; 1987b), CST (1984a), MITI (1988b) and Cabinet Council (1986). See also Irvine and Martin (1984, Chapter 5), Irvine (1988), Ishizaka (1986), Lynn (1986), Makino (1987), NSF (1988a) and Yoshimura (1987).
3. Representation of the Minister of Finance on the Council for Science and Technology enables decisions on broad priorities to be linked with discussions on the government R&D budget. The Ministry of Finance has been especially influential in research-policy decisions during the 1980s because of the strict cash limits placed on government expenditures.
4. While Monbusho is also formally responsible for servicing CST, its role is in practice considerably less important than that of STA.
5. A 1980 report by the CST Panel on Life Sciences helped establish a framework for coordinated national policy throughout the decade, identifying around thirty priority themes (CST, 1980; STJ, 1983, p.9). In 1981, the government asked CST to develop a long-term plan for the field, and this subsequently targeted twenty-nine research areas (e.g. development of novel cell-fusion technologies — see CST, 1984b; STJ, 1984, p.10). Among the inputs to the exercise were surveys of emerging scientific opportunities. A key problem, however, was the agreement of funding responsibilities among relevant ministries and agencies. MITI in particular wished to move into the life-sciences but this was resisted by other ministries traditionally responsible for such research, the conflict continuing during later discussions on the Human Frontier Science Programme.
6. Academic research is an exception in coming outside the aegis of STA. This is because Japanese universities have traditionally had 'freedom of ... activities [and] any coordination of university research by the government is understood [to be] against this freedom' (Yoshimura, 1987, p.3).
7. This followed a 1985 recommendation by the Ad Hoc Council on Administrative Reform — see NSF (1985a).
8. One result was that Monbusho had by the mid-1980s already established a programme on 'new superconducting materials' (NSF, 1987c). This financed the work of Tanaka at Tokyo University which was crucial in confirming the IBM discovery in late 1986, enabling him to beat the US multinational by three days in filing for the first Japanese patent on high-temperature superconductivity (Cross, 1988, p.30).
9. This remains the only version of the forecast to have been fully translated into English. However, since the approach employed has not changed significantly in subsequent surveys, the methodological description is still relevant. See also Irvine and Martin (1984, pp.107–14).
10. The possibility that industry might collectively finance a private-sector think-tank to carry out the survey was also considered. It was decided, however, that the level of commitment required from the research community would only be forthcoming if it had the official imprimatur of government.

11. The Steering Committee was chaired by Noboru Makino (Chairman of Mitsubishi Research Institute), who is highly respected within the Japanese research system. He has headed all four thirty-year STA forecasts.
12. The cost of the 1986-7 forecast was estimated by IFTECH officials to be over $0.5 million. This includes a $0.2 million contract from STA, honoraria for sectional group members, staff and administrative costs incurred by STA, and an institutional subsidy of $0.2 million provided by IFTECH.*
13. In the initial 1971 forecast, a postal survey was used to obtain judgements from 1,500 experts on: (i) the principal social and economic needs over the next ten and next thirty years; and (ii) the research developments necessary to meet those needs. Responses were then analysed by working groups — see Irvine and Martin (1984, p.109).
14. In estimating likely time-scales for commercialization, the views of industrial members of working groups were given greatest weight. In some cases, academics were found to have rather unrealistic expectations.*
15. These proportions varied widely across fields — 70 per cent of life-science topics were new, while those in energy and transportation remained largely unchanged. Where similar topics were proposed by several working groups, IFTECH had to eliminate duplication (in consultation with the Steering Committee) and gain agreement on modification to wording of survey questions.*
16. This entailed considerable work by IFTECH to check the research activities of potential participants (e.g. to determine whether they had published recently in the academic or professional literature).
17. The issue of a preliminary invitation to participate in the forecast means that the response rate of 80 per cent claimed by STA needs to be treated with caution (IFTECH, 1988a, p.2). Taking into account the initial refusal rate of 57 per cent (3,400 of the 6,000 experts invited) and subsequent losses in the two stages of the survey of 21 per cent and 19 per cent, respectively, the figure would more conventionally be calculated as 31 per cent.
18. Although of high calibre, the STA officials responsible for previous forecasts lacked the time and professional science-policy background to undertake comprehensive analyses of the factors underlying long-term trends. Devolving responsibility for the forecast to IFTECH has the added advantage of ensuring continuity. As in all government departments, STA officials are regularly moved to new duties and this often results in complete rotation of staff between one forecast and the next. The technical competence developed is therefore lost, and indeed IFTECH found it difficult to establish exactly how previous forecasts had been carried out.*
19. This provides further evidence for the self-confidence of Japanese researchers and their perception of being at or near the leading-edge in many fields. Cooperative international programmes were, however, seen as valuable for topics in life sciences and medicine, fields in which the United States and Europe still dominate.
20. Several companies have shown interest in exploring the reasons for such divergencies in opinion among experts from different disciplines, and this has formed part of the follow-up work on the forecast undertaken by IFTECH.*
21. Hitachi was only ten days behind IBM in applying for Japanese patents on several important breakthroughs in the manufacture of superconductors, while Sumitomo had by February 1987 filed for 600 patents including a number covering key technologies for producing thin films, wires, new devices and vapour processing techniques (Cross, 1988, p.30).
22. Among predictions made for information technology in the 1971 STA forecast was that 'all communication channels (transit trunks) will be digitised by 1987', which encouraged companies to move into integrated digital communications systems. There are, however, other examples of creative forecasts which have

proved less accurate such as 'the development of robot-fish which will round up fish and create shoals by 1988' (STA 1972).

23. The Chairman of the Steering Committee, Noboru Makinu, stressed this role: 'when the report was published, many newspapers, general economic and science journals, and even children's publications, featured the results extensively. I was also on TV a lot. The Japanese people are generally very interested in the future of science and technology, and so the survey has a great impact on public debate and understanding'.*
24. One of the report's recommendations was to establish a 'scientific assessment authority' to review university performance and establish appropriate levels of funding for different areas of science (SCJ, 1988).
25. The CST guidelines and recommendations from the Science Council of Japan also constitute important inputs to Monbusho policy formulation.
26. It should be stressed, however, that by the early 1980s Monbusho had already begun to give greater prominence to strategic research, and officials respond to criticism on the lack of targeted programmes by pointing to the highly successful initiative on superconducting materials over 1984–7 — see NSF (1987c).
27. The problems facing academic research in biotechnology are outlined in Tanaka (1988), while those in chemistry are described in Swinbanks (1988) and CSJ (1988).
28. Funding for the programme has been obtained in part by terminating two other support mechanisms: (i) 'special long-term research' (providing finance for cancer and fusion research); and (ii) 'special project research' (NSF, 1987a; Monbusho, 1987).
29. These were issues raised in an influential MITI report by the *ad hoc* Research Committee on Technological Innovations and the Potential for New Markets, entitled *Energise Yourself Japan* (MITI, 1987b). This aimed to establish a new '*raison d'être* for MITI's activities over the next decade'.*
30. 'Buzzwords' are culturally extremely important in Japan, with terms like 'frontier' and 'technological fusion' enjoying wide currency in official circles. Much effort is put into developing integrative concepts for research initiatives since this assists in gaining political and budgetary support. The practice is also common in industry, with NEC, for example, using the slogan 'C and C' (the integration of computers and communication) up to 1988, when corporate strategic priorities required a change to 'individuals, originality and group creativity'.*
31. ASPRONC has around eighty corporate members, primarily in the materials, chemicals and biotechnology sectors. Its activities include organizing workshops and symposia on developments in 'new chemistry', and a 'research executives forum' on 'the future outlook for relevant frontier industries and problems in R&D' (ASPRONC, 1987a). It is financed by interest on a $30 million endowment fund (50 per cent of which was contributed by MITI).
32. MITI consequently invested considerable effort in publishing the report of the round-table in a popular format so as to generate widespread public discussion.
33. This is also evident in the nationalist sentiments expressed in the forty-six technologies report. As one foreign science attaché in Tokyo noted: 'the report is interesting in that it portrays a basic Japanese insecurity. In spite of their advances in recent years, they feel that the US and Europe are now getting their act together and will use their leadership in fundamental science to regain technological superiority. This cynically could also be a theme likely to appeal to the Ministry of Finance and Japanese Cabinet'.†
34. Companies reported sometimes visiting Ministry of Finance officials to advocate the merits of new programmes and projects, as well as making their views known to government through politicians with whom they have links.*
35. As one industrial interviewee noted: 'much of the current work in bioscience is very fundamental. On the one hand, MITI has been worried about entering the area since Japanese industry and the universities are still rather weak. On the other,

it recognizes that the US, FRG, and the UK have many basic patents and is nervous about whether access to these technologies will continue to be provided so freely as in the past, given the concern over trade and technological friction. MITI is worried about bilateral friction placing barriers on technological imports from abroad, and has therefore concluded that it is necessary to stress basic research in these areas'.*

36. At one point, MITI suggested there should be a postgraduate training element to HFSP. Monbusho quickly made it clear that this would be unacceptable and the idea was dropped.*
37. The MITI committee was from an early stage aware of this problem. As one member stated: 'I was involved in the early Human Frontier Programme planning process, and we originally used the concept of science fusion in biology and electronics. But people in MITI were very worried it would cause foreigners to talk about the implications of fusing these fields since everyone knows we are strong in electronics. As a result, the fusion of life sciences with electronics was downplayed, and the documents given to foreigners focussed only on basic biological processes'.*
38. The National Institutes of Health (NIH) have been perhaps the strongest critic of the programme, arguing that Japan would demonstrate a more honest commitment to repaying its international scientific debts (and recycling its excessive trade surplus) if it financed a major programme on tropical diseases in less-developed countries. NIH regarded the failure even to countenance such a suggestion as reflecting a desire on the part of Japan to challenge the US lead in life sciences.*
39. Such studies were a major element in Mitsubishi Research Institute's income from government of around $25 million in 1986 (MRI, 1987).
40. JICST is a 'public corporation' funded by STA and through subscription income. It provides remote access to a variety of domestic and foreign data-bases on science and technology through JOIS (the JICST Online Information System), which attracts large numbers of users from companies, government agencies and universities. It also publishes several journals targeted at different audiences (e.g. a review of overseas technological developments of interest to smaller companies). The Japanese government has regarded the development of an efficient science information system as a national priority, and this is reflected in the financial support of $50 million given to JICST in 1988. The remainder of its annual budget of around $85 million is earned from subscriptions and other income. See JICST (1986), Miyakawa and Nasu (1987), and Sigurdson and Greatrex (1986).
41. One interesting finding concerns the increasing contribution of Japan to creating new 'seed technologies' for microelectronics. IFTECH identified 140 breakthroughs made during the three decades from 1950 to 1979, and classified them into eight types (e.g. 'novel theory/principle' or 'new production method'). The results showed the country had moved from a position of relative weakness in the 1950s to one of near parity with the rest of the world in the 1970s, its record being especially good in 'components and circuits' (NIRA, 1985a, p.66).
42. By aggregating the projected 'degree of necessity' (DN) and 'application rate' (AR), four categories of technology were distinguished: (i) 'priority technologies' (DN >0.7 and AR >10 per cent); (ii) 'selective technologies'(DN >0.7 and AR <10 per cent); (iii) 'transitional technologies' (DN <0.7 and AR >10 per cent); and (iv) 'non-priority technologies' (DN <0.7 and AR <10 per cent) – see NIRA (1985a) and Baba (1988).
43. During interview in 1983, researchers at Hitachi Research Institute expressed reasonable confidence about their predictions, stressing the main problem was likely to be conservatism. There was some concern that survey respondents in other companies would be reluctant to give away information on their long-term R&D strategies, with the result that judgements might possibly have been adjusted to suggest development targets would be achieved later than actually

expected (Irvine and Martin, 1984, p.134). This is an intrinsic problem when commissioning private-sector think-tanks with specific commercial interests to undertake public-knowledge projects. Against this has to be set the significant experience such organizations bring to research foresight compared with most of their public-sector equivalents.

44. A fifth firm, Suntory, was also visited in order to study industrial approaches to foresight in biotechnology, but insufficient information was obtained for any conclusions to be drawn.
45. A questionnaire-based approach had been attempted previously but proved unsatisfactory.
46. The latter is an important forum for discussing the long-term prospects for key technologies. In 1983, for example, it resulted in the establishment of four strategic 'project areas' in the fields of new fibres, biotechnology and electronics, which are now central to Toray's business activities.*
47. These have a typical investment horizon of between eight and ten years (e.g. to finance power-plant construction) and it is therefore necessary to engage in longer-term forecasting to adjust risk portfolios in line with projected changes in industrial structure. In the view of the Long-Term Credit Bank of Japan, research foresight is crucial in making investment decisions in high-technology sectors underpinned by rapid scientific change. The bank has a policy of disseminating the sector-based reviews of prospects for particular technologies produced by its industrial research division both for marketing purposes (requests for reports often generate contacts with potential customers), and to win the external recognition necessary to be routinely consulted by government during the planning process for research initiatives (thereby gaining early access to information as well as an opportunity to influence policy formulation).*
48. It is perhaps significant that these are similar to the core areas originally identified for the Human Frontier Science Programme. Hitachi believes, however, that its research is now far ahead of the public sector and the main benefits from any initiative in the area will be in training younger scientists.
49. Hirasawa was in 1988 conducting a study on science-policy processes in Japanese agencies and advisory councils. Case-studies were being undertaken of a variety of research initiatives, including the Human Frontier Science Programme and Basic Technology for Future Industries Programme. An assessment was also being made of the usage and effectiveness of different decision-making approaches and techniques. Preliminary results from a survey of officials in eight ministries and agencies show that greatest reliance is placed on brainstorming, direct consultation with researchers, and 'behind-the-scenes contacts' (the so-called *nemawashi*) (Hirasawa, 1988, pp.2–4).

REFERENCES

AIST (1981), *Toward New Research and Development*, Report of the Study Meeting for the Formulation of a Long-Term Industrial Technology Development Program, Tokyo, Agency of Industrial Science and Technology.

—— (1986a), *Basic Survey to Collect Systematically Information Concerning Technology Development Trends*, Tokyo, Agency of Industrial Science and Technology.

—— (1986b), *Technology Development and International Cooperation Toward the 21st Century*, Study Committee on Technology and International Cooperation, Tokyo, Agency of Industrial Science and Technology.

—— (1987a), *Agency of Industrial Science and Technology 1987*, Tokyo, Agency of Industrial Science and Technology.

—— (1987b), *Research and Development Project of Basic Technology for Future Industries*, Tokyo, Agency of Industrial Science and Technology.

—— (1987c), *Basic Technologies for the 21st Century: Japan's Role and Contribution to the Third Technological Revolution,* Tokyo, Tsusanchosakai.

ASPRONC (1987a), *ASPRONC: Objectives and Outline,* Tokyo, Association for the Progress of New Chemistry.

—— (1987b), *Creation of Frontier Industries through Chemistry: The First International Symposium on New Chemistry,* Tokyo, Association for the Progress of New Chemistry.

Baba, Y. (1988), 'Systemic Innovation: Its Nature and How to Benefit from It' (mimeo), Brighton, SPRU, University of Sussex.

Cabinet Council (1986), *General Guidelines for Science and Technology,* Tokyo, Cabinet Council.

Chiba, G. (1985), 'The Trends in Science and Technology and Problems of Internationalization in Japan', *Scientific Bulletin,* vol.10, no.4, pp.101–6.

Cross, M. (1988), 'Superconductors Surge to the Fore in Japan', *New Scientist,* 1 September, p.31.

CSJ (1988), *Research Environment Surrounding Japanese Chemistry: Investigation Concerning Funding and Equipment Needs,* Tokyo, Chemical Society of Japan.

CST (1980), *Views Concerning Promotion of Life Science,* Tokyo, Council for Science and Technology.

—— (1981), *Exploratory Research on Seeds of Innovative Technology by Means of the Dynamic Aggregation System,* Tokyo, Council for Science and Technology.

—— (1984a), *Comprehensive Fundamental Policy for Promotion of Science and Technology to Focus Current Changing Situations from the Long-Term View,* Tokyo, Council for Science and Technology.

—— (1984b), *Recommendation No. 10 on the Inquiry on Basic Programme for Research and Development on Leading and Fundamental Technology in Life Science,* Tokyo, Council for Science and Technology.

—— (1986), *Guidelines for Research Evaluation,* Tokyo, Council for Science and Technology.

—— (1987), *Current Developments in Research in Relation to the Principal Biological Functions and Future Focal Points,* Tokyo, Council for Science and Technology.

—— (1988a), *Council for Science and Technology,* Tokyo, Council for Science and Technology.

—— (1988b), *The Human Frontier Science Program,* Tokyo, Council for Science and Technology.

Dixon, B. (1988), 'Dilemmas of Evaluating Research', *British Medical Journal,* no.297, p.366.

Dore, R. (1983), *A Case Study of Technology Forecasting in Japan,* London, Technical Change Centre.

FAIS (1986), *The Long-Term International Impact of Japan's High-Technology Industries,* Tokyo, Foundation for the Advancement of International Science.

FTMCI (1985a), *Chemical Industry Towards the 21st Century: Its Role and Strategy in the Innovation Age,* Tokyo, Federation of Technological Management in Chemical Industries.

—— (1985b), *Technological Management Strategy of European and US Chemical Industries: Chemical Industries in the Year 2000,* Tokyo, Federation of Technological Management in Chemical Industries.

Giget, M. (1988), 'The Bonsai Trees of Japanese Industry', *Futures,* April, pp.147–54.

Hirasawa, R. (1988), 'Summary of Decision-Making Processes in the Science and Technology Policies of Japan' (mimeo), College of Arts and Sciences, Tokyo University.

Hirooka, M. (1984), 'Technological Innovation and Chemical Industry: Advanced Materials and Their Industrial Development', *Chemical Economy and Engineering Review,* January/February, pp.5-12.

Hitachi (1987), *Central Research Laboratory,* Tokyo, Hitachi Ltd.

—— (1988), *Advanced Research Laboratory,* Tokyo, Hitachi Ltd.
HRI (1983), *Long-Term Development Strategies for Electronics,* Tokyo, Hitachi Research Institute.
—— (1984), *In Search of Future Technology in Electronics,* Tokyo, Hitachi Research Institute.
IBJ (1986), *Light and Shadow: The Two Aspects of the New Materials Business,* Tokyo, Industrial Bank of Japan.
IDRI (1985), *Future Industrial Basic Technology Development and Social and Economic Effects,* Tokyo, Industrial Development Research Institute.
IFTECH (1979), *Survey on Forecast of Technology for Preservation of Environment and Prevention of Pollution,* Tokyo, Institute for Future Technology.
—— (1982), *Study on the Systematic Formulation of R&D Themes in Science and Technology from the Viewpoint of Social Needs,* Tokyo, Institute for Future Technology.
—— (1983), *An Analysis of the Development of Electronics,* Tokyo, Institute for Future Technology.
—— (1988a), *Future Technology in Japan: Forecast to the Year 2015,* Tokyo, Institute for Future Technology.
—— (1988b), 'Forecasting the State of Superconductive Materials' (mimeo), Tokyo, Institute for Future Technology.
Irvine, J. (1988), *Evaluating Applied Research: Lessons from Japan,* London, Pinter Publishers.
—— and Martin, B.R. (1984), *Foresight in Science: Picking the Winners,* London, Pinter Publishers.
Ishizaka, S. (1986), *Innovation Cycle and Japan's Role,* Tokyo, National Personnel Authority.
JATES (1982), *Social and International Influence of Microelectronics,* Tokyo, Japan Techno-Economics Society.
—— (1986), *Introduction to JATES Activities,* Tokyo, Japan Techno-Economics Society.
—— (1987), 'Questionnaire on Practical Application of Technology Forecasts' (mimeo), Tokyo, Japan Techno-Economics Society.
JICST (1986), *The Japan Information Center of Science and Technology,* Tokyo, Japan Information Center of Science and Technology.
JRDC (1988), *ERATO: Exploratory Research for Advanced Technology,* Tokyo, Research Development Corporation of Japan.
JSDRI (1987), *Survey on Conditions for the Promotion of Technology Development: Basic Study on Technological Trends Using Data-Bases,* Tokyo, Japan Systems Development Research Institute.
JSPS (1986), *Research Cooperation Between Universities and Industry: Encouragement of Creative Advanced Research in Japan,* Tokyo, Japan Society for the Promotion of Science.
Kodama, F. (1986), 'Japanese Innovation in Mechatronics Technology', *Science and Public Policy,* vol.13, no.1, pp.44–51.
Kondo, J. (1988), *Trend of Science in Japan,* Tokyo, Science Council of Japan.
Kono, Y. (1986), 'Basic Idea: Setup of Science and Technology for the 21st Century', *Science and Technology in Japan,* April/June, pp.52–5.
KPC (1983), *Mechatronics Innovation and Future Industrial and Labour Policies,* Kansai, Kansai Productivity Centre.
Kurachi, S. (1986), 'A Message from the President', in *Exploratory Research for Advanced Technology,* p.1, Tokyo, Research Development Corporation of Japan.
—— (n.d.), *A New Approach to the Development of Radically New Technologies,* Tokyo, Research Development Corporation of Japan.
LTCB (1984a), *Japan's High Technology Industries,* Tokyo, Long-Term Credit Bank of Japan.

—— (1984b), *Fine Ceramics Industry in Japan*, Tokyo, Long-Term Credit Bank of Japan.
—— (1986), *Ocean Development in Japan*, Tokyo, Long-Term Credit Bank of Japan.
Lynn, L. (1986), 'Japanese Research and Technology Policy', *Science*, vol. 233, pp.296–301.
Makino, N. (1987), *Decline and Prosperity: Corporate Innovation in Japan*, Tokyo and New York, Kodansha International.
Marsh, P. (1987), 'Why Research is in the Driving Seat', *Financial Times*, 8 June, p.20.
Martin, B. R., Irvine, J., Narin, F. and Sterritt, C. (1987), 'The Continuing Decline of British Science', *Nature*, vol.330, pp.123–6.
Maruta, Y. (1987), 'Tomorrow's Chemical Industries and the Importance of Basic Research', in *First International Symposium on New Chemistry*, Tokyo, Association for the Progress of New Chemistry.
Ministry of Finance (1987), 'Japan's FY 1987 Science and Technology Budget', *Scientific Monthly* (translation by National Science Foundation).
MITI (1980), *The Vision of MITI Policies in the 1980s*, Tokyo, Ministry of International Trade and Industry.
—— (1984), *A Future Course for Policy on the Development of Industrial Technology*, Tokyo, Ministry of International Trade and Industry.
—— (1986), *Basic Concept of Industrial Society as the 21st Century Approaches: Toward an Internationally Harmonious Industrial Structure with a Concentration on Creative Knowledge*, Tokyo, Ministry of International Trade and Industry.
—— (1987a), 'Ministry of International Trade and Industry R&D Budget for JFY 1987', *Scientific Monthly*, July (translation by National Science Foundation).
—— (1987b), *Energise Yourself Japan*, Tokyo, Ministry of International Trade and Industry.
—— (1988a), *Harmonization of International Economy and Society: Revitalization*, Tokyo, Ministry of International Trade and Industry.
—— (1988b), *Trends and Future Tasks in Industrial Technology*, Tokyo, Ministry of International Trade and Industry.
Miyakawa, T. and Nasu, Y. (1987), 'Japanese Database Industry — The Present and the Future', *MRI Research Review*, vol.2, no.1, pp.2–3.
MKI (1984), *Industrial Structure in the Information Age*, Tokyo, Mitsui Knowledge Industry.
Monbusho (1983), *Studies on Scientific Research Trends: (a) Electrical Engineering; (b) Linguistics; (c) Theoretical Economics and Econometrics*, Tokyo, Ministry of Education, Science and Culture.
—— (1984), *Studies on Scientific Research Trends: (a) Biology; (b) Chemistry; (c) Solid-State Physics*, Tokyo, Ministry of Education, Science and Culture.
—— (1986), *The University Research System in Japan*, Tokyo, Ministry of Education, Science and Culture.
—— (1987), 'Monbusho's R&D Budget for JFY 1987', *Scientific Monthly*, July (translation by National Science Foundation).
MRI (1977), *Progress in Science and Technology — Its Impact on Social and Economic Systems*, Tokyo, Mitsubishi Research Institute.
—— (1984), *The International Evolution of Japan's Technology Policy*, Tokyo, Mitsubishi Research Institute.
—— (1987), *Mitsubishi Research Institute in Brief*, Tokyo, Mitsubishi Research Institute.
MSC (1984), *On Basic Measures for Improvement of the System of Scientific Research*, Tokyo, Monbusho Science Council (translation by National Science Foundation).
—— (1986), *Recommendations for Promotion of Bioscientific Research in Universities*, Tokyo, Monbusho Science Council.
Narin, F. and Olivastro, D. (1988), *Identifying Areas of Leading Edge Japanese Technology*, Washington, DC, National Science Foundation.

Negishi, M. (1988), 'National Research Activities as Seen in Publication Data: A Survey Analysis', *Japanese Scientific Monthly*, vol.41, no.7, pp.40–7.

NIRA (1985a), 'Comprehensive Study of Microelectronics 1985', *NIRA Report*, vol.2, no.2, pp.5–144.

—— (1985b), *Social Impact of Electronics: A Summary of Discussions Regarding a Worldwide Questionnaire Survey*, Tokyo, National Institute for Research Advancement.

—— (1987a), *Think Tank Almanac 1986/87*, Tokyo, National Institute for Research Advancement.

—— (1987b), *National Institute for Research Advancement 1987*, Tokyo, National Institute for Research Advancement.

NRI (1977), *In Search of Comprehensive Policy Guidelines for Japan in the Changing World: National Priorities for the 21st Century*, Tokyo, Nomura Research Institute.

—— (1983), *Studies on the Social Impact of Electronics*, Tokyo, Nomura Research Institute.

NSF (1985a), 'Reorganization of Japan's Science and Technology Agency', Report Memorandum No.85, Tokyo, National Science Foundation.

—— (1985b), 'Educational Reform in Japan: Some Controversial Recommendations', Report Memorandum No.67, Tokyo, National Science Foundation.

—— (1985c), 'STA's Resources Institute to Review Future Directions of Basic Research to Induce Innovative Technologies', Report Memorandum No.87, Tokyo, National Science Foundation.

—— (1986a), 'Recommendations of Monbusho's Science Council for Promotion of Bioscientific Research', Report Memorandum No.99, Tokyo, National Science Foundation.

—— (1986b), 'The Human Frontier Program: MITI's Initial Think Piece on the Nature of the HFP', Report Memorandum No.111, Tokyo, National Science Foundation.

—— (1987a), 'Monbusho Launches New Category for Grants-in-Aid for Research on Priority Areas', Tokyo, National Science Foundation.

—— (1987b), 'Monbusho to Establish University-Industry Joint Research Centers', Tokyo, National Science Foundation.

—— (1987c), 'Monbusho Support for Superconducting Materials Research JFY 1984-1986', Tokyo, National Science Foundation.

—— (1987d), 'Science and Technology Agency to Propose New Initiative for R&D on New Superconducting Materials', Report Memorandum No.129, Tokyo, National Science Foundation.

—— (1988a), *The Science and Technology Resources of Japan: A Comparison with the United States*, Washington, DC, National Science Foundation.

—— (1988b), 'Japanese Government R&D Programs with Industry: MITI and Ministry of Education', Report Memorandum No.158, Tokyo, National Science Foundation.

RIKEN (1986), *Frontier Research Programme*, Saitama, Institute of Physical and Chemical Research.

SCJ (1975), *Science Council of Japan: The First 25 Years (1949-1974)*, Tokyo, Science Council of Japan.

—— (1986), *Science Council of Japan*, Tokyo, Science Council of Japan.

—— (1988), *Trends in Scientific Research in Japan*, Tokyo, Science Council of Japan.

Sigurdson, J. and Greatrex, R. (1986), *Electronic Databases in Japan*, Research Policy Institute, University of Lund.

—— (1986), *Industry and State Partnership in Japan: The Very Large Scale Integrated Circuits (VLSI) Project*, Research Policy Institute, University of Lund.

STA (1972), *Science and Technology Developments up to AD 2000*, Tokyo, Japan Techno-Economics Society.

—— (1982), *Technology Development Forecast up to 2010*, Tokyo, Science and Technology Agency.
—— (1983), *Japan Science and Technology Outlook*, Tokyo, Fuji Corporation.
—— (1985), *White Paper on Science and Technology 1985: New Developments in R&D and the Era of Cooperation (Summary)*, Tokyo, Foreign Press Center.
—— (1986), *Science and Technology White Paper 1986: Toward a Better Environment for Man (Summary)*, Tokyo, Foreign Press Center.
—— (1987a), *STA: Its Roles and Activities*, Tokyo, Science and Technology Agency.
—— (1987b), *White Paper on Science and Technology 1987: Toward the Internationalization of Japan's Science and Technology (Summary)*, Tokyo, Foreign Press Center.
—— (1987c), *Process of Formulating Policies and Science and Technology and TA in Japan*, Tokyo, Science and Technology Agency.
—— (1988), *National Institute of Science and Technology Policy*, Tokyo, Science and Technology Agency.
Statistics Bureau (1988), *Survey of Research and Development in Japan*, Tokyo, Management and Coordination Agency, Prime Minister's Office.
STJ (1983), 'Research and Development of Life Sciences in Japan', *Science and Technology in Japan*, April/June, pp.8–16.
—— (1984), 'Report on Inquiry No.10 on Basic Plan for R&D of Pioneering and Basic Technologies in Life Sciences', *Science and Technology in Japan*, July/September, pp.8–10.
—— (1985), 'Survey of Trends in Basic Research Directed Towards Innovative Technology', *Science and Technology in Japan*, October/December, pp.33–5.
—— (1987a), '6th Science and Technology Forum', *Science and Technology in Japan*, April/June, p.45.
—— (1987b), 'Special Coordination Funds for Promoting Science and Technology', *Science and Technology in Japan*, April/June, pp.18–19.
—— (1987c), 'Survey on the Direction of Japan's Technological Development', *Science and Technology in Japan*, November/December, pp.34–8.
—— (1987d), 'ERATO's New Research Themes', *Science and Technology in Japan*, August, pp.72–3.
Suzuki, E. (1987), 'Chemistry Challenges Frontier Industry', in *First International Symposium on New Chemistry*, Tokyo, Association for the Progress of New Chemistry.
Swinbanks, D. (1988), 'Japan's Research Predicament Revealed in Society's Survey', *Nature*, vol. 336, p.197.
Tanaka, M. (1987), 'Evaluation of Innovation Policies in Japan', paper presented at Conference on Evaluation of Programmes Promoting Technological Innovation, December (Ref DSTI/SPR/87.51), Paris, OECD.
Tanaka, M. (1988), 'Industry-University Relations in the Case of the New Biotechnology in Japan' (mimeo), University of Saitama, Japan.
Toray (1987), *R&D Activities at Toray Industries*, Tokyo, Toray Industries.
Veitch, A. (1986), 'How to disappear into the great blue yonder and still end up, probably, a winner', *The Guardian*, 7 March.
Yamauchi, I. (1983), 'Long-Range Strategic Planning in Japanese R&D' (mimeo), Tokyo, Nomura Research Institute.
Yoshimura, H. (1987), 'The Organization of Science and Technology in Japan', paper presented at International Conference on the Organization of Science and Technology in Western Industrialized Countries, Bonn, Tokyo, Science and Technology Agency.

7 Research foresight in Australia

1. INTRODUCTION

In the first of our case-studies of small to medium-sized economies, we examine the role of foresight in attempts to modernize Australia's research infrastructure during the 1980s. Falling prices for traditional agricultural and mineral products have imposed strains on the balance of payments and highlighted the need for high-technology industries which are innovative and internationally competitive. This is seen as requiring greater targeting of the R&D resources deployed within the public sector, and has created a climate in which 'selectivity, concentration and accountability are the bywords' (Franklin, 1988a, p.34).

A central goal of recent government policy[1] has therefore been to strengthen the research system so that it can contribute more effectively to generating increased added value to primary sector exports, while at the same time reducing the substantial trade deficit in high-technology and manufactured goods. Tax incentives have contributed to a rise in corporate R&D expenditure (up from a very low 0.25 per cent of GDP in 1981–2 to 0.39 per cent in 1985–6 — DITAC, 1987i, p.25). In addition, the foreign multinationals dominating many strategic industrial sectors have been encouraged to expand their domestic R&D activities, and to exploit more intensively the nation's considerable scientific and technological skills by establishing central laboratories serving the entire Pacific-rim region.

While total R&D spending increased in real terms by 27 per cent between 1981–2 and 1985–6, the figure of A$2,750 million still represented only 1.15 per cent of GDP — rather less than the OECD norm. With nearly 70 per cent of research concentrated in the public sector, government has placed restructuring high on the political agenda following the realization that 'Australian science is high quality, but marginal to the economic system, [and] locked into a separate matrix to that of application and productive activity' (ibid., p.5). One aim has consequently been to achieve improved integration between policies for R&D support and industrial development, leading to the Department of Science (DOS) being disbanded in 1987 and the transfer of many of its functions to an expanded Department of Industry, Technology and Commerce (DITAC). The Commonwealth Scientific and Industrial Research Organisation (CSIRO) has also been reorganized to give greater priority to strategic programmes oriented to meeting longer-term industrial needs, and it now reports to the Minister for Industry, Technology and Commerce. This can be seen from Figure 7.1 which shows the principal

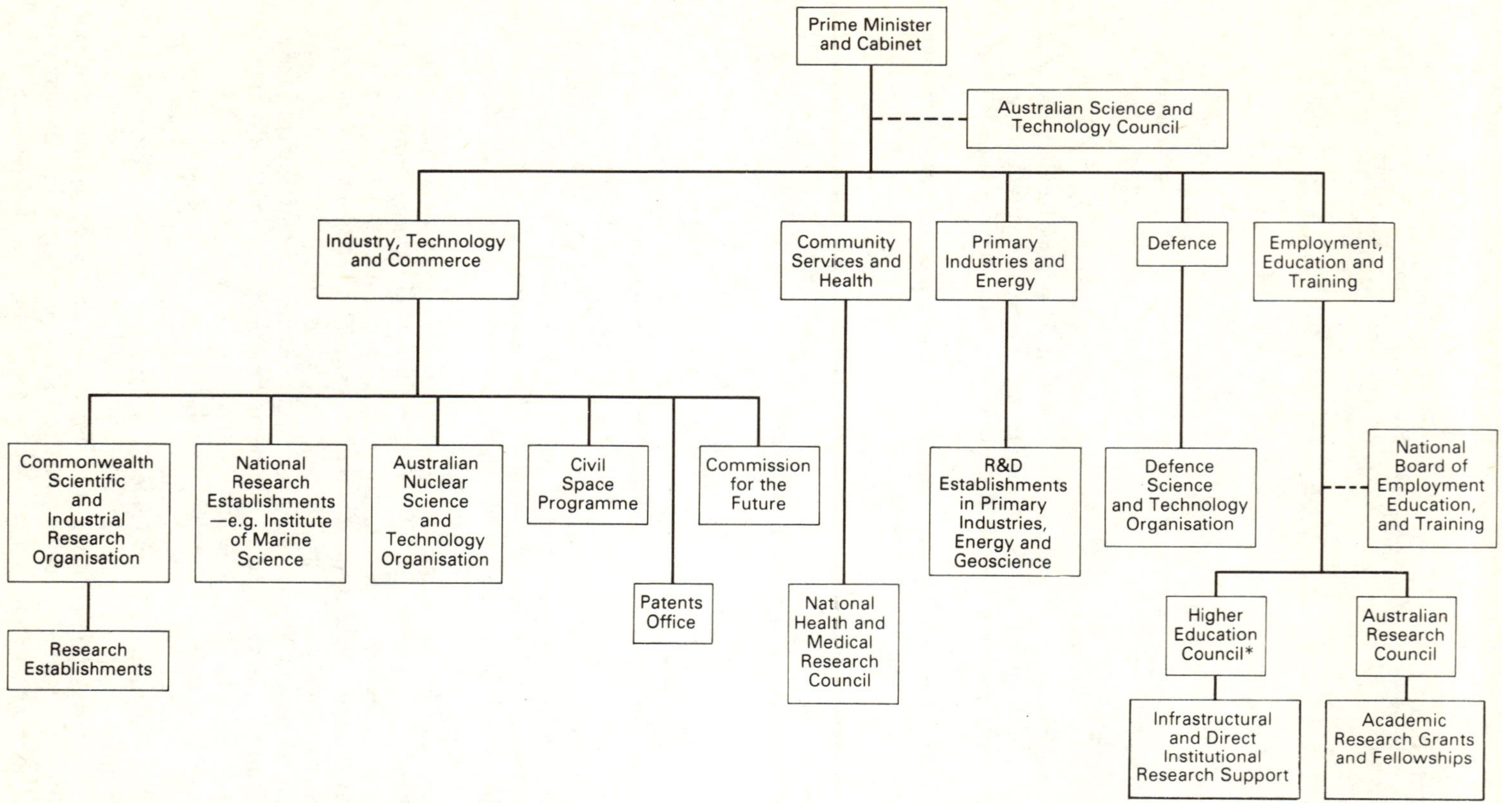

Source: Constructed from various official documents. Recent restructuring has resulted in existing organizational charts being superseded.
* Previously Commonwealth Tertiary Education Commission.

Figure 7.1 Outline of the Australian federal R&D system

features of the federal R&D system (but excludes regionally-focussed agencies and research programmes of the six Australian states).

Support for academic research has similarly been restructured. The Department of Employment, Education and Training (DEET) was given responsibility for the Higher Education Council, which provides core funding to universities for research infrastructure and designated centres (formerly the task of the Commonwealth Tertiary Education Commission). DEET also oversees the Australian Research Council (ARC) which has brought together within one agency various grant schemes for non-medical research. This centralization of policy functions previously under two ministries (DOS and the Department for Education and Youth Affairs) reflects a desire to set clearer priorities for academic research (Aitkin, 1988), and is being accompanied by institutional changes in the universities paralleling those in the United Kingdom (DEET, 1987; 1988a). Policy-makers have consequently been exploring the merits of introducing more systematic approaches to planning academic research and graduate training.

Overall, we shall see that, although a wide range of foresight activities has been undertaken in Australia, the impact on decision-making has been limited. Among the reasons are a political culture relying upon negotiated compromise among major stakeholders and intuitively preferring 'rule-of-thumb' judgements to those based on 'objective' analysis; a pluralist approach to research support with no real mechanism for establishing national R&D priorities; the lack of any long-term planning tradition; a free-market ideology inimical to attempts by funding agencies to pick 'winners' (whether these be industrial sectors, technologies or research fields); and strongly autonomous scientific institutions which have defended the well-established peer-review system. The emphasis now being placed on central coordination and planning of R&D is, however, resulting in an environment where foresight seems certain to make a greater contribution to research policy.

In what follows, we first examine the role of foresight in establishing national priorities for research. Several initiatives within the higher education sector are then considered, followed by a discussion of how DITAC and CSIRO have deployed prospective techniques in helping plan programmes oriented towards strategic research and basic technology. Finally, we draw a number of general conclusions arising from Australian experience with foresight.[2]

2. FORESIGHT IN NATIONAL PRIORITY-SETTING

Systematic attempts to identify research areas likely to be important to the nation over the longer term have been made by the Australian Science and Technology Council and the Department of Science. Two especially interesting exercises merit consideration along with several other foresight activities undertaken by these organizations.

(a) Australian Science and Technology Council

ASTEC serves as a high-level advisory body to the Prime Minister, and has a membership of fifteen drawn from the business community, universities and the trade-union movement. Its statutory tasks include furnishing information and advice on: (i) the development and application of science and technology in relation to the furtherance of national well-being; (ii) the identification and support of promising new research areas; and (iii) the relative priorities that should be attached to the budget proposals submitted annually by ministries with science and technology responsibilities (ASTEC, 1987d, pp.4–5). However, while ASTEC's institutional location enables it to provide independent advice to government (a function resented to some extent in other parts of the R&D system), the distributed nature of power in federal decision-making means that its recommendations are unlikely to be implemented unless there is sufficient political support among relevant agencies. This is well illustrated by the limited impact of a study initiated in 1980 which aimed to identify priorities for Australian research to the end of the decade.

The exercise stemmed primarily from ASTEC's desire to develop a more systematic basis for determining future national funding policy. The Council had earlier produced the first in a planned series of reports on the state of Australian science (ASTEC, 1979), which surveyed developments in twenty-five key fields and distinguished four categories of priority (from 'urgent high priority' down to 'recommended'). Although not repeated (ostensibly because of the effort involved), it was decided to continue this line of work by exploring the method of identifying research priorities developed by UNESCO.

The approach adopted in the ASTEC study was sophisticated, explicitly combining science-push and demand-pull factors, as well as taking into account the likely impact of advances in one discipline on progress in others. Several stages were involved:

(1) Two academic consultants first reviewed forecasting and priority-setting exercises undertaken overseas (Tisdell, 1981; Ronayne, 1984). In addition to drawing lessons on the factors determining success and failure in foresight, they identified the UNESCO approach as particularly promising.[3]

(2) A 'pilot workshop' was next held in which a modified version of the UNESCO methodology was tested in two fields (minerals and energy research). After favourable review, a 'priorities workshop' was organized which brought together nearly eighty senior scientists, technologists, business leaders and government officials.[4] Participants represented both 'users' of scientific information (who concentrated on the national objectives to which R&D might contribute), and 'providers' of that information (who focussed on identifying research areas important in relation to both achieving those objectives and advancing knowledge).[5] The 'user' and 'provider' groups were each divided into ten panels, based either on industrial sectors (e.g. primary industries) or fields of science and technology (e.g. earth and space sciences).

(3) Prior to the workshop, 'users' (the O-group) agreed a set of national objectives which could realistically be achieved through concerted R&D efforts over the next five to ten years (e.g. a reduction in land degradation from agricultural and other uses). After screening by ASTEC officials to eliminate duplication, and consideration of comments by O-group participants, a revised list of objectives was produced which constituted

one starting point for the workshop. In parallel, 'providers' (the S-group) were sent the UNESCO research classification scheme, and asked to suggest changes to disciplinary categories which would facilitate analysis of emerging areas showing promise for meeting Australian needs. After two rounds of comment, a modified scheme was prepared by ASTEC officials and this formed a second input to the workshop.

(4) The first task undertaken by the O-panels was to review the draft lists of objectives in the ten sectors, and categorize them into four levels of priority for future government funding. At the same time, the S-panels identified the disciplines in each field most likely to contribute to meeting the various objectives. After some revision of the research classification scheme, S-panels presented their results to the O-group for comment.

(5) A central element in the exercise was to establish the extent of cross-impacts and dependencies among both objectives and disciplines. Judgements were made by each O-panel as to whether the attainment of specific objectives was necessary for, or dependent upon, other objectives. The judgements were analysed using a matrix to establish which linkages were most critical (using a four-point scale). O-panels then agreed the strengths of the relationships among their respective sector-based objectives, while S-panels assessed the impact and dependence of disciplines in their field on other R&D areas. Negotiations also took place between panels in cases where interests overlapped. This yielded a list of 166 areas ranked in terms of pervasiveness of impact.

(6) The final step was to develop consensus between the two sets of panels on the relative importance of R&D disciplines to national objectives. O-panels evaluated the extent to which realization of each objective would depend on advances in specific areas, while S-panels assessed the contributions likely to be made by their respective disciplines.[6] This provided the basis for negotiations between each O-panel and S-panel to agree an overall ranking.

(7) Following the workshop, an ASTEC panel reviewed the results for consistency and made a number of minor revisions. It then analysed the results in more detail, employing a set of indicators to represent the linkages among objectives, and between disciplines and objectives.[7]

Illustrative results from the exercise are shown in Figure 7.2, which lists the disciplines judged to be 'critical' for realizing especially important national objectives (see left-hand column) together with those of lower priority. Disciplines are also ranked in terms of their relevance to the entire set of objectives in each category. As can be seen, applied mathematics achieved the highest relevance figure (25) for the top priority objectives, in large part because of its critical importance for the construction, energy and environmental sectors. In contrast, the results in Figure 7.3 show the extent specified disciplines were regarded as likely to contribute to progress in other R&D areas. Again, applied mathematics scored highest overall.

Further analysis by the ASTEC panel compared the priorities determined during the workshop with those implicit in the pattern of government funding for research, and in the employment of trained scientists and engineers broken

Relevance Index	Critical Support for Priority 1 Objectives	Critical Support for Priority 2 Objectives	Critical Support for Priority 3 Objectives	Critical Support for Priority 4 Objectives	No Support for Objectives at Critical Level
100–30					
30–20	25 Applied Maths. 23 Lower Atmos. Sciences 20 Computer Technology	21 Chem. Eng. & Technology			20 Mech. Eng. & Technology
20–10	18 Plant Science 17 Hydrology 16 Ecology, Forestry Science, Animal Science 15 Physical Chemistry 14 Occupational Medicine, Genetics 13 Geophysics, Soil Science, Clinical Sciences 12 Medical Neurosciences, Pharma. and Clinical Pharmacy 11 Cancer, Epidemiology Geology, Organic Chemistry, Plant Physiology and Biochemistry 10 Oceanography, Biochem. Technology, Non-Linear Maths, Degenerative Vascular Diseases, Geochemistry	17 Agricultural Systems 15 Construction Technology Solid State Physics 14 Nuclear Technology, Materials Technology 12 Electronic Technology Metallurgical Technology	12 Acoustics 11 Toxicology, Optics 10 Molecular Biology		17 Instrument Technology 15 Urban Planning 14 Electronics 12 Telecom, Technology 11 Microbiology & Virology 10 Physics of Fluids, Biochemistry, Petrol. and Coal Technology
10–0	9 Preventive Medicine, Mining Technology 8 Entomology 7 Power Technology	7 Fishery Science	8 Transport Systems Technology 7 Macromolecular Chemistry 6 Medical Technology		8 Analytical Chemistry, Inorganic Chemistry 7 Electromagnetism, Human Physiology 6 Nuclear Physics 5 Animal Physiology, Neurosciences, Motor Vehicle Technology 4 Developmental Biology 3 Instrumental Astronomy, Immunology, Molecular Physics 2 Nucleonics

Source: ASTEC (1981b, p. 23).
Note: The four categories of national objectives range from 'extremely important' (priority 1) down to 'limited importance' (priority 4).

Figure 7.2 Relevance of R&D disciplines to Australian national objectives (priority profile)

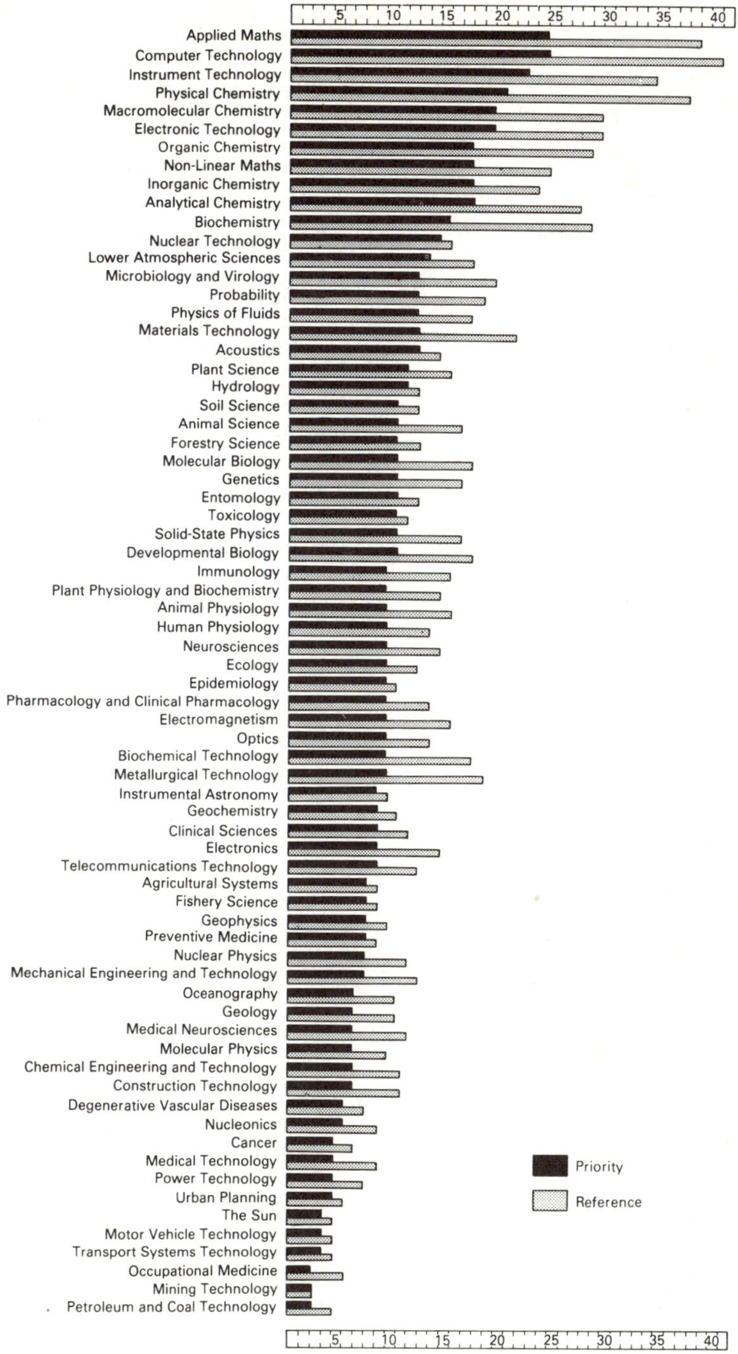

Source: ASTEC (1981b, p. 110).
Note: Cross-support profiles' give an indication of the overall relevance of each discipline for progress in all other research areas. The reference index assumes all disciplines are equally important, while the priority index provides a weighted measure of 'relevance' taking into account the aggregate rankings agreed by panels. Scales go up to 100 in both cases.

Figure 7.3 Interrelations between disciplines: cross-support profiles (priority and reference) indentified in 1981 ASTEC study

down by discipline. Several important mismatches were identified, and the panel therefore recommended increased support for medical, environmental and physical sciences, and a reduction for agricultural research. It also suggested that the list of disciplines most relevant to national objectives be used in determining future priorities for graduate training programmes, while areas with the greatest potential impact on other research were seen as forming an input to discussions on targeting of basic science. Also useful here would be a further list of more than sixty specialties singled out by the panel as 'most important' for the advance of scientific knowledge — for example, condensed-matter physics (because of developments in non-crystalline states, solid-state devices, superconductors and surfaces — ASTEC, 1981b, p.119).

Despite the undoubted methodological sophistication of the exercise, the immediate response of the research community was overwhelmingly negative. Criticism from certain ASTEC Council members resulted in the findings appearing only as an unofficial 'working paper' (ASTEC, 1982a), while a later government review concluded that 'the report ... appears to have had a minimal impact on the identification of research areas warranting selective ... funding' (DOS, 1987a, p.8). According to one official, 'the report got canned for a whole variety of reasons':*

(1) Systematic approaches to science policy were at the time relatively new in Australia and there was an inherent suspicion among researchers about the exercise. There was no foresight or evaluation culture to provide a receptive context for the results.
(2) The study was initiated by the newly appointed ASTEC Secretariat which was in the process of establishing a role for the organization, and there was insufficient prior consultation with researchers, scientific institutions and other government agencies. As a result, little consensus existed on the need for the exercise.
(3) The reliance upon senior researchers and industrialists in panel discussions provoked fears, especially among scientists in newer subfields or those without representation, about the possible repercussions for future funding. Criticism focussed especially on the high ranking gained by applied mathematics, which was explained in terms of the pervasive definition of the field employed, and led to accusations that the exercise was overly influenced by mathematicians.* Moreover, by commencing with the UNESCO classification scheme, the exercise had an inbuilt bias towards traditional disciplines. Coupled with the lower familiarity of older researchers[8] with emerging areas, it was perhaps not surprising that certain specialties already recognized as strategically important in other industrial nations — for example, neurosciences, molecular electronics, molecular biology and optics — were not given greater priority. (In marked contrast, however, these areas figured, along with superconductivity, artificial intelligence and laser devices, as especially promising in the list characterized by high intrinsic scientific merit!)
(4) Finally, the study was carried out in a political vacuum since ASTEC had no institutional machinery for implementing the recommendations, while failure to secure the wider involvement of funding agencies meant there was no interface with the budget-setting process or with established peer-review procedures. The inevitable result, according to one

interviewee, was that 'there was never any possibility of the study impacting on policy — with hindsight it was a theoretical and fanciful exercise'.* This conclusion is shared by Johnston (1987, p.4):

> the lesson that could be drawn from this failed exercise was that there was no attempt at (or certainly no success in) establishing a commitment to develop and improve a process of priority-setting. The participants approached the exercise with a perspective of determining whether the exercise was accurate, and secondly whether *their* field was likely to benefit, and more importantly to be penalised, by the adoption of such a procedure. Once some inadequacies were discovered in the methodology, the approach could be discarded and the status quo re-established.

Although accepting much of this criticism in its own review of the exercise, ASTEC cautioned against judging the results solely on the basis of immediate policy impact. Perhaps just as important over the longer term would be:

> the extensive debate within and outside ASTEC which the 'priorities workshop' has engendered ... It has increased the knowledge of most ASTEC members concerning both practice and theory in the difficult area of priority assessment ... It has exposed members, some for the first time, to a major experiment in science policy, serv[ing] to alert ASTEC to some of the difficulties in seeking to be objective about matters involving complex judgements. [ASTEC, 1982a, p.8]

This has indeed turned out to be the case. However, in other countries, such a negative initial response is by no means uncommon and to some extent should be seen as part of the lengthy learning process typically involved in establishing a role for foresight in national priority-setting.

ASTEC subsequently concentrated on studies dealing with more general policy issues and (through its Technological Change Committee) with needs and opportunities in specific fields.[9] The latter have covered such areas as interactive information services, computer-related manufacturing and artificial intelligence (ASTEC, 1983; 1985a; 1987c), and include a foresight element in that evidence is presented on technological change and its likely socio-economic impact. Although scenario techniques were employed in the interactive information services report, trend-analysis methods are more commonly used. Given Australia's dependence on imported technology, significant attention is given to surveys of R&D in other industrial nations, which are normally undertaken by visiting delegations but occasionally by specialist consultants.[10]

In recent years, however, the trend towards increasing coordination of R&D by government, coupled with the growing institutional strength of ASTEC, has seen a return to initiatives aiming to set a national agenda for basic research. These include, first, an influential review of the higher education sector notable, among other things, for its recommendation that the Australian Research Council be established to take a lead in 'the identification and selective support of emerging or underdeveloped fields of research which are likely to be important in the national interest' (ASTEC, 1987b, p.36).

Second, a study was launched in late 1987 to develop a detailed 'profile of Australian science'. The main objectives were: (i) to carry out an audit of expenditure, scientific expertise and infrastructure (e.g. experimental facilities) across all fields of basic research to identify 'strengths, weaknesses, gaps and opportunities'; (ii) to assess current national research performance relative to both historical achievements and the contributions being made by other countries; (iii) to determine what implications existing priorities have for the country's likely research performance over future years; and (iv) to draw conclusions concerning potentially promising areas of science, the redirection of resources and the overall level of support for basic research (ASTEC, 1987e, p.1).

The remit for the study recognized that peer-review procedures, although important for programme-level decisions, are limited by the reluctance of the scientific community to be seen as picking winners or losers, a reluctance reflecting the desire to avoid generating self-fulfilling prophecies which may result in funds for less favoured areas being cut.[11] Consequently, the intention was to construct the scientific profile with the aid of a range of input and output indicators — for example, data on specialties in which particularly large numbers of funding requests are being made, or information on the demographic characteristics of the research community (to determine the disciplines most attractive to younger scientists). Particular emphasis was placed on exploring the utility of bibliometric techniques for mapping areas of national scientific strength and weakness, and for identifying rapidly growing and declining fields. Interviews conducted with a cross-section of leading researchers from different disciplines showed relatively good agreement with the bibliometric results, as did submissions made by professional scientific institutions. The ASTEC report was thus able to conclude that formal foresight techniques provided a potentially valuable tool for helping set national research priorities. Further development and testing of such techniques in a policy context was therefore proposed (ASTEC, 1989).

While it is too early to make any assessment of the likely policy impact of the study, the institutional environment is today considerably more receptive than when formal committee-based procedures were employed in 1981. Success will depend to a great extent on endorsement of the findings by the Australian Research Council, and on the response of the scientific community more generally. ASTEC is aware of the need for broader involvement in decisions on longer-term research policy, and has taken steps to institute new consensus-seeking mechanisms which enable the scientific community to participate in shaping national R&D priorities. November 1986 saw the first annual Science and Technology in the Budget Forum, convened jointly with the principal scientific and engineering academies. The intention was to promote wider debate on the priorities established in the national budget, with officials from government and the funding agencies making presentations.[12] The forum was repeated in 1987 (see ASTEC, 1987d, p.13; Tegart, 1987a).

Finally, mention should be made of a major exercise to assess the potential impact of science and technology on economic development during the 1990s. This was initiated by ASTEC in 1988 following various reports to government suggesting that stablilization of the nation's external debt will require a growth in exports of 5–6 per cent per annum. Research is focussing on three issues: (i) longer-term commercial opportunities and market trends; (ii) the likely impact of promising technologies on industrial competitiveness; and (iii) constraints on exploiting emerging opportunities (for example, inadequate research infrastructure or stock of R&D expertise). One output from the study will be 'advice to government on the policies most likely to maximise the contribution of science and technology to Australia's economic growth and export performance' (ASTEC, 1988a, p.2).

(b) Department of Science

The second exercise aiming to identify national research priorities in a systematic manner was undertaken by a DOS Research Capability Assessment Task-Force in the period immediately prior to absorption of the department by DITAC. The objectives set for the study were to 'highlight those areas of basic research which are capable of being exploited to the economic advantage of Australia through: (a) development of a methodology to identify strong areas of research activity in Australia; and (b) characterizing their relationship to our natural, cultural, social, economic and infrastructural advantages' (DOS, 1987a, vol.1, p.1). Although the resulting report on *Research Strength and Economic Benefit* has no official status and seems likely to remain unpublished, it is worth reviewing here since the methodological approach employed is relatively novel.

The initial step was to review experience with research foresight both overseas (especially in the EEC, Japan and the United States) and in Australia (e.g. the 1981 ASTEC exercise). This provided the basis for constructing the analytical framework adopted in the study (see Figure 7.4), which emphasized matching up areas of national scientific strength to commercial opportunities in technology sectors where industry is economically competitive.

In operationalizing this approach, it was first necessary to develop measures of *research strength*. This entailed analysing: (i) various expenditure and personnel statistics to gain an overview of the fields in which existing efforts were concentrated (e.g. astronomy and ecology), and how the distribution of activities differed from the international norm;[13] and (ii) data on research outputs to identify areas in which Australia had made significant contributions (e.g. plant genetics and certain medical specialties). The latter involved using two sets of bibliometric indicators constructed from the *Science Citation Index* — national shares of publications and citations broken down by field (drawn from the Science Literature Indicators Data-Base compiled by CHI Research), and a bibliometric model of Australian research activity based on clusters of co-cited papers (described below).

The next task was to develop a framework for assessing the potential *commercial exploitability* of particular fields over the period to the mid-1990s. Three sets of factors were considered: (i) characteristics of the area — intrinsic potential for commercial exploitation, willingness and ability of existing research groups to participate, need for further work in related 'enabling' fields, links of research groups with industry and with overseas institutions, and the preferred roles of government and the universities in developing new activities; (ii) the prospects for industry to capitalize on the commercial opportunities within each field — extent of consensus among government and industry on potential for exploitation, additional resources required for R&D infrastructure, likelihood of adequate investment by companies to develop emerging technologies, and relationship to national patterns of economic activity and competitive strength (e.g. manufacturing sectors based on natural resources); and (iii) projected economic benefits from investment in the area

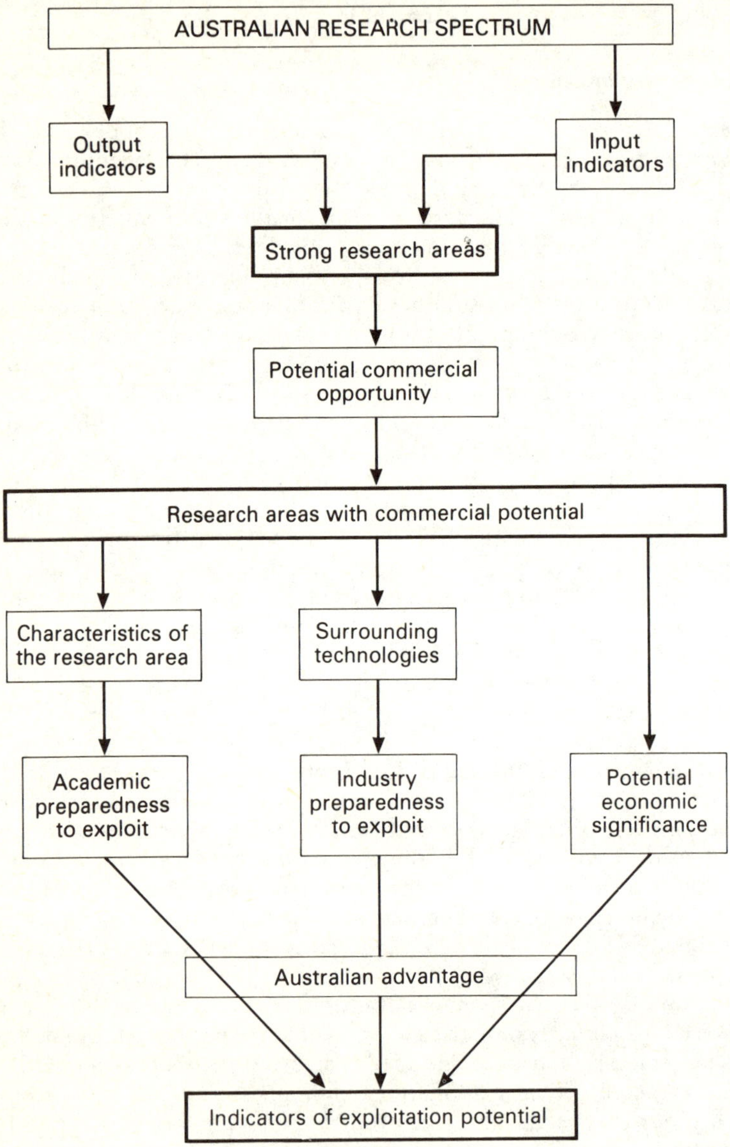

Source: DOS (1987a, vol. 1, p. 12).

Figure 7.4 *Outline of Australian Department of Science task-force approach to identifying exploitable areas of science*

— likely domestic and foreign markets associated with new technologies, and external factors constraining commercial impact (e.g. trade friction or major foreign R&D initiatives in the field). Experts from government, academia and industry were then invited to participate in a survey requiring judgements to be made (on a five-point scale) on the potential exploitability of certain fields. Background statistical data were supplied to assist respondents, including not just input and output indicators but also information on linkages across scientific fields, and among related areas of basic science and technology.

The third task was to conduct pilot studies assessing the exploitability of three strong areas — plant molecular genetics, polymer chemistry, and astronomy and astrophysics. These drew upon consultations with experts on likely scientific developments and the scope for application, one aim being to validate results from earlier stages in the exercise. The task-force was thus able to make a convincing case that plant molecular genetics would be central over both the short and longer term in increasing the competitiveness of the agricultural and forestry sectors (particularly through the impact of genetic engineering and tissue-culture research). Polymer technology was also judged promising, with major benefits likely to arise from R&D on 'very large polymer products' and on advanced optical applications. However, the opportunities associated with astronomy and astrophysics (stemming principally from technological advances in gallium arsenide-doped detectors and charge-coupled devices) were seen as more limited, and commercial exploitation would involve entering the highly competitive international market for aerospace and telecommunications products.

At the end of the study, considerable effort was devoted to assessing the utility of the approach for science-policy purposes. The overall conclusion was that, although in principle providing a viable method for assessing exploitability, extensive inputs need to be made in a structured fashion by experts if meaningful results are to be obtained. The dangers of officials relying solely on formal foresight exercises were particularly stressed: 'The importance of expert advice cannot be overemphasized. Each step along the process of research assessment — including research evaluation and the identification of gaps, preparedness for exploitation, industry attitudes, economics — requires special expertise' (ibid., p.61). The report recommended that DOS undertake a comprehensive exercise to assess existing departmental research profiles and identify areas of future national importance. However, this first required significant improvements in the science and technology indicators produced by government (especially regarding the disaggregation of research expenditures and personnel by field, and the range of patent statistics collected). Moreover, an effective advisory committee mechanism would be essential, with industry being given ample opportunity to evaluate the merits of the proposed framework.

Although representing an important addition to the range of foresight techniques available, the DOS approach can be criticized for its conservative science-push perspective. By starting from existing research strengths, there was an inbuilt tendency to concentrate on established disciplines even though an important task in priority-setting is to identify emerging cross-disciplinary areas. These are normally hidden from view in expenditure and personnel data (which are invariably based on traditional field definitions). Nor is it easy to throw light on growth points in science with conventional bibliometric techniques (the CHI data are again based on established disciplinary categories), although the task-force was perhaps unfortunate in not being able to benefit as much as had been hoped from the claimed advantages of co-citation analysis (which was found to require further technical development before its potential policy utility might be realized).

As regards the impact of the exercise, several officials commented that departmental restructuring had made the task of implementation more difficult.[14*] One problem is that the approach developed by the task-force relates to establishing priorities for basic science, but responsibility for funding such research lies with DEET and ARC rather than DITAC (which took over the DOS policy group undertaking the study). Initially at least, scope to apply the approach for in-house policy purposes was limited, and follow-up studies have therefore concentrated more on analysing the development and commercial utilization of generic technologies through patent indicator data.

Another problem is that acceptance of the need for greater selectivity has been much slower than in DEET. DITAC primarily supports industrially oriented research, where the *laisser-faire* view that priorities should be determined in a responsive mode according to market needs remains strong. However, the climate of opinion has been changing rapidly, and this was reflected in the establishment of an interdepartmental committee in late 1988 'to make immediate recommendations on the most effective way to strengthen research and development in science and technology' with a view to 'planning now for the industries of the 21st century' (DITAC, 1988b).

3. FORESIGHT IN BASIC RESEARCH

(a) Higher education sector

The institutional locus of basic science has shifted over the last decade as the sixty-five or so universities and higher education colleges have taken over many of CSIRO's former functions in the national research system. Between 1978 and 1985, spending on academic R&D rose by 18 per cent in real terms to reach A$730 million (remaining a constant 0.33 per cent of GDP — DITAC, 1987i, p.48), and further growth is planned up to the early 1990s. Finance has traditionally come in three main forms: (a) 'general university funds' (GUF) from the Commonwealth Tertiary Education Commission (CTEC); (b) direct support for designated centres; and (c) fellowships and peer-reviewed grants, principally from the Australian Research Grants Committee (ARGC) and National Health and Medical Research Council (NHMRC). While GUF still constituted 64 per cent of overall expenditure in 1985, its relative decline is set to accelerate following policies adopted by DEET, the Higher Education Council and ARC since their establishment in 1987.[15] Before considering this more recent experience, it is instructive to examine the role of foresight in CTEC and ARGC.

(i) Commonwealth Tertiary Education Commission

The nature of university funding, coupled with a long-established peer-review tradition, has been responsible for a rather low level of foresight in the higher education sector. Distribution of GUF by CTEC was primarily based on student numbers (mediated by the continuing demands of state governments for new universities), with institutions receiving 20 per cent of their grants-in-aid in the form of unconditional funding. Priorities therefore implicitly tended to reflect the popularity of disciplines among undergraduate students.

In the mid-1980s, however, growing demands for public accountability and government criticism that universities were not playing a full part in economic restructuring resulted in two developments within CTEC. First, exploratory studies were carried out on the possible introduction of formal evaluation methods (including bibliometric techniques) to assess research performance (Linke *et al.*, 1984; Bourke, 1986). Second, and of more relevance here, a procedure was established for reviewing the current status and future needs of individual disciplines. In the first of the reviews, an expert panel considered current and projected imbalances in the supply and demand of graduates and researchers in the field of law. The resulting report recommending greater concentration of effort (including closure of some departments), and more stress on areas of national need (CTEC, 1987), predictably generated considerable controversy. It was followed by a review of engineering focussing on the objectives for academic courses (both undergraduate and postgraduate) up to the end of the century (CTEC, 1986b). An attempt was also made to identify future growth points within different engineering specialties, as well as to develop indicators of national and institutional research performance. The results had not yet been published in 1988, but were again expected to provoke reaction from the universities if major changes were proposed.* These disciplinary reviews represent the first attempt to introduce external criteria into the distribution of GUF.

CTEC has also been responsible for two schemes providing direct research support to designated institutions. Launched in 1982, the larger of the initiatives concerns Special Research Centres, which each receive around A$0.6 million per annum for a six-year programme in an agreed field (e.g. clinical immunology or policy studies). Centres are selected by a panel of leading researchers using a competitive proposal-based procedure, with scientific excellence being the main criterion.[16] This is not the case, however, in decisions to establish Key Centres in Teaching and Research, which are given annual funding of A$0.2 million for mixed programmes of research and teaching. Evaluation of proposals for new centres, again conducted by an expert panel, takes into account not only scientific excellence but also future industrial needs for researchers in more applied specialties. Yet, while five 'likely' areas for funding were indicated when the initiative was announced in 1985 (e.g. industrial design), no systematic foresight appears to have been undertaken by CTEC in identifying these fields. As with the Special Research Centres, the policy has rather been to adopt a responsive approach in line with academic preference.[17]

(ii) Australian Research Grants Committee

Until 1987, ARGC represented the principal source of support for academic projects in fields other than medicine and health (which was, as now, the responsibility of NHMRC). The criteria for awarding grants were based exclusively on scientific excellence as judged by peer-review (ARGS, 1987), and foresight played no role. Following encouragement by DOS, however,

ARGC in 1985 began a study to explore the utility of bibliometric methods for identifying current and emerging Australian scientific strengths. Co-citation studies of three fields were commissioned from the Centre for Technology and Social Change. As noted below, the experiment was not wholly successful, with the panels responsible for evaluating the bibliometric models tending to concentrate on assessing their accuracy rather than what contribution they might make to expert decision-making.

Criteria other than scientific merit did, in contrast, play a part in establishment of the National Research Fellowship Scheme in 1984. Many of the fellowships awarded by ARGC have been in eight priority areas (such as materials technology) selected through informal consultation with ministries and R&D agencies. External criteria have also featured in other academic funding schemes: in particular, four levels of mission-oriented priorities were identified for the National Energy Research Development and Demonstration Program (DPIE, 1988), while the Marine Science and Technologies Grant Scheme has since the early 1980s supported work in nine target areas (Ronayne, 1984). In both cases, priorities were set by expert committees following national workshops which considered future research opportunities and needs. Neither made use of formal foresight techniques.

(iii) Department of Employment, Education and Training

The relative autonomy accorded to ARGC and CTEC enabled the universities to resist any suggestion that peer-review be complemented by other decision-making mechanisms. However, this situation has changed markedly since DEET published a 'green paper' outlining proposals for a fundamental restructuring of the higher education system (DEET, 1987). These include: (a) the negotiation of individual contracts with universities specifying the fields in which teaching and research should be concentrated; (b) a gradual decline in GUF to release funds for an expansion of the direct support mechanisms administered by ARC; and (c) greater emphasis on research in fields important for the nation's future, especially 'exploitable areas' of natural science and engineering.[18]

The changes would involve DEET assuming a more active role in longer-term planning, with a consequent reduction in the autonomy of the research community (although peer-review would continue to be used in allocating funds within broad priority areas). Not surprisingly, the proposals caused consternation in academia[19] and resulted in a concerted campaign to overturn the more radical recommendations. However, the government seems unlikely to accede to this pressure, since the reorganization of higher education is central to its policy of ensuring the national R&D infrastructure is able to meet demands for the scientific and technological expertise required to guarantee Australia's economic future.*

A subsequent white paper attempted to stem the criticism by acknowledging 'the relevance of all disciplines to our current economic circumstances', and stressing that priority support for 'those fields of study of greatest relevance to the national goals of industrial development and economic restructuring ... will be implemented gradually as part of a significant expansion of total higher education effort' (DEET, 1988a, p.8). However, academic resistance has continued, fuelled by suspicion of recommendations which, on the one hand, leave institutions 'free to establish their own priorities and develop their strengths', but, on the other, limit freedom of choice through targeting of research. Universities are now required to produce rolling medium-term strategic plans which identify future priorities, taking into account institutional strengths and national needs such as those for graduates in specific fields (DEET, 1988d). They must also include data on research publications, postgraduate student output, and level of peer-reviewed grants by field. This has enabled DEET to compile an overview of institutional strengths and weaknesses, useful both for its own decision-making and to assist universities in preparing their strategic plans (DEET, 1988b).

The government's intention is that ARC will become increasingly influential in 'steering' academic research. In addition to overseeing several funding schemes, it was given responsibilities for framing national research policy.[20] This includes implementing the guidelines set out in the 'green paper' that more emphasis on competitive funding mechanisms is necessary if resources are to be targeted flexibly on 'areas of ... national economic and social importance' (DEET, 1987, p.68). According to the ARC Chairman, the key problem is how to reach agreement on procedures for setting academic research priorities:

> The understandable urge by governments to establish priorities has ... been met by reluctance and even hostility on the part of universities ... because of the fact that the universities' funding includes a large but not precisely identified element for research: discussions about research priorities, when the university world is involved, quickly become discussions about university funding and university management. University people tend to take up intransigent positions on these matters: teaching cannot be separated from research, even conceptually; performance indicators for research are hopelessly flawed; national priorities require national goals and national goals are intrinsically vacuous; if priorities are to be set the mechanism must be 'bottom-up' because only researchers know where the possibilities lie; and so on. The consistent message is 'touch me not'. Yet touched the universities are, and will continue to be, so long as they depend on the government for the greater part of their income. A sensible ordering of priorities in research is not beyond the wit of man, and I have no doubt that we will achieve one before very long ... We need new mechanisms for paying attention to the future, making decisions about areas, and backing those decisions with funds. [Aitkin, 1988, pp.2–3][21]

As a result, ARC quickly recognized the need to develop a capacity for undertaking systematic research foresight, and began to plan studies exploring the strengths and weaknesses of different approaches. At the same time, reviews would be carried out of experience with the two national basic research programmes (especially that on marine sciences)[22] to ensure selective policies do not endanger academic creativity. In the first instance, therefore, no more than 20 per cent of overall ARC research funding is being allocated to five targeted areas (which in 1990 will include cognitive science and materials research). One task of a Research Review Committee established in late 1988 is to 'provide a statement of national research priorities to guide the Australian Research Council in recommending on the competitive allocation of research funds to higher education institutions' (DEET, 1988e, p.1).

The existence of a clear need for priority-setting, linked to specific decisions on selective budgetary support, seems likely to provide an important opportunity to test systematic foresight techniques in a policy context. The growing consensus on research targeting — with activities in DEET and ARC being matched by efforts in DITAC to set priorities for basic technologies — also appears certain to provide a more fertile environment for ASTEC in its quest to link broad decisions on R&D funding explicitly with future socio-economic needs.

(b) Centre for Technology and Social Change

The Centre for Technology and Social Change (TASC) at the University of Wollongong is Australia's largest academic science-policy institute, and works mainly in the fields of R&D evaluation, technology assessment and research foresight. Of special interest is a series of studies undertaken for four state governments to help develop policies for strengthening the regional infrastructure for R&D and targeting support for research. For example, a 'science and technology profile' constructed for the Queensland Science and Technology Council aimed not only to identify the capabilities and limitations of the state research system and its links to industry, but to establish a technology identification process in six key industries (TASC, 1987, p.8). The results have reportedly been well received both by policy-makers and the industrial and public-sector research communities.*

TASC has also been engaged in a long-term programme of basic research on science and technology indicators, focussing in recent years on bibliometric techniques for mapping emerging specialties (Franklin and Johnston, 1988). In particular, its contribution to the DOS exercise to identify promising research areas involved constructing a co-citation model of Australian science (i.e. a model of national participation in clusters of highly co-cited papers published world-wide) for the period 1983–4 (Franklin, 1987a; 1987b).[23] This was complemented by detailed case-studies of national strengths and weaknesses in astronomy, acoustics and computer science (e.g. Franklin, 1986), with panels of academic researchers nominated by the Australian Academy of Science being used in validating the results. Reaction was apparently somewhat mixed; the astronomy model was found to be of some potential utility, but less positive judgements were made about the other two.* Apart from the technical complexity involved in understanding and analysing the co-citation clusters,[24] the models were criticized for omitting certain areas of research, for including specialties of limited relevance, and for not being sufficiently specific to allow a detailed assessment of Australian efforts relative to other countries.[25] Such problems arose partly because co-citation analysis is less suited to more applied fields like acoustics, and partly because of difficulties with using the technique in diffuse areas like computer science.* There was also a tendency for the panels to see the technique as a threat to peer-review[26] rather than as a means of providing additional management information, a common occurrence when bibliometric approaches are first introduced.

Despite these problems, it is generally accepted that the TASC study has been valuable in building a national competence in early-warning indicators (e.g. ASTEC, 1988b). Together with more traditional bibliometric data on publication and patent world-shares broken down by field or sector,[27] the results have been drawn upon by ASTEC and DITAC in initiatives to explore new approaches to identifying opportunities for Australian research. In addition, DITAC has commissioned a further study from TASC to assess the utility of patent analysis (including citations to basic research) for strategic planning. This builds on recent work to map the science-technology interface by studying the linkage between co-citation clusters and patents (Coward and Franklin, 1989). As with similar institutes overseas (e.g. the Lund Research Policy Institute in Sweden — see Chapter 9), one overall lesson here is the value of long-term academic research in developing basic expertise in foresight. TASC is thus now in a position to make an effective contribution to priority-setting initiatives within both government and funding agencies.

4. STRATEGIC RESEARCH AND BASIC TECHNOLOGY

(a) *Department of Industry, Technology and Commerce*

The demise of DOS left DITAC with responsibility for longer-term strategic research. As well as developing a closer working relationship with CSIRO, the department has put in place a comprehensive set of support mechanisms for industrially-oriented research under the Grants for Industrial R&D Scheme (DITAC, 1988c). This has three elements: (a) 'discretionary grants' aimed at start-up firms and larger companies involved in restructuring; (b) 'generic technology grants' to develop enabling technologies likely to improve industrial competitiveness over the next decade; and (c) 'national interest agreements' providing support for promising R&D projects which companies would otherwise not undertake because of technical uncertainty or the diffuse nature of likely commercial benefits.[28]

Although DITAC has been exploring the utility of foresight techniques, their impact on policy has been limited by a widespread ambivalence towards longer-term planning. This is reflected in the continuing struggle between those in government (principally officials in R&D agencies) advocating a strongly focussed national technology policy based on 'rational' analysis of emerging industrial opportunities and needs, and their opponents (in the private sector, the Treasury and the Bureau of Industry Economics) who maintain the traditional *laisser-faire* position that it is dangerous for the state to take over functions best left to the market (Johnston, 1987). The case for selective technology support has, however, been gaining ground, resulting in several planning initiatives with a foresight element. Although largely focussing on individual research fields or industrial sectors, two have involved attempts to establish macro-level priorities — the National Technology Strategy and a consultative exercise on Selecting Technologies for the Future.

(i) National technology priorities

The early 1980s saw officials in the former Department of Science and Technology (DST) taking up the notion increasingly popular overseas that government should target resources on research areas likely to form the basis of future science-based industries. In part, this was a consequence of the 1981 ASTEC exercise, but important political lessons had also been learned about the need to move towards selective support for R&D in an incremental fashion. A document outlining DST proposals for a 'key technologies' policy stressed that:

> Support for key technologies [is] always seen to be too much like government 'picking-the-winners' ... Governments can support key technology areas without

unduly trespassing on commercial aspects. Indeed, it is an essential part of the ... approach to ensure that industry is left to itself to develop the 'market niches' based on new capability ... The key technology approach is compatible with an industry development policy which recognizes that industry must make its own decisions. [Tegart, 1983, pp.4–5]

The election of a Labor government and the appointment of Barry Jones as Minister for Science and Technology provided an opportunity to pursue this policy.[29] Jones felt strongly that Australia was being left behind in the global technology race, and that a 'shock of awareness' was necessary to stimulate a national debate on R&D policies for the future.[30] This took the form of a three-day National Technology Conference held in 1983, bringing together government, industry and the scientific community to discuss emerging threats and opportunities.[31] The results provided an input to the National Technology Strategy published the following year, which identified five generic areas (e.g. biotechnology and industrial microbiology) likely to impact across a wide range of industrial sectors (DST, 1984). Despite being supported by detailed analyses of trends in R&D and market growth, the document was strongly attacked by large sections of industry, the trade unions and the public-sector research community, which for varying reasons regarded targeting as a threat to their interests. It therefore had to be substantially revised to give greater stress to infrastructural R&D support, and to define 'key technologies' more broadly so that the emphasis was shifted away from generic areas (DOS/DITAC, 1985). Publication was then delayed by transfer of the 'technology division' of DST to DITAC, whose industry and commerce sections had from the outset been among the leading opponents of any move to introduce targeting. The report which finally appeared was consequently accorded little importance in government circles and made no direct policy impact.[32*]

In retrospect, proponents of the key technology approach see the experience with the 'national strategy' as having been valuable in at least opening up a public debate. It also made clear the constraints on research foresight and priority-setting, especially the scope for government departments to launch *dirigiste* initiatives lacking wider political support. Finally, it highlighted the need to emphasize 'process' factors in foresight.

These lessons were taken into account in the second macro-level exercise, Selecting Technologies for the Future, launched by DITAC in 1986.[33] The approach employed was based on the assumption that sufficient consensus existed on the need to strengthen support for nationally important generic technologies to allow agreement to be reached on priorities through a structured consultative procedure. The first step was therefore to establish 'a process which leads to an agreed mechanism for selecting technologies and a set of priorities regarding generic technology evolution, and to enhance the legitimacy of making choices of such technologies in an environment which generally shies away from "picking the winners"' (DITAC, 1987a, p.ix).

A report was prepared assessing experience with priority-setting in Australia, Canada, the EEC, Sweden and the United States. As well as evaluating the methods used, it summarized the results of a number of studies identifying emerging strategic fields (including several reviewed in this and other chapters). On this basis, a five-stage approach to technology selection was proposed: (a) define the broad-brush (generic) technology categories (level 1) most relevant for Australian industry (e.g. advanced materials); (b) identify the associated 'enabling technologies' (level 2) without which only limited improvements in critical areas of industrial innovation will be possible (e.g. high-performance ceramics); (c) determine whether these technologies are sufficiently developed to allow market appraisals to be made of possible new products, processes or services; or, alternatively, identify 'linking technologies' (level 3) necessary to commercialize the products of enabling technologies (e.g. ceramic fabrication methods); (d) view all these technology elements in a framework spanning basic research to development, and agree appropriate mechanisms for collaborative R&D between the public and private sectors; and (e) construct scenarios for marketing, improving design quality, and other continuing support of new products, processes and services (DITAC, 1987a, p.41). Guidelines were also set out for operationalizing the approach, including a series of questions to help map the structure of individual technologies (with illustrative case-studies provided for fields like advanced materials).

While DITAC officials would have an important role in compiling background material to aid discussion on priorities, the report stressed that 'the preferred process by which the future technologies for Australia will be identified must involve *wide consultation* among the appropriate expert community, with a view to *generating a consensus approach* to the directions and priorities the community deems most desirable to achieve our economic objectives' (ibid., p.45, emphasis added). A national workshop was subsequently hosted by the Australian Academy of Science in an attempt to generate such consensus, attracting 180 participants representing a wide range of producers and users of R&D. They were divided into six syndicates, which were asked to use the DITAC framework to identify priority generic and enabling technologies within various industrial sectors over the next ten to twenty years. To help structure discussion, a set of twenty questions was provided (e.g. on pervasiveness of technologies, or national strength in the relevant research fields).

According to several participants, the conference failed to live up to expectations.[34] Scientific interest groups concentrated on ensuring *their* fields were represented in any list of identified priorities, while many industrialists reacted strongly against what they saw as DITAC again trying to steer national technology policy, arguing instead that the 'market' should decide and emphasis should therefore continue to be placed on infrastructural support.* The result was that:

> the workshop suffered from many of the same problems as the National Technology Conference and the 1981 ASTEC exercise. This is the difficulty of trying to gain agreement on priorities in a diverse pluralistic society. People excluded from the list of priorities kick up trouble with the result that you can only get the lowest common denominator accepted. Basically, there are too many special interests.*

In short, despite an explicit attempt to avoid the impasse reached in developing the National Technology Strategy, it still proved difficult to agree on priorities.

DITAC officials were less critical, claiming the exercise not only brought people together from academia, government and industry, but also resulted in far more companies coming to accept the need for a targeted long-term R&D strategy.* It was originally planned that the workshop discussions would be used 'to distil some major priority areas and a general strategic plan for the future' (DITAC, 1987a, p.xi), which would then be subject to further consultation. Whether this will still take place is unclear.

(ii) Research field priorities

Despite the lack of success in agreeing an overall research strategy through a consultative process, implicit national priorities for targeted programmes do nevertheless now exist in practice (DITAC, 1988d). These have emerged in an incremental and *ad hoc* fashion through a series of 'search conferences' held to identify priorities for new programmes in specific technology areas.

In developing the mechanism of the search conference, DITAC has drawn upon experience from a number of previous initiatives regarding the conditions under which consensus-seeking approaches can effectively be deployed in foresight. Lessons from the establishment of three national programmes have been particularly important. While priorities in the first of these (on marine science and technology) essentially reflected the recommendations made by an ASTEC working party in 1980,[35] the process employed to frame the second (on biotechnology) provided an opportunity for wider involvement of the scientific community. The political case for the initiative was again made by ASTEC (1982b), following a conference held by DST to discuss opportunities in genetic engineering, but identification of promising lines of research was carried out in a specialist workshop. This made use of a structured selection process in which experts from industry, government and the universities rated forty-two areas of biotechnology (e.g. immunoproteins) in terms of: (a) relative national research strength; (b) potential for exploitation of results; and (c) likely time-horizon for commercialization. Fourteen priority areas were thus identified (e.g. animal-breeding techniques) and these were incorporated in 1983 in the National Biotechnology Programme (DITAC, 1987a, pp.91–103). This approach to agreeing priorities for strategic programmes was soon recognized within DST as providing a model for future efforts (Tegart, 1983, pp.15–16). Moreover, the results of the initiative were later judged to have had a significant impact on industry (Tegart, 1987, p.15).

Experience with information technology (IT) provides an interesting contrast. Here, DST approached the task of identifying research needs and opportunities by commissioning a major study from the consultants, W. D. Scott and Arthur D. Little. This identified various priority fields in which competence was required to develop a future IT industry and to guarantee a national capacity to act as an intelligent user of emerging technologies (Scott and Little, 1984; DITAC, 1987a, p.104). The report was criticized by interviewees, in particular for failing to tailor its recommendations to the limited resources likely to be available. Furthermore, the absence of any consensus-seeking mechanism denied the IT community an opportunity to demonstrate to government the commercial importance of the priorities it saw for the future. The overall effect was that the study had limited impact, and it took longer to establish a national programme than in biotechnology or marine science.*

As a result, DITAC has since placed greater emphasis on workshops as a means of enabling researchers in the public and private sectors to discuss needs and opportunities in more narrowly defined technology sectors — for example, lasers, expert systems, and membrane technology (DITAC, 1985, 1987b, 1987c). However, although proving reasonably successful in generating consensus on future R&D priorities (as well as a basis for initiating the more comprehensive exercise on Selecting Technologies for the Future), these workshops had only an indirect influence on policy since they were not tied to any specific research initiative.

DITAC now recognizes that search conferences are most effective when organized to establish priorities for defined programmes on which political agreement already exists for budgetary support. This allows discussion to focus on specific choices in the field (rather than the case for a programme in one area relative to others, or the respective merits of targeted versus infrastructural support). A recent example is the national workshop, IT Search 88, the aims of which were: (a) to agree a broad long-term vision for R&D in information and communications technologies; (b) to establish priorities for precompetitive generic research; and (c) to suggest ways of ensuring R&D activities in target areas are integrated with Australia's Information Industries Strategy (DITAC, 1988a; 1987h).[36] The meeting brought together sixty experts from industry, government and the universities, who, through structured discussions in five syndicate groups,[37] agreed detailed priorities in the fields of hardware, communications, information systems, and systems and applications software. A report on the workshop was then circulated within the IT community.*

Such an approach to establishing priorities for the generic technologies programme was first used in 1985. At two workshops convened by DOS, ten target areas for longer-term materials research were identified (e.g. zirconia-toughened structural ceramics) and these subsequently became programme priorities. Further search conferences are planned in the materials field and in biotechnology (DITAC, 1987d; 1987e; 1987f). One feature of this experience has been the willingness of industry to participate in setting priorities without complaining about government interference. In part, this is due to the fact that the generic technologies programme still represents only a small element in DITAC support for R&D, but it is perhaps more a function of the growing recognition by companies of the long-term need to develop competence in strategic fields. Another lesson stressed by DITAC is the valuable role which can be played by national advisory committees in providing a specific audience for priority-setting workshops, and a mechanism for ensuring that recommendations have high-level impact.*

(iii) Infrastructure for research foresight

Prior to being disbanded, DOS was devoting considerable effort to studying the dynamics of emerging fields and the implications for future research policy. One aim was to follow up the work on 'exploitable areas of science' (which resulted in the *Research Strength and Economic Benefit* report discussed earlier) with a series of sectoral reviews using an approach piloted in the field of molecular electronics (DOS, 1987b). This evaluated current and projected activities in Australia and abroad in rapidly developing areas such as electroactive organic materials and molecular computing. It also assessed the longer-term commercial prospects for identified areas in which concentrated effort might capitalize upon existing research strengths and technological capabilities in industry.

Although DITAC initially gave less emphasis to science-push factors, the review of 'national science capacity' (DITAC, 1988b) undertaken in collaboration with ASTEC and ARC reflects a return to the DOS position on research policy. Work has also continued on studying the growth characteristics of basic technologies (e.g. by exploring the links between patent and bibliometric indicators), and extending the coverage of official science and technology indicator series (e.g. DITAC, 1987i).

(b) Commonwealth Scientific and Industrial Research Organisation

CSIRO previously occupied a more central position within the R&D system,[38] its five institutes undertaking basic and applied research in a broad spectrum of fields including agriculture, environmental science, manufacturing technology, minerals and energy, and radio astronomy (CSIRO, 1986a). Its policy of carrying out research only within intramural laboratories contributed to growing institutional ossification,* as well as being the focus of conflict with the universities as they developed a stronger research capacity and were able to respond (in a way CSIRO was not) to the needs of the economy for trained scientists.[39] Government recognition of these problems culminated in a series of reviews of CSIRO,[40] the last of which was carried out by ASTEC in 1985 and led to extensive restructuring and a curtailment of its basic research activities.[41] One consequence has been to weaken previous opposition within CSIRO to systematic approaches to planning and evaluation of research. Analysis of the introduction of foresight is interesting in that it highlights the time-consuming nature of the process, the tensions between formalized strategic planning and traditional decision-making processes, and the extent to which success depends upon an effective institutional management structure and a conducive political environment.

(i) Divisional and subject reviews

Like similar organizations overseas, CSIRO has since the late 1960s developed improved mechanisms for internal review and priority-setting. Reliance has mainly been placed upon evaluations of each of the forty or so research divisions, carried out every five to seven years by expert committees which include both industrial and foreign scientists. These assess existing activities, decide whether directors should be reappointed, and, increasingly, give advice on future lines of research. For example, the remit for a recent review of the Division of Chemical Physics was to make recommendations on 'appropriate research for CSIRO in the general area of materials science and technology, including identification of high priority research needs and opportunities, and suitable organizational arrangements for that research' (CSIRO, 1986d, p.14).

Submissions are generally invited from a wide constituency of researchers and organizations in the academic, governmental and industrial sectors, as well as from CSIRO staff. These form an input to the review committee which conducts a programme of site-visits and consultations over a two-day period, and then prepares a report. Recommendations can be quite critical (as in the chemical physics review), although more commonly they advocate the need for increased levels of activity.[42] However, no use has been made of systematic foresight techniques in this or the other main planning mechanism — subject reviews. Carried out in a similar fashion, these focus on fields likely to have a broad economic impact, or on emerging areas which cross-cut the activities of divisions and units — for example, recombinant DNA techniques (CSIRO, 1981) and space research (DITAC, 1987a, pp.103–4). In the latter case, several key technologies like digital holography were identified, together with an assessment of their likely applications in space science (e.g. solar physics) and industry (e.g. in avionics).

(ii) Planning and Evaluation Advisory Unit

The first attempt to introduce systematic foresight in CSIRO began in 1979 with the establishment of a Planning and Evaluation Advisory Unit (PEAU) attached to corporate headquarters. This followed recommendations by an independent committee that the organization's strategic planning capability should be upgraded (Ronayne, 1984, p.171). In particular, divisional and subject reviews needed to be complemented by other mechanisms capable of tackling 'such issues as making cuts in resources to divisions or programs on a selective basis' (CSIRO, 1984, p.12).

To overcome these problems, PEAU was given the task of developing more formal approaches to priority-setting. Two main tasks were envisaged: contributing systematic analyses of scientific and socio-economic data to the policy-making process; and building an in-house competence in planning and evaluation techniques. Somewhat ambitiously, PEAU set out to review CSIRO's main lines of research in terms of their potential contribution to social and economic development, and thus to identify priorities which more accurately reflected national needs. This involved a four-stage procedure: (a) determining the fields likely to yield significant pay-off to future research; (b) agreeing broad priorities within each field; (c) identifying projects in each priority area; and (d) ranking projects in order of importance (Johnston and Girdleston, 1983, p.2). As a preliminary step, PEAU prepared a matrix linking major economic sectors — for example, manufacturing, water resources and 'community and the environment' — with relevant areas of research. This then formed the basis for a series of sector reviews mapping out research priorities over the next decade and beyond.

Among the more successful of these reviews was one on agriculture conducted jointly with the Bureau of Agricultural Economics (BAE). This developed a socio-economic scenario for Australian research, drawing on a wide range of data to set out trends within the agricultural sector, as well as to identify commercial, technical and social constraints on improving farm productivity. Agricultural and industrial 'users' were also consulted, together with knowledgeable scientists, to determine broad areas where new programmes would benefit farmers and the wider research community. The report was regarded as providing a useful overview of economically significant R&D options.*

A more difficult task was to translate the options into ranked priorities for R&D. In conjunction with an internal CSIRO working party, PEAU first prepared a document outlining: (a) results from the joint study with BAE; (b) the extent to which the organization's existing portfolio of research addressed agricultural needs (programmes were divided into 'near-term' and 'long-term' according to whether benefits were likely in under or over ten years) and overlapped with activities of other organizations; and (c) promising research initiatives judged by CSIRO staff to have a reasonable chance of success (in both technical terms and the acceptability of new products or techniques to farmers). The resulting document was circulated widely for comment, especially on 'gaps and imbalances' in the proposed research. After analysing the responses, the working party identified seven priority areas (e.g. increasing the efficiency of fertilizer usage), along with thirty-seven sub-priorities (e.g. biological nitrogen fixation) ranked in terms of their scientific and economic potential (CSIRO, 1982).

According to CSIRO officials, neither the agricultural review nor most other sector studies had much impact on policy,[43] and the overall planning exercise was therefore not judged successful. Among the reasons given were:

(1) it proved difficult for PEAU to recruit experienced staff (who were extremely scarce in Australia at the time);
(2) studies took considerably longer than planned, causing resentment among scientists because of the effect on their research;
(3) the participation of all interested institute and division directors in the working groups responsible for priority-setting had two predictable results — the areas selected were rather general and of limited value for strategic planning, and established institutional interests could ensure most of their preferred priorities were included in the final list;
(4) it was unclear how the results of reviews were to be used apart from providing general guidance to the CSIRO Executive — little prior consideration had been given to how the target areas identified by individual studies might be combined to yield an overall set of priorities, and no links were established at the outset with the budget-planning process; and
(5) insufficient attention was given to the *process* for carrying out the exercise,[44] in particular on how to secure the participation of researchers who would inevitably react against the introduction of socio-economic criteria to determine priorities given the previous emphasis on scientific excellence.*

While some of these difficulties might have been avoided, the main problem, according to a member of the CSIRO Executive during this period, was 'the resistance to the concept of forward planning in the senior administration and among senior scientists'. Despite attempts to support PEAU in what he termed 'the struggle against the forces of darkness',† the unit was disbanded in 1983 when its head retired. One lesson to be drawn is that any centralized foresight activity is likely to be ineffective in a research organization with a directorate dominated by entrenched institutional interests. This was the situation in CSIRO prior to restructuring and there was consequently little scope for strategic decision-making.

(iii) CSIRO Corporate Planning

The role of research foresight in CSIRO has increased appreciably since closure of PEAU. Indeed, the Chief Executive now regards the introduction of effective research planning as a central task for the future since Australia 'cannot afford to spread its limited resources too thinly' (CSIRO, 1987a, p.10). This concern underlies the reorganization of CSIRO which has taken place since government accepted the recommendations of the 1985 ASTEC review. The main changes are, first, the introduction of a new management structure with broad strategic decisions on longer-term priorities vested in an independent board. Second, the five discipline-based institutes (e.g. the Institute of Physical Sciences) have been replaced by six with a sectoral orientation (e.g. the Industrial Technologies Institute), enabling previously distinct research areas like animal production and wool technology to be brought together.[45] Third, a budgetary system has been set in place which reflects the refocussing of CSIRO's mission towards applications-oriented work and technology transfer, with industry and other 'customers' expected to pay a greater share of research costs. Although this is resulting in reduced support for fundamental research, priority continues to be given to longer-term strategic programmes.[46]

In framing its new corporate planning system, CSIRO has been guided by the experience with PEAU, recommendations by an independent review committee (CSIRO, 1984),[47] and the results of a previous initiative by senior management to develop an overall research strategy (CSIRO, 1986a, pp.50–2).[48] A fundamental principle adopted at the outset is that foresight activities should be distributed throughout the organization. This was regarded as essential in order to avoid the difficulties associated with both the 'guru-at-the-centre approach' epitomized by PEAU, and the decentralized tradition in CSIRO (through which established scientific groups could too easily defend their interests).*

Another feature of the system is that a link has been established with the budget-setting process, CSIRO beginning work in 1988 on its first rolling 3–5 year corporate plan, which also sets out projections for R&D over a further fifteen years. Inputs are made by an overall strategic plan prepared by the CSIRO Board, while each of the thirty-five divisions is responsible for producing their own more detailed plan in association with newly appointed industrial advisory committees and institute directors. All six institutes now have planning officers who assist directors in this function. A Corporate Planning Organization (CPO) reporting to the Chief Executive is responsible for synthesizing the various inputs into a draft corporate plan, as well as preparing background policy studies, undertaking data analysis, and developing an in-house capacity in the use of foresight techniques.

In this way, significant freedom has been accorded to divisions (and institutes) to define their own priorities, but within the context of an overall strategy and taking into account their capacity to raise external income. Research foresight has thus been built into the management structure, with responsibility for identifying promising projects and programmes devolved to scientists. As the Chairman of CSIRO has stressed: 'The Board will not be picking winners ... Scientists will argue for the funds they want and they will decide how they should be spent. The difference now is that they will have to give greater attention to priorities. In short, it is the scientists who will be trying to pick winners' (Wran, cited in Sherwell, 1988).

Implementing the new system is, however, proving difficult, especially in regard to developing a workable division of labour between CPO and the planning officials in institutes. Problems have also arisen as a result of the lack of experience in CSIRO with prospective analysis, and CPO has therefore been appraising different methods for identifying promising research areas and aims to link its activities with foresight exercises in DITAC, ASTEC and ARC.*

At the programme and project level, however, considerable progress has been made following an exploratory study carried out by McKinsey (in collaboration with four senior CSIRO staff).[49] This developed a systematic approach for estimating the economic benefits from prospective research projects, employing a four-stage approach based on commercial practice: (a) identify the business systems through which the results of successful R&D are to be commercialized (i.e. potential customers and markets need to exist); (b) determine the magnitude of the possible economic benefits stemming from successful research, and the capacity of Australian firms to capitalize upon likely opportunities; (c) assess the technical and human factors which will determine the probabilities and costs of technical success, setting time-frames for the work together with objectives and appropriate mid-term goals; and (d) determine suitable arrangements for industrial collaboration (CSIRO, 1987b). Pilot studies were then undertaken of rare-earth processing and rare-earth magnet production, two areas where R&D might lead to significant commercial opportunities for the nation. After the results were evaluated positively by experts, further testing was carried out in a review of polymer technology and other McKinsey-type studies are therefore planned. The approach is, however, clearly more appropriate for applied and development work than longer-term strategic research where the impacts are diffuse and difficult to estimate. As a result, a degree of ambivalence exists among CSIRO research staff (Oldham, 1988).* If used in a comprehensive manner across the organization, the approach might well result in an over-emphasis on short-term research best carried out in companies.

Despite these limitations, the McKinsey exercise is seen within CSIRO as having been valuable in encouraging researchers to broaden their science-push visions of economic impact to include a market-pull dimension. Most importantly, it has helped develop the evaluation culture so necessary if scientists are to participate willingly in foresight. As a senior planning official stressed: 'It has taken a decade in CSIRO to get to a situation where the McKinsey suggestions are accepted. Ten years ago they would have been laughed out of court. Ultimately they will probably play the role of a catalyst in developing an internal methodology for prospective studies.'*

(c) Commission for the Future

The Commission for the Future was established in 1985 by the Minister of Science with a remit to 'promote public understanding of developments in science and technology and their implications for economic and social planning ... so that people are able to ... participate with government, business, industry, education systems and the unions in planning for the future and determining priorities for ... research and development' (CFF, 1987b, p.1). Although a federal agency under DITAC since 1987, the Commission has no direct advisory role *vis-à-vis* government policy.

From the outset, the Commission has concentrated on developing an Australian version of the participatory model of decision-making in science and technology popular in the Netherlands and Scandinavia. Its annual budget of A$1 million supports four main programmes: (a) analyses of futures issues (including an evaluation of the likely effects of telecommunications technologies and a survey for UNESCO of forecasting studies in Australia); (b) public dissemination of information on emerging trends (in particular, through the magazine, *In Future*, which claims a readership of 100,000);[50] (c) organization of conferences on such topics as 'skilling for the future'; and (d) various 'public involvement' initiatives like operating a 'clearing house on futures issues' and 'science and technology shops',[51] as well as a project on the 'greenhouse effect' carried out jointly with the CSIRO Division of Atmospheric Research (CFF, 1987a).

The work of the Commission has so far been more concerned with technology assessment than with research foresight, but it hopes to redress the balance in the future. Although enjoying support among citizens groups and trade unions, it has encountered hostility from both government agencies and political parties because of a widespread perception that its activities are more concerned with fostering critical assessment of science and technology than with encouraging the redirection of research to meet long-term industrial and social needs. This has consequently restricted the influence of the Commission on government policy.*

5. CONCLUSIONS

Australian experience with research foresight has undoubtedly been mixed, the relatively large number of initiatives having achieved only limited impact. While it would be easy to overemphasize the negative lessons, this would miss the more important point that the setbacks have not led to a wholesale rejection of foresight but rather to a determination to learn from mistakes and improve the techniques and processes employed. Such tenacity has ultimately begun to pay off in the growing acceptance that foresight techniques are useful, indeed often indispensable, in guiding the difficult decisions which need to be made on future scientific priorities. Several conclusions can thus be drawn:

(a) A precondition for the acceptance of foresight was the recognition of a need for more systematic approaches to research policy among key stakeholders in the R&D system. This emerged gradually as it became clear that the nation's longer-term survival was at risk if science and technology were not harnessed more effectively, and that existing priority-setting mechanisms were inadequate. Resistance to change from entrenched interests required significantly greater coordination of research policy, leading to the establishment of ARC, the growing political power of ASTEC, and a new role for DITAC in setting priorities for basic technology programmes. This has ultimately created a set of 'users' of foresight with the ability to think and plan long-term and strategically.

(b) Also important has been the recognition that an incremental approach is vital when introducing foresight into a research community lacking a well-developed evaluation culture. Initial attempts to deploy foresight in decision-making, especially more quantitative approaches, were invariably resisted by scientists worried about the threat to their autonomy. This was evident in the 1981 ASTEC exercise and the attempt by DOS to use co-citation techniques in mapping Australian research strengths. In contrast, scientists have shown much greater willingness to participate in the search conferences organized by DOS and DITAC, even when prior inputs are made in the form of quantitative analysis of emerging research trends. The general lesson is that the long learning-curve with foresight can best be approached by beginning with relatively focussed exercises relating to initiatives which have no implications for existing funding, leaving more sophisticated exercises for later.

(c) As in other countries, a strong free-market ideology has constrained government efforts to target support for R&D. It has therefore been important to demonstrate that foresight is not to be equated with inflexible bureaucratic planning, a task again requiring an incremental approach. The search conferences have been especially useful in changing the climate of industrial opinion in that they provide an alternative model of how foresight can facilitate the agreement of priorities, rather than merely serving as a mechanism for government to impose its wishes on the market. As a result, more firms now accept the need for strategic planning of public R&D support, and this may ultimately make it possible for industry to set aside specific sectoral interests to arrive at consensus on national priorities. Until then, macro-level exercises like Selecting Technologies for the Future seem likely to prove fruitless.

(d) It has been recognized that implementation structures are central to the success of foresight and need to be built in from the outset. This requires the involvement of those with the power to act upon the results — for example, through their influence in the budget-setting process (as in the search conferences) or through high-level advisory committees which provide an interface with government (a mechanism deployed in the national biotechnology programme). Conversely, when foresight has been carried out in a political vacuum, most notably in the 1981 ASTEC exercise, the impact has generally been limited. As ASTEC itself has stressed:

> The factors influencing the impact of foresight activities on policy are likely to be the same as those affecting the impact of any advice, including timeliness, pertinence, acceptance by government ... and *easy coordination of the advice with current or proposed policy initiatives*. Acceptance of the results is probably a function of the standing of the institution responsible in the eyes of policy-makers. The methods used and quality of forecasts are not important in comparison, unless the results are extensively publicized.†

(e) While Australian agencies understand that state-of-the-art methodology is itself no guarantee of successful foresight (as is clear from the 1981 ASTEC study), they have continued to explore the merits of different approaches. The DOS exercise on Research Strength and Economic Benefit demonstrated the potential utility of various quantitative methods, while early-warning indicators are being investigated by DITAC and ASTEC as an aid to decision-making on strategic areas. Although pilot studies with co-citation analysis were not wholly successful, one result is that Australia now has a capacity to use these and related data-based approaches. It has also been recognized that the existence of high quality science and technology indicators is a precondition for effective research foresight, and significant improvements have been made by DITAC to the national R&D statistics.

(f) Careful design of the foresight process is essential if an appropriate balance is to be achieved between supply-side (science and technology push) and demand-side (user needs and market pull) factors, and if top-down and bottom-up inputs are to be married effectively. A continuing problem in Australia has been how to tap the collective wisdom of scientists and engineers without allowing their interests to become dominant. The experience of CSIRO is particularly illuminating on the problems associated with top-down planning and the hostility engendered in the research community, while the organization's new policy of long-term strategic planning coupled with a distributed process of establishing budgetary priorities may well prove useful as a model for other agencies. Where too great an emphasis has been placed on 'bottom-up' approaches — as in CSIRO previously and in the DITAC exercise on Selecting Technologies for the Future — existing interests have been vigorously defended with highly conservative results. The restructuring of higher education research currently under way reflects a recognition that strategic planning by the universities themselves is an essential component of any move to effective priority-setting at a national level.

(g) Another conclusion in respect of the science-push dimension of foresight concerns the dangers of conservatism stemming from use of traditional field categories. Initial assumptions about the structure of science made, for example, in the Research Strength and Economic Benefit study, limited the identification of promising areas to the fields selected at the outset. Even when procedures are built in for assessing cross-disciplinary impacts, as ASTEC did in 1981, problems can arise because diffuse fields like applied mathematics will almost by definition be categorized as pervasive and therefore meriting priority. A vital element missing in the Australian foresight methods employed so far has been a mechanism for identifying interdisciplinary areas.

(h) Finally, Australian experience with foresight highlights the need to take into account cultural, political and institutional factors. Hence, no universal prescriptions on method are possible, and appropriate approaches can change over time even within a single country. Until now, limited initiatives like the search conferences have often been the only way of introducing foresight into priority-setting. Comprehensive studies have been feasible solely for background information purposes (i.e. anticipatory intelligence) since even the slightest budgetary implications have provoked largely irreconcilable disagreement among different interest groups. Whether this will continue is doubtful given the mood of realism in the country, and the recognition by the scientific community that strategic planning and priority-setting are unavoidable.

NOTES

* Refers to material collected during interviews with Australian officials in February 1988, while † denotes information drawn from subsequent correspondence.

1. Developments in Australian R&D policy are reviewed in Ronayne (1984) and OECD (1986). See also ASTEC (1989), DITAC (1987i), *Nature* (1986), Franklin (1988a) and Dodgson (1988).
2. Australia was not initially included in the study but an invitation to visit the country coincided with the period of fieldwork, and it was felt that sufficiently interesting lessons could be drawn to justify extending the coverage. However, time constraints unfortunately precluded any attempt to study foresight in industry. Nor was it possible to consider initiatives by state governments which have increasingly regarded the creation of regional infrastructure for R&D as a central element in their support for industrial development. This can be seen, for example, in the activities of the New South Wales Science and Technology Council (NSWSTC, 1987a; 1987b; 1987c). State policies for science and technology are outlined in OECD (1986, pp.73–83), while the findings of a recent Victoria government report identifying various strategic technologies are summarized in DITAC (1987i, p.98).
3. The UNESCO method was based on techniques developed by the US Department of Defense and the French Delegation for Scientific and Technological Research (UNESCO, 1978). Since it was intended primarily for application in less-developed countries, certain modifications needed to be made by ASTEC following discussions with UNESCO officials.
4. The priorities workshop is described in ASTEC (1981a; 1981b; 1982).
5. The Australian Academy of Science was involved in identifying experts to serve on panels.
6. This is probably the most sophisticated use for policy purposes of the criteria proposed by Weinberg (1963) for assessing the relative merits of competing claims on scientific resources, in particular his third 'intrinsic criterion' concerning the extent to which research in one discipline may contribute to progress in neighbouring fields.
7. Indices were contructed to depict both *cross-support* (or *relevance* in the case of R&D disciplines) and *dependency* (on a scale of 100), according to whether objectives and disciplines were treated as having equal priority (defined as a *reference index*) or were presented in terms of the relative priorities determined in the workshop (*priority index*).
8. Few younger scientists participated. Most of the twenty-four industrial panelists, for example, were senior executives (often in general management rather than chief scientist positions), while many academic and institute participants held professorial or equivalent posts. This led to criticism that too much reliance was placed on broad knowledge of a field, and too little on specialist disciplinary expertise.†
9. The main criterion underlying choice of areas is pertinence to the current or required development of national research policy. After possible areas have been identified by the Secretariat or Chair, they are discussed and narrowed down by the Council. Although major initiatives require considerable investment of resources, studies on average cost around A$50,000 and involve eight person-months of effort. Among the more influential was a review of marine science and technology carried out shortly before the 1981 priority-setting exercise (see ASTEC, 1980).
10. Unlike Sweden, for example, the country does not have a particularly well-developed system of foreign technical attachés, a situation explained in part by decisions taken during the 1970s to contract, rather than expand, the existing 'scientific liaison service'. The recent OECD report on Australian science policy recommended a substantial increase in monitoring of overseas R&D (OECD, 1986).
11. ASTEC has pointed out, however, that 'redirection of resources from the least impressive research to areas of intellectual and/or strategic promise can only

benefit the standing of basic research and strengthen arguments for more favourable funding treatment' (ASTEC, 1987e, p.4).

12. The aim was to replicate the annual American Association for the Advancement of Science forum which brings together the research community, funding agencies and industry to examine the priorities and policies implicit in the federal R&D budget (see Chapter 5).
13. Severe difficulties were encountered in obtaining similarly disaggregated data for other countries, a problem discussed in Irvine and Martin (1986).
14. The Secretary of the Department of Science, W. J. McG. Tegart, moved to the equivalent position in ASTEC in 1987. Having been influential in initiating foresight activities in DOS, he was in a position to ensure that ASTEC exploited the expertise developed in these earlier studies.
15. It is planned to transfer A$125 million from 'general university funds' to schemes supporting competitively-awarded research projects over the period 1989–91 (DEET, 1988c).
16. Some of the fifteen centres operating in 1988 were working in emerging strategic fields — e.g. plant cell biology. However, no prior restrictions have been placed on fields in which proposals for centres can be made. The Special Research Centres scheme was favourably reviewed in the mid-1980s (CTEC, 1986a).
17. In 1983, CTEC carried out an exploratory questionnaire survey in which academics were invited to identify promising areas of research. This provoked a great deal of criticism from scientists worried both about the imposition of priorities from above and the funding implications for their fields.*
18. A new category of direct support has been established in the form of Special Research Funding for Technological Institutions. This provides seed money for technical research in the forty-five colleges of advanced education, institutions previously excluded from full participation in the academic system.
19. Academic objections to the proposals in the 'green paper' are discussed in Bourke (1987).
20. This is the task of the new ARC Research Policy Consultative Committee, whose members include officials from the main research-oriented ministries as well as from agencies like CSIRO, NHMRC and the Defence Science and Technology Organisation. This in principle ensures the wide involvement needed to fulfil its task of coordinating national policy. However, there is some concern about the wisdom of ARC reporting to the Minister for Employment, Education and Training* given that its recommendations will often have implications for other departments.
21. In order to create the flexibility to introduce new priority programmes, it is first necessary to withdraw resources from academic institutions not making contributions to science commensurate with the costs involved. Consequently, the introduction of formal techniques for assessing research performance is also seen as a priority, since peer-review is regarded as inappropriate for making the hard decisions necessary in reallocating funds from established institutions and fields (Aitkin, 1988).
22. The other national programme on biotechnology has had a more strategic focus (see OECD, 1986, pp.68–9).
23. This was supplied by the Center for Research Planning, a US consultancy company specializing in the use of bibliometric data for science-policy purposes.
24. See also DOS (1987a, vol.1, pp.61–2).
25. An outline of the problems involved in using co-citation techniques for policy purposes is given in Hicks (1987). See also the exchange between Franklin (1988b) and Hicks (1988).
26. This was probably not helped by the fact that consideration was reportedly give to employing the acoustics model in an evaluation of the National Acoustics Laboratory.*

27. These were provided to DOS (and later to DITAC) by CHI Research, with TASC being commissioned to analyse the computer tapes. It is perhaps significant that Australia is now one of the few countries to report bibliometric indicators of research activity in its official statistical series (DITAC, 1987, pp.7–8 and 183–202).
28. In addition, incentives were introduced in 1985 giving companies a 150 per cent tax allowance on R&D expenditure. For an assessment of the impact on industry, see Dodgson (1988).
29. It should, however, be noted that the previous Minister for Science and Technology had publicly made the case for key technologies to be supported, distinguishing this from the unacceptable policy (to a centre-right government at least) of targeting resources on specific industries. Earlier still in 1980, the Myers Report had identified six generic technologies of potential importance to Australia but its recommendations had little policy impact (Johnston, 1987).
30. Soon after election of the Hawke government, a National Summit Conference was held to discuss future policy directions, with trade unions being centrally involved. According to one participant, Jones was angry that science and technology were not even on the agenda and therefore decided to hold his own forum. The initial plan was that Australia should target resources on sixteen sunrise industries, but this was soon dropped when it became evident that such an interventionist policy was unlikely to prove politically acceptable.*
31. According to DITAC (1987a, p.94), the conference 'did not aim at consensus but rather fostered "creative tension" between interest groups working in parallel areas, encouraging analysis of overall goals, appropriate measurement of performance and specific areas of agreement and disagreement'.
32. An indication of the reduced status of the revised report is given by the absence of the original forward by the Prime Minister (DOS/DITAC, 1985).
33. The initiative was largely driven by the enthusiasm of an experienced DITAC official (A. J. Jones) who had previously convened several national workshops aiming to identify research opportunities in specific strategic fields (e.g. lasers).
34. One participant reported that the extent of disagreement in his syndicate was so great that it proved impossible to move beyond the second question in the time available.*
35. This distinguished nine fields of national importance which merited priority support (for example, geochemical studies). Following acceptance of the ASTEC recommendations by government, the Marine Sciences and Technologies Grant Scheme was established.
36. Details of the Australian Information Industries Strategy are given in DITAC (1987g).
37. Other inputs were made by specialist reviews circulated prior to the workshop, and lectures given by leading authorities in the IT field.
38. Although declining in relative importance, its 1987 R&D budget of around A$450 million was still equivalent to more than half the total research expenditure within the Australian higher education system.
39. The origins of the of conflict between CSIRO and the universities are described in Ronayne (1984, pp.163–7).
40. In recent years, CSIRO has been subject to numerous internal and external reviews of both its organizational structure and mission orientation – see ASTEC (1985b), Ronayne (1984, pp.163–7), Flood (1984), CSIRO (1984) and Oldham (1988).
41. Oldham (1988) documents the process by which the ASTEC (1985b) review was carried out, and its subsequent policy impact. CSIRO (1986a, pp.25–33) outlines the organization's response to the recommendations.
42. See, for example, the review of the Division of Protein Chemistry and Molecular and Cell Biology Unit (CSIRO, 1983a), and the subsequent evaluation of the

Division of Molecular Biology (CSIRO, 1986c). Recommendations often involve institutional restructuring (e.g. upgrading a unit to a division).

43. One exception was the review of energy research carried out in 1980. This examined the lead-time for implementation of a variety of technologies, and set out projections for when research was likely to be exploited. According to a member of the CSIRO Executive at the time, the exercise 'had a considerable influence on the setting of national priorities carried out later by the National Energy Research Development and Demonstration Program as a basis for allocating its research funding'.†

44. A related problem seems to have been a lack of sensitivity in acknowledging the contributions to studies, one researcher complaining about 'putting a lot of effort into identifying priorities, but then finding that they were integrated into the PEAU document without recognition — it thus became solely *their* product'.*

45. The management consultants, McKinsey, were the architects of CSIRO's new organizational structure, having been commissioned in 1987 to carry out what became known as the 'model institute study' (CSIRO, 1987a, pp.8-11). Implementation of their recommendations was accompanied by the replacement of almost all institute directors and a reduction in staff numbers (primarily through early retirement). Oldham (1988) discusses the policy background to restructuring, in particular the part played by the ASTEC review.

46. In one division, the number of projects was cut from around a hundred to twelve (Sherwell, 1988).

47. This was carried out by a committee set up by the CSIRO Executive in 1983 to review the entire range of planning activities in the organization, its membership including two industrialists with considerable expertise on the subject. The main recommendations made in its report — greater stress on research foresight and priority-setting, coupling strategic planning at the centre with decentralized planning at institute and division levels, and the integration of planning with budget-setting — formed the basis of the corporate planning system subsequently introduced. The committee was, however, clearly aware of the problems likely to be faced by the existing CSIRO organizational structure in operating such a system: 'constant vigilance on the part of top management will be needed to make the scheme work well. If top management is not committed to the system it could easily become a paper exercise for legitimizing whatever proposals come to the top as a result of internal political forces competing for a share of the available resources, with insufficient regard to the ultimate beneficiaries of the research' (CSIRO, 1984, p.26). The committee's report provides a comprehensive overview of the problems involved in setting up a corporate planning system in a public sector R&D agency, and merits wider distribution than it has so far received.

48. Although this was in part a defensive move by CSIRO to limit the extent of future restructuring, it did pinpoint areas where urgent action was necessary (e.g. improving the interface with industry and staff training). It also set down criteria for selecting areas of R&D growth, recommended establishing an Executive Sub-Committee on Research Evaluation (which initiated work on methods for assessing the commercial prospects for research), and identified for the first time a list of priority fields — information technology, biotechnology, generic manufacturing technology, advanced materials, oceanography and waters and solids. However, these priorities made little impact on funding decisions within CSIRO since, as Johnston (1987, p.4) notes, 'it was decided at the time that this focus would not be achieved at the expense of the traditional focus on rural research, suggesting that the hard decisions, i.e. discontinuing other research, were still to be made'.

49. The study was initiated in 1985 by the Research Evaluation Committee established following the completion of the 1984–5 exercise to develop an overall strategy for CSIRO (see previous note). The remit set for the committee included making

recommendations on the techniques institutes and divisions might use to assess prospective benefits from planned projects (CSIRO, 1986a, p.51).

50. See, for example, the special issue on 'The Information Society: Challenge and Change' (CFF, 1987c).
51. This is modelled on the Dutch science shops described in Netherlands Ministry of Education and Science (1987).

REFERENCES

Aitkin, D. (1988), 'The Establishment of Research Priorities', paper presented at 14th ACU Quinquennial Congress, Canberra, Australian Research Council.

ARGS (1987), *The Case for Research: Funding Requirements for 1988*, Canberra, Australian Research Grants Committee.

ASTEC (1979), *Science and Technology in Australia 1977–78*, Canberra, Australian Government Publishing Service.

—— (1980), *Marine Sciences and Technologies in Australia — Priorities for Additional Research and Development 1980–81*, Canberra, Australian Science and Technology Council.

—— (1981a), *Identification of Priorities for Australian Science and Technology Policy in the 1980s*, Canberra, Australian Science and Technology Council.

—— (1981b), *National Objectives and Research Priorities*, Canberra, Australian Science and Technology Council.

—— (1982a), *Report on the ASTEC Workshop on National Objectives and Research Priorities*, Canberra, Australian Science and Technology Council.

—— (1982b), *Biotechnology in Australia*, Canberra, Australian Science and Technology Council.

—— (1983), *Videotex in Australia: Interactive Information Services*, Canberra, Australian Science and Technology Council.

—— (1985a), *Computer-Related Technologies in the Metal Trades Industry*, Canberra, Australian Government Publishing Service.

—— (1985b), *Future Directions for CSIRO*, Canberra, Australian Government Publishing Service.

—— (1987a), *Education and National Needs*, Canberra, Australian Government Publishing Service.

—— (1987b), *Improving the Research Performance of Australia's Universities and Other Higher Education Institutions*, Canberra, Australian Government Publishing Service.

—— (1987c), *Computerised Assistants — New Tools for Society*, Canberra, Australian Government Publishing Service.

—— (1987d), *ASTEC Annual Report 1986–87*, Canberra, Australian Government Publishing Service.

—— (1987e), 'Profile of Australian Science', Council Document 8/7, Canberra, Australian Science and Technology Council.

—— (1988a), *The Contribution of Science and Technology to Australia's Balance of Payments to the Year 2000*, Canberra, Australian Science and Technology Council.

—— (1988b), 'Summary of the Co-Citation Analysis of Australian Acoustics Research ', Canberra, Australian Science and Technology Council.

—— (1989), *Profile of Australian Science: A Study of the Current State and Potential of Basic Scientific Research*, Canberra, Australian Science and Technology Council (forthcoming).

Bourke, P. (1986), *Quality Measures in Universities*, Canberra, Commonwealth Tertiary Education Commission.

—— (1987), 'The national need: who knows what it is or how to find it', *Canberra Times*, 5 October, p.3.

CFF (1987a), *Towards a Productive Culture: Annual Report 1986-87*, Canberra, Australian Government Publishing Service.
—— (1987b), *Draft Corporate Plan for 1987-88*, Carlton, Commission for the Future.
—— (1987c), 'The Information Society: Challenge and Change', special edition of *In Future*, December, Carlton, Commission for the Future.
Coward, H. R. and Franklin, J. J. (1989), 'Identifying the Science-Technology Interface: Matching Patent Data to a Bibliometric Model', *Science and Technology Studies*, vol.14, no.1, pp.50–77.
CSIRO (1981), *CSIRO Energy Research: Planning and Policy*, Canberra, Commonwealth Scientific and Industrial Research Organisation.
—— (1982), *Agricultural Research — Planning Review*, Canberra, Commonwealth Scientific and Industrial Research Organisation.
—— (1983a), *Review of the Division of Protein Chemistry and the Molecular and Cellular Biology Unit*, Canberra, Commonwealth Scientific and Industrial Research Organisation.
—— (1984), *Corporate Planning in CSIRO: A Review of CSIRO's Strategic Research Planning Activities*, Canberra, Commonwealth Scientific and Industrial Research Organisation.
—— (1986a), *Annual Report 1985/86*, Canberra, Commonwealth Scientific and Industrial Research Organisation.
—— (1986b), *CSIRO in Brief*, Canberra, Commonwealth Scientific and Industrial Research Organisation.
—— (1986c), *Review of the Scientific Programs of the Division of Molecular Biology*, Canberra, Commonwealth Scientific and Industrial Research Organisation.
—— (1986d), *Review of Materials Science and Technology in CSIRO and Review of CSIRO Division of Chemical Physics*, Canberra, Commonwealth Scientific and Industrial Research Organisation.
—— (1987a), *Annual Report 1986/87*, Canberra, Commonwealth Scientific and Industrial Research Organisation.
—— (1987b), *Assessing the Commercial Prospects for Research*, Canberra, Commonwealth Scientific and Industrial Research Organisation.
CTEC (1986a), *The Report of the Steering Committee to Review Special Research Centres and the Future of the Programme*, Canberra, Commonwealth Tertiary Education Commission.
—— (1986b), 'Discipline Assessment of Engineering' (mimeo), Canberra, Commonwealth Tertiary Education Commission.
—— (1987), *Australian Law Schools: A Discipline Assessment*, Canberra, Australian Government Publishing Service.
DEET (1987), *Higher Education: A Policy Discussion Paper*, Canberra, Australian Government Publishing Service.
—— (1988a), *Higher Education: A Policy Statement*, Canberra, Australian Government Publishing Service.
—— (1988b), *Profile of Research Activities*, Canberra, Department of Employment, Education and Training.
—— (1988c), *Review of Higher Education Research Policy*, Canberra, Department of Employment, Education and Training.
—— (1988d), *Research Management Plan*, Canberra, Department of Education, Employment and Training.
—— (1988e), 'Research Review Committee Announced', Canberra, Department of Employment, Education and Training.
DITAC (1985), *Lasers — II Australia: Workshop Report and Recommendations*, Canberra, Department of Industry, Technology and Commerce.
—— (1987a), *Selecting Technologies for the Future: A Discussion Paper*, Canberra, Department of Industry, Technology and Commerce.

—— (1987b), *Expert Systems: Proceedings of the National Workshop on Expert Systems*, Canberra, Australian Government Publishing Service.

—— (1987c), *Membrane and Separation Technology: The Australian Perspective*, Canberra, Australian Government Publishing Service.

—— (1987d), *Supporting Australian Innovation: Industry Research and Development Board*, Canberra, Department of Industry, Technology and Commerce.

—— (1987e), *Supporting New Materials Technology*, Canberra, Department of Industry, Technology and Commerce.

—— (1987f), *Grants for Industry Research and Development: The Generic Technologies Program*, Canberra, Department of Industry, Technology and Commerce.

—— (1987g), 'An Information Industries Strategy', special issue of *Australian Technology Magazine*, Canberra, Department of Industry, Technology and Commerce.

—— (1987h), *Generic Technology Grants: Information and Communications Technologies*, Canberra, Department of Industry, Technology and Commerce.

—— (1987i), *Measures of Science and Innovation: Australian Science and Technology Indicators Report 1987*, Canberra, Department of Industry, Technology and Commerce.

—— (1988a), *IT Search 88: Workshop on Research Priorities in Information Technology*, Canberra, Department of Industry, Technology and Commerce.

—— (1988b), *Review of Science Capacity*, Canberra, Department of Industry, Technology and Commerce.

—— (1988c), *Annual Report 1987–88*, Canberra, Department of Industry, Technology and Commerce.

—— (1988d), *Corporate Plan 1988–89*, Canberra, Department of Industry, Technology and Commerce.

Dodgson, M. (1988), 'Research and Technology Policy in Australia: Legitimacy in Intervention' (mimeo), Brighton, Science Policy Research Unit.

DOS (1987a), *Research Strength and Economic Benefit: An Examination of the Strong Areas of Australian Basic Research and Their Potential for Economic Exploitation, Volume 1 (Report) and Volume 2 (Appendices)*, Canberra, Department of Science.

—— (1987b), *Molecular Electronics: A Sectoral Review*, Canberra, Department of Science.

—— /DITAC (1985), *National Technology Strategy — Revised Discussion Draft*, Canberra, Department of Science and Department of Industry, Technology and Commerce.

DPIE (1988), *National Energy Research Development Program, Guidelines and Priorities for Grants*, Canberra, Department of Primary Industries and Energy.

DST (1984), *National Technology Strategy*, Canberra, Department of Science and Technology.

Flood, J. (1984), 'The Advent of Strategic Management in CSIRO: A History of Change', *Prometheus*, vol.2, no.1, pp.38–72.

Franklin, J. J. (1986), *The State of Computer Science Research in Australia: An Analysis Based on the Australian 1983–84 Bibliometric Model*, report to Department of Science, Wollongong, Centre for Technology and Social Change.

—— (1987a), *Introduction to the Policy Application of Bibliometric Data: Highlights of the Australian Research Activity in the 1983-84 Model*, Wollongong, Centre for Technology and Social Change.

—— (1987b), *Highlights of Australian Research 'Strengths' and 'Weaknesses' in the 1983–84 Bibliometric Model*, report to Department of Science, Wollongong, Centre for Technology and Social Change.

—— (1988a), 'Selectivity in Funding: Evaluation of Research in Australia', *Prometheus*, vol.6, no.1, pp.34–60.

—— (1988b), 'Testing and Using Quantitative Methods in Science Policy Contexts', *Social Studies of Science*, vol.18, no.2, pp.365–75.

—— and Johnston, R. (1988), 'Cocitation Bibliometric Modelling as a Tool for S&T Policy and R&D Management: Issues, Applications and Developments', in A. F. J. van Raan (ed.), *Handbook of the Quantitative Study of Science and Technology*, Amsterdam, Elsevier.

Hicks, D. (1987), 'Limitations of Co-Citation Analysis as a Tool for Science Policy', *Social Studies of Science*, vol.17, no.2, pp.295–316.

—— (1988), 'Limitations and More Limitations of Co-Citation Analysis/Bibliometric Modeling', *Social Studies of Science*, vol.18, no.2, pp.375–84.

Irvine, J. (1988), *Evaluating Applied Research — Lessons From Japan*, London, Pinter Publishers.

—— and Martin, B. R. (1986), 'Is Britain Spending Enough on Science?', *Nature*, vol.323, pp.591–4.

Johnston, R. (1987), 'Selecting Technologies for the Future: An Overview of Australian Experience', Wollongong, Centre for Technology and Social Change.

Johnston, B. and Girdleston, J. (eds) (1983), *Implications for Future Research of Recent Developments and Trends in Agriculture*, Canberra, Bureau of Agricultural Economics and Commonwealth Scientific and Industrial Research Organisation.

Linke, R. D., Birt, L. M., Fensham, P. J., Scott, R. D. and Smith, B. W. (1984), *Report of a Study Group on the Measurement of Quality and Efficiency in Higher Education*, Canberra, Commonwealth Tertiary Education Commission.

Nature (1986), 'Editorial Comment on Australia's Basic Science', vol. 326, p.426.

Netherlands Ministry of Education and Science (1987), *Technology Assessment: An Opportunity for Europe — Netherlands*, The Hague, Government Printing Office.

NSWSTC (1987a), *Research and Development in New South Wales — An Overview Report*, Sydney, New South Wales Science and Technology Council.

—— (1987b), *Tertiary Education Research and Development in New South Wales*, Sydney, New South Wales Science and Technology Council.

—— (1987c), *New South Wales Government R&D*, Sydney, New South Wales Science and Technology Council.

OECD (1986), *Reviews of National Science and Technology Policy — Australia*, Paris, Organisation for Economic Cooperation and Development.

Oldham, G. (1988), *Commentary on the ASTEC Review of CSIRO*, ASTEC Occasional Paper No. 2, Canberra, Australia Government Printing Office.

Ronayne, J. (1984), *Science in Government*, London, Edward Arnold.

Scott, W. D. and Little, A. D. Inc. (1984), *Information Technology in Australia: Capabilities and Opportunities*, Canberra, Department of Science and Technology.

Sherwell, C. (1988), 'The good, the bad and the ugly of Australian science', *Financial Times*, 9 March, p.22.

TASC (1987), *Annual Report 1986*, Wollongong, Centre for Technology and Social Change.

Tegart, W. J. McG. (1983), 'The Key Technology Approach to Technology Policy', Address to Eighteenth Meeting of Australian Manufacturing Council, July, Canberra, Department of Science and Technology.

—— (1987a), 'The Development of Science and Technology in Australia: The Framework of the Forum' (mimeo), Canberra, Australian Science and Technology Council.

—— (1987b), 'Biotechnology — A Success Story for Selectivity in Australian Science', paper presented at Annual General Meeting of Australian Academy of Science, Canberra, Department of Science.

Tisdell, C. A. (1981), *Science and Technology Policy: Priorities of Governments*, London, Chapman and Hall.

UNESCO (1978), *Method for Priority Determination in Science and Technology*, Paris, UNESCO.

Weinberg, A. M. (1963), 'Criteria for Scientific Choice', *Minerva*, vol.1, pp.159–71.
Wilson, G. (1987), 'From now, CSIRO will try to pick winners — Wran', *National Farmer*, 2 October, pp.1–2.

8 Research foresight in Canada

1. INTRODUCTION

A major issue in recent Canadian discussions on research policy has been how to marshal national resources to meet the challenge from existing and emerging industrial world powers.[1] There was a time when the specialized nature of the economy (based on natural resources, transportation engineering and telecommunications) and the technological dependence on American multinationals were not perceived as major problems. However, the desire to secure a more independent future has resulted in a period of national debate on strategies for strengthening domestic R&D and improving the capacity to access commercially promising foreign technology (e.g. MOSST, 1983a; NABST, 1988a; de la Mothe, 1989). Although R&D expenditure has historically been relatively low, representing only 1.38 per cent of GDP in 1987 (Statistics Canada, 1988b, p.2), both federal and provincial governments now appear to be committed to an overall expansion in research budgets. Particular emphasis has been given to stimulating industrial R&D which rose by 50 per cent in real terms during the six years to 1987 (Statistics Canada, 1988a, p.2).[2]

The stress on economic and technological regeneration has also had implications for academic and government research (NABST, 1988a–e). Policies have been introduced to improve university–industry links, to target research through earmarked programmes and national centres of excellence, and to reorient postgraduate training towards fields with strategic potential (MOSST, 1987b). In addition, the federal government has encouraged the National Research Council to reduce the priority given to fundamental science and focus the work of its laboratories more on emerging industrial technologies.

Accompanying these changes has been increased coordination and planning of research, which was previously limited by lack of agreement between federal and provincial governments over their respective responsibilities. The centrality of R&D in industrial development policy is reflected in the fact that all ten provinces have formulated their own longer-term plans and priorities (Dufour, 1985), paving the way for the historic agreement of a National Science and Technology Policy (CSTM, 1987). Parallel to this has been the establishment, again for the first time, of a National Strategy for Science and Technology (InnovAction) which set out a federal agenda: (i) to promote a 'science-oriented culture'; (ii) to stimulate innovation and technology transfer; (iii) to increase the country's capacity to exploit emerging technologies; and (iv) to manage the deployment of R&D resources to meet societal priorities more effectively (MOSST, 1987e). Another key development in 1987 was the creation of the National Advisory Board on Science and Technology (NABST). This advises the Prime Minister on overall guidelines for federal R&D, as can be seen from Figure 8.1 which outlines the main elements of Canada's science-policy apparatus.

The strategic importance attached to R&D also underlay a restructuring of ministries in late 1988, when Industry, Science and Technology Canada (ISTC) was set up to integrate the industry portfolio with that for research (which was previously under a separate Ministry of State for Science and Technology — MOSST). ISTC inherited responsibility for administering a large fraction of federal R&D (38 per cent), with the National Research Council and Science Council of Canada reporting to Parliament through the Minister of State for Science (one of two junior ministers within the department). ISTC's other activities include framing policy for federal R&D and servicing NABST. However, its influence on academic research remains limited. Finance for the universities is provided by a dual-support system in which institutional core-funding (transferred to provincial governments through the Secretary of State) is complemented by project-based grants from the Natural Sciences and Engineering Research Council (NSERC), the Social Sciences and Humanities Research Council and the Medical Research Council. A measure of pluralism in academic support has been maintained in that different ministries continue to be responsible for each of the research councils (e.g. NSERC comes under ISTC), and this is reinforced by the existence of a parallel network of provincial funding agencies.

In what follows, we first examine the role of foresight in federal initiatives to establish overall research priorities, especially those by NABST and MOSST. Prospective activities of the Science Council of Canada are also reviewed, in particular its programme on Emerging Science and Technologies and the associated National Consultation on Emerging Technology. Next, we discuss foresight in academic basic research, focussing on two of the research councils and the university system more generally. Analysis of strategic and applied government R&D centres upon the National Research Council, while private-sector approaches to foresight are outlined using case-studies of Bell Northern Research and the Canadian Advanced Technology Association. Finally, we summarize the lessons for research policy which can be drawn from Canadian experience with foresight.[3]

2. FORESIGHT IN NATIONAL PRIORITY-SETTING

(a) Federal government

The election in 1984 of the Conservative government under Mulroney led to a major shift in science policy. The election manifesto included a pledge to increase federal support for R&D, especially in areas with high economic potential.[4] However, before making available the promised additional funds, the new government concentrated on establishing administrative structures appropriate for building consensus on research priorities and implementing selective funding of targeted areas. This involved four related initiatives:[5]

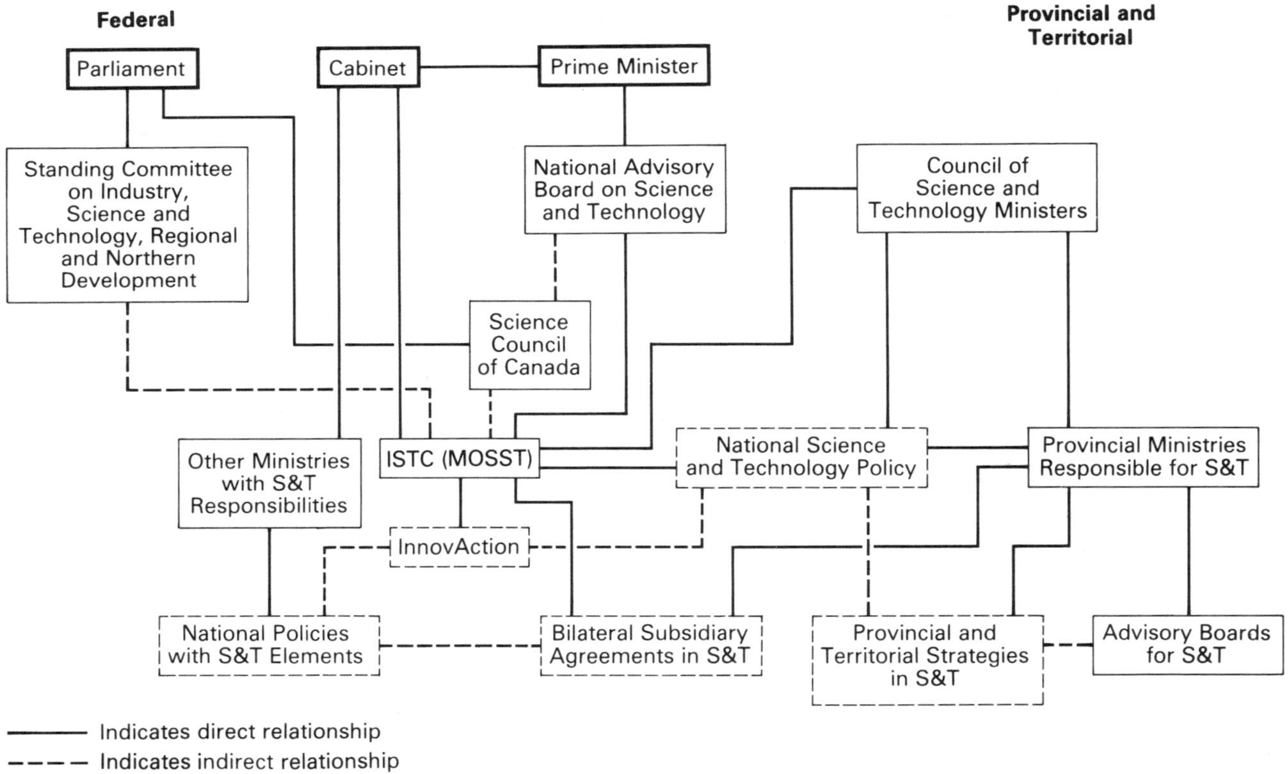

Source: Constructed by P. Dufour, Canadian Department of External Affairs.

Figure 8.1 *Key elements within Canada's national science and technology policy apparatus*

(1) NABST was given a remit to 'develop national science and technology goals and policies, and their application to Canada's economy' (NABST, 1987). Chaired by the Prime Minister, its main task is to advise government on R&D trends, on improving the links between public and private sector research, and on mechanisms for strengthening industrial R&D and reaching national consensus on research priorities. The creation of this high-level body was seen as one way to focus discussion on the difficult but vital decisions which need to be made on selective restructuring of R&D support — for example, whether 'government should rely more on universities and less on federal laboratories for the performance of research' (NABST, 1988b, p.58).

(2) The Council of Science and Technology Ministers was also established in 1987 in an attempt to bridge the gap between federal and provincial decision-making. With a membership comprising the federal Minister of State for Science and relevant ministers from provincial governments, it functions as a consultative forum on issues requiring coordination of R&D policy. Given previous conflict between Ottawa and certain provinces, efforts have been made to develop consensus on how federal initiatives might better complement regional R&D priorities. Guidelines for federal and provincial cooperation were formally set out in the National Science and Technology Policy (CSTM, 1987).

(3) The Decision Framework for Science and Technology represents an attempt to ensure that federal R&D priorities are taken into account in departmental policies and programmes (MOSST, 1987d). It has three main components, the first of which is a set of criteria for deciding whether proposed R&D merits public support (the criteria relate to objectives, mission relevance and appropriate sector of performance — e.g. in industry or government laboratories), together with guidelines for programme selection. Second, federal departments and agencies involved in R&D have, since 1987, had to prepare annual strategic plans outlining priorities for the next few years (including an assessment of anticipated research outcomes and the likely time-scale of any socio-economic impacts). In addition, potential beneficiaries must be identified and encouraged to participate in priority-setting and joint financing, thereby building in an early commitment to exploit future results. Third, the Minister of State for Science was made responsible for preparing an *Annual Overview of Federal R&D* which synthesizes departmental and agency plans, and identifies gaps, duplication and key issues. The overview provides a basis for government discussions on strategic directions for R&D.[6]

(4) The Canadian Strategy for Science and Technology (InnovAction) set in place a coordinated framework for targeting resources on fields with economic and regional development potential, or those likely to contribute to the advancement of knowledge and training of researchers in areas of national need. Clear federal R&D priorities now exist in three strategic fields — biotechnology, microelectronics/information technology and advanced industrial materials — and a large increase in federal support is scheduled over 1988–92 (MOSST, 1988).

These initiatives together reflect a move towards a more coordinated approach to decision-making in science and technology, and have provided a context for greater use of systematic foresight techniques. However, given the need to secure agreement on R&D policies among federal and provincial governments, efforts have primarily concentrated on determining priorities through consensus-seeking workshops and committee discussions. For example, the Science Council of Canada organized a National Science and Technology Forum in 1986 at which 200 invited representatives from industry, trade unions, government and universities discussed R&D issues relevant to the country's future.[7] Workshops were held focussing on three themes — 'development and acquisition of new knowledge', 'putting knowledge to work and creating opportunities' and 'involving all Canadians and adapting to change'. Participants were provided with relevant background material and a set of questions to structure debate and facilitate the drawing of conclusions. The report on the forum (SCC, 1986a) represented a major input to a meeting of federal and provincial ministers later that year to agree the basic framework for the National Science and Technology Policy. It also had an impact on two subsequent conferences organized under the aegis of InnovAction: the National Forum on Post-Secondary Education (MOSST, 1987g) and the National Conference on Technology and Innovation held in 1988.

While forging consensus on a national policy framework was an essential first step, federal and provincial governments have also had to undertake detailed analysis of R&D options. A growing role has been played by NABST[8] and its various committees (particularly those on industry, government and the universities) which have outlined R&D strategies for each sector (NABST, 1988a–f). The University Committee has been especially active, in line with the widespread view that restructuring of higher education is a precondition for modernizing the R&D system to meet future societal needs. Its advice has been that universities should focus their research and postgraduate training more on nationally important fields like mathematics and systems technology (where industrial demand for trained scientists is projected to rise by 200 per cent in the period to 1995, compared to 25-40 per cent in engineering and only 10 per cent in management science — NABST, 1988b, p.22).[9] To aid reorientation of priorities, the Committee proposed that 'a long-term plan for publicly sponsored research be developed' to ensure priority support for: (i) individuals and groups at the forefront of their disciplines; (ii) state-of-the-art equipment and central facilities; and (iii) targeted research of national interest (ibid., p.57). The Government Committee has also highlighted the need for improved foresight, recommending that greater emphasis be given to monitoring overseas technological activities, and to determining the most promising lines of research within each national priority area (e.g. opportunities arising from cold-region technologies — NABST, 1988d, pp.5–6).

ISTC is formally responsible for conducting background policy research for NABST and has (through MOSST) attempted to expand its in-house capacity to undertake prospective analysis. This followed the recognition that:

strategic, or enabling, technologies are based on multidisciplinary contributions of science and they interact synergistically. Their impacts are generic, and they will be critical to enhancing competitiveness in all sectors of the economy. This changing environment is placing new demands on Canada, and the Ministry, for technology monitoring and strategic planning in science and technology. [MOSST, 1987e, pp.15–16]

MOSST first took on responsibilities for futures research in the mid-1970s, providing advice on long-term trends in science and technology and their implications for social and economic policy (MOSST, 1982, p.9; 1983b). However, this function gained in significance as government exerted pressure on the Ministry to fulfil its mandate 'to establish priorities for science and technology ... and their application in furthering national objectives' (MOSST, 1987b, p.7). Also important was the 1983 decision to designate the Secretary of MOSST as Chief Science Advisor with a remit explicitly including the integration of a long-range perspective into federal R&D policy, and identifying active research areas showing particular promise for Canadian industry (MOSST, 1987e, pp.7–8).

Prospective studies were initially contracted out entirely to consultants like SRI International, but this did not prove wholly satisfactory. A greater proportion of foresight is now carried out in-house, primarily by the Strategic Technologies Branch which has become increasingly active in developing procedures for setting national programme priorities since the early 1980s. This followed a task-force proposal that a coordinated federal strategy be developed to capitalize upon emerging commercial opportunities in biotechnology (MOSST, 1981). Representatives of industry, government and the universities were subsequently brought together in a National Biotechnology Advisory Committee to frame R&D priorities consistent with Canada's longer-term socio-economic needs. These were agreed at a series of national workshops organized by *ad hoc* working groups, drawing upon background studies by MOSST and other agencies of developments overseas as well as upon foreign visits by committee members. Four technologies were identified as especially promising (e.g. plant-strain development), together with six related fields of basic research (e.g. cellular genetics — MOSST, 1986a). The biotechnology programme succeeded in drawing attention to areas of commercial opportunity (e.g. in advanced forestry products) and catalysing R&D initiatives in an important field where Canadian efforts were previously weak.[10]

A similar consultative process was employed in creating the national programme on microelectronics (MOSST, 1987h), with MOSST officials coordinating discussions to agree an outline strategy. Since this is a field in which Canada has a leading world presence through companies like Northern Telecom, the approach was to invite industry and the scientific community to identify potentially important gaps (such as the need for training programmes or for fundamental research in specific areas). A National Microelectronics Advisory Committee then established detailed priorities and administered a programme budget earmarked under InnovAction. The Interdepartmental Working Group on Advanced Industrial Materials later employed a similar planning procedure to develop a programme for the third priority field, the main goal of which is to strengthen the nation's technological capacity for adding value to metal and mineral exports. Among the longer-term research requirements identified was work in materials characterization and properties (MOSST, 1986b; 1987i).

The Strategic Technologies Branch has also carried out more specific surveys of opportunities in areas like photovoltaics (MOSST, 1982, p.9), while recent unpublished studies have analysed the prospects for promising technologies in five resource-based sectors (e.g. the impact of biotechnology on aquaculture) and outlined federal strategies for supporting research in artificial intelligence and high-temperature superconductivity.* These have again involved widespread consultation with industry and researchers, the aim being to set in place programmes tailored to Canada's future needs rather than simply jumping on successive international bandwagons. Officials stressed that each emerging technology has its own specificities relating to indigenous technical strength, the structure of domestic industrial activity and the growth dynamics of the underlying scientific specialties. This was emphasized in the NABST (1988d) report on InnovAction, and in a scenario-based report on opportunities in hydrogen technology which proposed building upon Canadian expertise to develop alternative energy and propulsion systems without the environmental problems associated with fossil fuels (RAGHO, 1987).

Of the main problems facing ISTC/MOSST, one has been a shortage of staff with a background in science and technology policy. This has limited its use of systematic foresight techniques, although the hope is to strengthen in-house expertise in this area over future years.* Second, and more fundamentally, the recent departmental restructuring failed to address the pressing need to establish a clearer link between advice on research priorities and decisions on the overall government R&D budget.[11] In particular, it is questionable whether ISTC should have been made responsible for servicing NABST given the potential conflict of interest arising from the fact that it also oversees a considerable proportion of federal R&D. If advice on R&D priorities is to be tied to decisions on resource allocation across departments, the only solution may be to transfer the national research-policy functions of ISTC to a new central agency (NABST, 1988c, p.10).

(b) Science Council of Canada

Established in 1966 to provide independent advice on science and technology, SCC has traditionally operated at arms-length from government, initiating and publishing its own studies as well as undertaking commissioned work. Its official responsibilities are : (i) to analyse science-policy issues and make recommendations to government; (ii) to alert Canadians to the likely impact of science and technology; and (iii) to stimulate debate on research policy among federal and provincial governments, industry and the universities (SCC, 1987a, p.3). SCC studies are carried out by consultants or in-house staff, often in association with external experts, and are overseen by *ad hoc* steering committees. Although normally based on literature review, interviews and case-studies, they have sometimes employed formal foresight techniques like Delphi exercises.

(i) National Consultation on Emerging Technologies

SCC has long advocated a more systematic approach to setting research priorities, and has periodically carried out reviews of national R&D needs and opportunities (e.g. SCC, 1972; 1979).[12] In 1984, the Minister of State for Science and Technology invited the Council to undertake a survey of promising areas of strategic research. After negotiations on the resources required, he awarded a contract for the National Consultation on Emerging Technologies. This was conducted within the context of a broader programme on Emerging Science and Technologies (EST),[13] the terms of reference of which were: (a) to ask Canadians at the leading-edge of technology to identify fields likely to be important for economic and social development; (b) to establish which sectors would probably be most affected; (c) to assess the prospects for commercializing relevant ESTs over the next decade; (d) to bring together research and business experts to discuss opportunities and constraints, and generate consensus on future action; and (e) to raise public awareness of the potential impact of emerging technologies (Steed and Tiffin, 1986, p.13).

SCC had earlier initiated a review of international experience with research forecasting. In this, Elzinga (1983) analysed the conceptual and operational problems that had been encountered,[14] and advocated a foresight approach combining systematic monitoring of scientific trends with consensus-seeking expert discussions. Three alternative foresight initiatives were proposed depending on the resources available. The medium-sized option, and the one eventually adopted, was structured around a national consultation modelled on the French exercises held in 1979 and 1981 (see Chapter 3). Although the involvement of MOSST meant that generic technologies were stressed more than initially envisaged by Elzinga, the basic philosophy underlying his approach was accepted: 'The process must serve a broader purpose than simply producing a forecast. Identifying a range of possible developments can help stimulate public discussion of desirable futures and of the role of government in such futures' (Steed and Tiffin, 1986, pp.12–13). SCC decided to establish a high-level advisory committee for the programme, and this exercised a strong influence on the methodology employed.

One lesson from the international review was the need to combine 'science and technology push' and 'end-user pull' perspectives, and these were built into the National Consultation from the outset. The advisory committee recommended that this be done by means of a matrix of R&D fields versus socio-economic sectors of impact, although some SCC staff involved expressed reservations about the likely technical difficulties.* The aim was to employ the matrix as the basis for a preliminary survey, first of leading researchers to obtain an overview of scientific opportunities, and then of industrial R&D managers to establish likely technological needs. It was planned that clustering algorithms would subsequently be used to identify cross-impacts and linkages, and much effort was devoted during the initial stages to constructing a suitable survey instrument.[15] However, the replies from a pilot study suggested that the proposed forty-page questionnaire was unlikely to yield a sufficient response rate, and the advisory committee therefore agreed to a more modest survey. This posed four questions: (a) which ESTs are likely to be significant for Canadian development over the next decade? (b) What sectors will be most affected? (c) What is the probability of each EST being commercialized during the next decade? (d) And what concerns exist about the country's response to important ESTs?

Telephone interviews were conducted by consultants with over 200 leading scientists, engineers and R&D managers from companies, government laboratories and academia.[16] However, the open-ended questionnaire (sent beforehand by post) elicited rather general responses which were difficult to analyse, with interviewees differing in their understanding of what constituted an EST and the level of aggregation of technology fields and industrial sectors to employ. In addition, some respondents (especially academics) showed a lack of realism about the time-scale for possible commercialization, while certain industrialists refused to provide detailed information because of concern about commercial secrecy.*

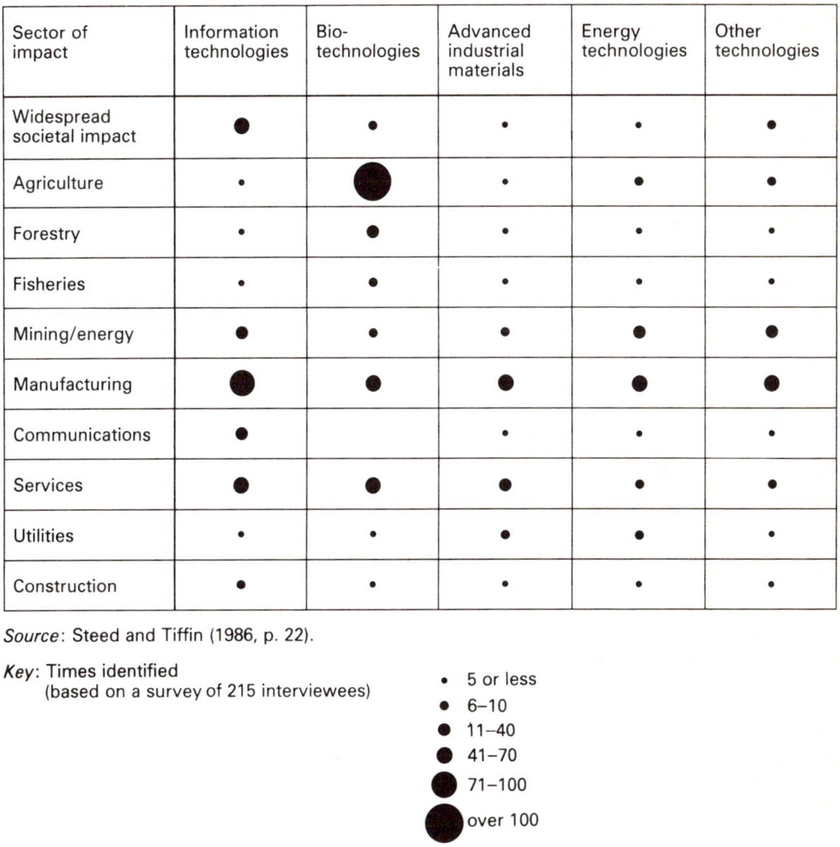

Figure 8.2 Matrix of emerging technologies and sectors of impact on Canada

The survey confirmed the relevance to Canada of the fields internationally regarded as important. Figure 8.2 shows that information technology was judged likely to have the widest impact, closely followed by biotechnology. Altogether, thirty ESTs (e.g. gallium-arsenide technology) were identified as having potential for significant sectoral impact, and a further fifteen, often at a more specific level of aggregation, were viewed as promising (e.g. natural language translation and superconducting technologies — ibid., pp.23–5). As regards the factors likely to constrain Canada's capacity to exploit future opportunities in these fields, respondents expressed concern over the poor links between academic and industrial R&D, shortages of researchers, inadequate government funding and the lack of a well-developed entrepreneurial culture. There was also agreement that R&D policy needed to recognize explicitly the synergy between EST fields, and that increased targeting of R&D was required to capitalize upon national technological and industrial strengths.

In an effort to identify technologies in which the country had a comparative advantage, a survey was carried out of 163 members of the Canadian Research Management Association (65 per cent of whom were industrialists). The questionnaire listed the fifty ESTs identified in the telephone survey, and required respondents first to indicate the fields in which their organization had R&D expertise. Those with relevant experience then ranked Canada's standing in each field ('leader', 'fair' or 'weak'), and in cases of technological leadership specified the companies or laboratories concerned. Areas of greatest strength included telecommunications, oil-recovery technologies and remote-sensing (ibid., pp.34–5).

The second phase of the National Consultation was designed to produce more detailed information from a broad cross-section of the science and technology community, especially active researchers and practising engineers. Among the objectives were developing an inventory of emerging technologies considered important for Canada, and helping establish networks of experts in fields of special strategic significance (e.g. sensor technologies). The views of over 11,000 researchers and managers in government agencies, universities and companies were sought using a postal questionnaire based on questions similar to those in the telephone survey, with care being taken to define the terms employed more precisely.[17]

A higher than expected response rate of 31 per cent resulted in immense data-processing problems given the decision to rely again on open-ended questions. Coupled with a 50 per cent cut in SCC's budget imposed by government in 1986, this contributed to a situation where only very preliminary analysis had been completed by early 1988.[18] Nevertheless, large numbers of technologies and sectoral impacts were found to be clustered within a few related areas. Three core technologies (microelectronics, biotechnology and advanced industrial materials) accounted for more than 50 per cent of all important ESTs, while most sectoral impacts could be grouped into thirteen categories (e.g. food and transportation). Further evaluation of interlinked core technologies and sectors suggested several potentially exploitable niche areas where Canada has the research capacity and domestic market to support future commercial development (e.g. hydrogen fuel-technology).

Although these findings were made available in the form of a discussion paper (Steed and Tiffin, 1986),[19] the failure to publish a detailed report on the second phase of the National Consultation led to some dissatisfaction within

MOSST and other agencies.* This was not helped by the lack of any national workshop to consider the implications of the results for future R&D, a problem stemming in part from insufficient thought being given at the outset by SCC and MOSST as to how the exercise would be used to inform policy debate and priority-setting.*

(ii) Field and sector studies

The other component of the EST programme involved holding workshops aiming (i) to develop consensus between the public and private sectors on R&D initiatives in core technology areas, and (ii) to explore the prospects for more specific ESTs and raise public awareness on potential threats and opportunities. Although the original intention was that most would be held after the postal survey results became available, plans for three workshops on advanced materials were brought forward after the advisory committee agreed that a more rapid response was desirable.†

SCC already had some experience in assessing the prospects for individual fields such as biotechnology (SCC, 1985a), as well as from a pilot project for the EST programme on aquaculture (SCC, 1985b). It therefore decided to adopt an approach combining consensus-seeking workshops with expert committee deliberations to explore national priorities for advanced industrial materials. Attended by 275 participants from government, academia and companies, the workshops identified opportunities in ceramics, composites, new metallic materials and polymers, resulting from rapid progress in such areas as microstructures. Besides revealing that little work was under way, they highlighted the fact that an interdisciplinary perspective is required to exploit opportunities arising from the synergy across fields, as well as improved linkages among firms in different sectors and between industrial, academic and government laboratories (Elzinga and McCutcheon, 1985). They also helped forewarn the mining and metal industries of potential threats.*

Financial constraints during 1986 precluded similar workshops being convened for the two other core technology areas. However, SCC had the previous year sponsored a conference on the impact of biotechnology on forestry research (SCC, 1986c, p.26; 1987b).[20] It had also commissioned a Delphi study on the future for IT from consultants but this ran into technical problems and received a poor response from international experts in commercially sensitive areas. Nor was it possible to initiate in-depth studies in more than a few of the thirty ESTs initially planned — for example, emerging sensor technology (Katz, 1987) and new plastics technologies. The latter was among the more influential SCC foresight exercises, in part because the Canadian Plastics Institute (which helped organize the consultative process) was able to bring together the main research performers and users. Some seventy Canadian and foreign experts were first surveyed to obtain ratings of the commercial prospects for specific ESTs (e.g. controlled molecular architecture) and, in the case of domestic respondents, the industrial and socio-economic sectors most likely to benefit. The results were then discussed at a workshop attended by thirty representatives from industry and the public sector. Table 8.1 shows a matrix of the emerging plastics technologies identified against the sectors judged likely to benefit from research in each

Plastics Technologies / Contribution to 1983 GDP ($10⁹)	Agriculture & Food	Fisheries	Forests & Products	Mines & Metals	Liquid Fuels	Automobiles & Parts	Other Transport Equipment	Software-Based Systems	Telecom. Equip. & Systems	Other Manufacturing	Construction	Entertainment & Recreation	Education	Health & Safety	Other Service Industries
Plastics Technologies	18	1	13	10	4	8	2	16	15	33	19	4	17	19	140
1. Surface Modifications	3					1	2				2				
2. Crosslinking Thermoplastics						1	3		2		4				
3. Coldworking	4					1	2				3				
4. Foams						2	3			4	1				
5. Joining Techniques			4			1	3				2				
6. Reactive Processing						1	2			4	3				
7. Composites						1	2			4	3				
8. Dosage, Membranes, Barriers	1	3	4											2	
9. Controlled Molecular Architecture			1						2	4	3				
10. Conductive Polymers						2	4	3	1						
11. Powder						2	4			1	3				
12. Process Equipment						1	2		4	3					

Note: The four sectors judged most likely by experts to benefit from technologies listed are indicated by 1 (highest likelihood down to 4 (lowest).
Source: Anderson *et al.* (1985, p. 12).

Table 8.1 Major impacts of new plastics technologies on various industrial and socio-economic sectors in Canada*

field. Future priorities were seen as lying in such fields as composites and conductive polymers (Anderson *et al.*, 1985).

A range of studies on other important science and technology questions has also been carried out by SCC (see Steed, 1988). One of the main themes addressed has been the future needs of the science and technology infrastructure, and recent reports have recommended strengthening the network of science counsellors responsible for monitoring overseas R&D (SCC, 1984) and measures to improve the long-term health and policy relevance of the social sciences (SCC, 1985c). Given the importance of upgrading the technological capacity of Canadian science-based industry, a second theme has been how to improve the supply of researchers in fields like engineering (Mascolo *et al.*, 1985). Finally, work on emerging environmental problems has resulted in a series of influential reports calling for longer-term research on such issues as water usage (Mitchell and McBean, 1985; SCC, 1988b) and the global ecological crisis (SCC 1988c). As in the EST programme, workshops involving experts from the research-user and producer communities have generally been employed. The study on social sciences, for example, drew upon the results of a two-day meeting in which some twenty academics and government officials discussed proposals for improving the relevance of the field to solving problems likely to develop over the next decade.

As regards the overall impact of the prospective studies undertaken by SCC, the budget cuts of 1986 are one indication of the mixed views held by federal government departments. Apart from the dissatisfaction in MOSST over inadequate progress in analysing survey data from the National Consultation, there was concern within the Treasury Board and elsewhere about the uneven quality of study reports and about what was seen as a failure to respond satisfactorily to the new science-policy climate following the 1984 election, in particular to the need for selectivity and concentration in research funding.[21*] Although this stemmed in part from the fact that SCC is an autonomous public body, it may also be explained by the previous involvement of its former Chairman (Stuart Smith) in opposition politics and the consequent opportunity this gave critics to interpret as partisan the stress placed by the Council on maintaining its role in providing independent scientific advice.* There was also greater scope for conflict with MOSST as it developed its own analytical and policy capability,[22] and this contributed to limiting SCC's influence on federal decision-making.

An alternative view holds that SCC was harshly judged in 1986 by a government committed to a more coordinated approach to research planning, and therefore unlikely to be sympathetic to an organization producing reports sometimes questioning the R&D policies of particular departments (especially in regard to the resource and environmental sectors). Much of the criticism of SCC is seen as misplaced in failing to appreciate the diffuse process by which its foresight activities have influenced decision-making. Certainly, there are few cases where direct links can be identified between SCC reports and subsequent policy action by government or industry, but this reflects the fact that studies have rarely been carried out to inform specific decisions. Rather, the impact of SCC has been in: (a) challenging and overturning orthodoxy in research policy (which is necessarily a long-term process); (b) providing a wide range of analytical material useful in a general way for framing policy positions or solving problems; (c) educating the general public, government and industry about emerging concerns and opportunities; and (d) catalysing the development of national consensus on key science and technology issues (Steed, 1988). Given that one of its roles is to 'act as an S&T gadfly and alert Canadians to perhaps misdirected public policy',* it is questionnable whether departmental responsibility for SCC should rest with the same minister overseeing MOSST (and now ISTC) and administrative support for NABST.[23]

Whatever the merits of these differing views on SCC, they highlight the point that wider organizational and political factors are often crucial in the success of foresight. For its part, SCC has become more aware of the need to design policy-implementation structures into foresight from the outset, and to ensure that methodological expertise is coupled with effective consensus-seeking mechanisms.

3. FORESIGHT IN ACADEMIC BASIC RESEARCH

A high proportion of Canadian basic research is carried out within the higher education sector. In 1986, overall spending of C$1,600 million on academic research represented 23 per cent of domestic R&D (NABST, 1988b, p.38), with just under 40 per cent coming in the form of 'general university funds' distributed by provincial governments. The provinces also provide around 20 per cent of the finance for academic separately budgeted research, although this is primarily the responsibility of the federal research council system. We concentrate here on foresight initiatives within the Natural Sciences and Engineering Research Council (NSERC) and the Social Sciences and Humanities Research Council (SSHRC), but also consider developments in university research policy.

(a) *The university system*

After a lengthy period of debate on the future of the universities, the federal government, provinces and industry all now agree that academic research must be strengthened considerably.[24] Many of the current problems stem from the heavy reliance upon infrastructural support distributed on the basis of student numbers.[25] Coupled with the policy of federal agencies to cover only direct costs in research grants, this means that no mechanisms have been established for priority-setting either within institutions or at provincial and national levels. While a set of elite universities can be distinguished,[26] the overall result has been fragmentation and duplication of effort and an inability to develop a critical research mass or to mount high-quality postgraduate programmes.* This is a particular problem in interdisciplinary fields (MOSST, have been strongly recommended by NABST (1988b).

These concerns featured prominently in the 1987 National Forum on Post-Secondary Education which sought to establish a joint federal and provincial strategy for higher education. It was agreed that improved management and planning of research were needed at all levels, but it remains to be seen whether this will lead to greater use of foresight. For example, while it was accepted as useful in principle to monitor emerging industrial needs for researchers, the dangers of 'mechanical' forecasting were stressed since in rapidly advancing fields the demand for skills often follows supply rather than vice-versa. The conclusions of the forum contributed to a federal decision to provide additional funding of C$240 million over five years (under InnovAction) for a network of national centres of excellence (IPCD, 1988). Since the criteria for selecting centres include relevance to industrial competitiveness, there may well be pressure on universities and cooperating companies to carry out prospective analysis when framing proposals.*

The centres of excellence initiative was largely modelled on the innovative plans drawn up by the Ontario Premier's Council for strengthening the province's R&D infrastructure in order to meet the future needs of companies competing in 'the new global economy' (OPC, 1986). The Council was created in 1986 to forge consensus on priorities for a long-term development strategy for the province. The cornerstone in this is a C$1,000 million technology fund, 20 per cent of which was allocated to a five-year programme to establish joint university-industry centres of excellence in precompetitive research areas. As well as building an advanced scientific capability, centres have the role of training front-line researchers and transferring new technology to companies. Following consultation with industry and a study of emerging R&D trends in countries like Japan (to which inputs were made by the Canadian Institute for Advanced Research), proposals for twenty-eight centres were reviewed by a panel of domestic and foreign experts and seven chosen in fields like laser and lightwave research (OPC, 1987). The establishment of similar centres on campuses by other provinces seems likely to encourage increased long-term planning by universities and more selective funding of academic research.*

(b) Natural Sciences and Engineering Research Council

Established in 1978, NSERC inherited most of the former responsibilities of the National Research Council for funding academic science and engineering research. Its budget of C$330 million in 1987 was distributed through several support mechanisms: (i) finance for the academic 'research base' (including project and programme grants); (ii) studentships and fellowships; (iii) targeted research ('strategic grants' and collaborative university–industry programmes); and (iv) general research support (e.g. for conferences — NSERC, 1987a). Constrained by level funding, NSERC has placed greater stress on effective research management and has recently completed evaluations of its 'operating' and 'strategic' grants programmes (DPA Group, 1986; 1987). The range of foresight activities carried out has also been expanded. These now cover: (i) preparation of the NSERC five-year plan; (ii) disciplinary reviews; and (iii) selection of priority topics for strategic research.

(i) NSERC five-year plan

Like other federal agencies, NSERC is required to submit a corporate plan to assist Cabinet decisions on the quinquennial budget allocations. This is drafted by the President, together with senior officials, using outline projections prepared by advisory and programme committees, and is then modified after discussion by the NSERC Council. The first plan covered 1980–5 and identified four priorities (e.g. expanding targeted research) within the framework of three budgetary scenarios. The exercise helped structure discussions on ensuring greater continuity in the organization's finances, as well as providing an opportunity to make a case to government for increased funding.*

The next plan for 1985–90 emphasized developing the national stock of intellectual capital and repairing deficiencies in academic research equipment and infrastructure resulting from the period of level funding. More than half the report was devoted to assessing the extent to which the goals set for the previous quinquennium had been realized, but some attempt was made to forecast the likely supply and demand for researchers up to 1990. Various scenarios were constructed assuming different R&D growth rates relative to GNP, per capita research expenditure and average career lengths. The results showed that, even under conservative assumptions, severe skill shortages seemed unavoidable in certain applied science and engineering disciplines (NSERC, 1985, p.132).

While this latter plan proved useful in defending NSERC's budget, it was criticized by some officials for taking insufficient account of the activities of other agencies and of scientific and user needs, and for generally failing to provide a strategic lead. This was linked to a wider unresolved conflict in the agency over basic policy and funding for targeted programmes compared with support for the 'research base'. Such critics thought future plans should draw more heavily upon systematic studies of scientific opportunities in the case of fundamental research, with potential impact on different industrial and socio-economic sectors being used as an important criterion in selecting strategic programme areas. This would enable broad priorities to be identified across fields rather than being framed primarily in terms of type of funding mechanism as has tended to be the case in the past.*

(ii) Disciplinary reviews

NSERC has funded various foresight activities by the scientific community, including workshops on the potential impact of new R&D developments and planning initiatives in fields presenting special opportunities (NSERC, 1987b, pp.39 and 52). The first disciplinary reviews were carried out in 1980 as part of the process of deciding priorities for the increased funds then made available to the Council. Scientists had voiced concern about a top-down approach and emphasis on external criteria in NSERC's first five-year plan, and saw the field reviews as a means of counterbalancing a perceived trend towards centralization. By 1988, reviews had been completed in four fields — earth sciences, physics, civil engineering and chemistry — with a fifth on subatomic physics under way. Each had a similar remit: (a) providing an overview of national strengths and weaknesses; (b) identifying the likely demand for scientific expertise over the medium- and longer-term; (c) defining research directions for both the next five years and the period beyond; (d) developing a plan for the field taking into account scientific and economic factors as well as potential international developments; (e) agreeing specific research priorities; and (f) making recommendations on the most effective funding mechanisms.

The first review on earth sciences had by far the greatest policy impact.* Carried out by an *ad hoc* committee established by the Geoscience Council of Canada, it was very much internally driven, researchers feeling the field to be seriously underfunded because 'many of the issues that concern the earth science community ... are long range and have not yet reached a sufficient level of crisis to become government priorities' (Wheeler, 1981, p.105). Given the fragmented disciplinary structure of earth sciences, it was recognized that increased support would only be forthcoming if consensus could be secured on a high-profile 'mega-project' promising 'state of the art conclusions on problems of vital concern to our science and to national objectives in mineral resources' (Fyfe and Rust, 1981, p.113). Moreover, such a multi-million dollar project needed to combine scientific merit with demonstrable long-term commercial potential.

With support from NSERC, a workshop was organized on research opportunities and needs over the next decade. This attracted over 100 participants from academia, government laboratories and industry. Although several potential programmes (e.g. on marine geoscience) were discussed by ten discipline-based committees, most attention focussed on an examination of the respective merits of two large-scale initiatives. Following plenary discussions, a committee was established to draft a report recommending that the Lithoprobe Project be given urgent priority. The project involved interdisciplinary research on the dynamics and evolution of the lithosphere, with emphasis on the long-term exploitation of Canadian earth resources and the development of an improved understanding of natural geological hazards (Fyfe and Rust, 1981). Strongly endorsed by researchers, the proposal was subsequently funded on a multi-agency basis (CGC, 1986).

NSERC officials attributed the policy impact of this initiative to the politically astute leadership of a few well-respected researchers who were able to marshal the community to agree a project combining high scientific and commercial promise. Most importantly, they succeeded in securing consensus among government and university researchers, as well as attracting active interest from the private sector. As a result, 'people were made to feel they had a voice and had participated in the process of selecting priorities even if their views were not taken into account. Industry and everyone else thought the proposed programme was superb ... We just had to reward this success.'* The Lithoprobe Project has since helped to integrate work in the different earth science specialties, so there is now greater cohesion and a shared vision of the future. However, concern was surfacing in 1988 that no follow-up workshop had yet been held and that the community might therefore again fragment.*

The review of physics was initiated in 1980 by the Canadian Association of Physicists, with NSERC once more providing financial support. Given the breadth of the field, it was decided to begin with surveys of individual specialties and then integrate the findings. Committees were established to ascertain the views of researchers on scientific opportunities and budgetary priorities in five main subfields. These drew on reviews of similar exercises abroad, information from postal surveys, and overseas visits to consult experts and funding agencies. The resultant reports all included summaries of recent research developments, together with international comparisons of expenditure, personnel and equipment which pointed (inevitably) to the parlous state of Canadian efforts in each subfield. A further common feature was the attempt to argue that, because the areas of special technological promise identified by industry during the exercise differed little from the lines of research selected by peer-review, non-targeted support should continue to be given priority since it allows greater scope for creativity and flexibility.

The quality and policy utility of the reports varied considerably, with the review of optics including information on areas judged by companies to have long-term commercial potential (e.g. optical information transmission and storage — CAP, 1981e). Overall, NSERC found the survey of condensed-matter physics of most use since the data reported on scientific opportunities and needs were relatively comprehensive. They included cross-national statistics on numbers of researchers disaggregated by subfield, as well as bibliometric data on the output and impact of different institutions (which were used to evaluate the relative importance to scientific success of such factors as size of group, access to research assistance, equipment and level of funding). In addition, a survey of researchers yielded a detailed list of currently active areas and those likely to become important over the next five years (e.g. superconductivity and disordered/amorphous materials), while a parallel review of industrial opinion identified subfields with particular commercial promise (e.g. lasers and cryogenics — CAP, 1981d). However, as in the other reports (e.g. CAP, 1981a–b), any attempt to agree ranked priorities was studiously avoided.

Although NSERC had stressed that a 'long shopping list of major new projects ... without any indication of rank, priority or preference, will serve only a very limited purpose', this is precisely what the five physics committees produced. All argued that *their* particular subfield had special scientific merit, and that no specific priorities could be identified other than the need for sharply increased funding, especially for non-targeted research and infrastructural support. Moreover, even after a review meeting to discuss the findings held at the Royal Society of Canada (CAP, 1982), the physics community proved unable to agree an overall set of priorities. Indeed, the nuclear physics committee submitted separate reports for low-, intermediate- and high-energy research, each recommending costly new facilities (CAP, 1981c). As a result, the physics review was widely regarded as a failure.* The use of a committee structure based on traditional subfields meant that insufficient attention was given to cross-disciplinary impacts, while the emphasis on panel discussions at the expense of consultative workshops resulted in the important function of consensus-building being neglected. Furthermore, the obvious shortcomings of the three nuclear physics reports did little to advance the case for big science, highlighting a more general need to improve the planning process for new facilities.[27]

The other field surveys on civil engineering and chemistry reportedly had even less impact. The former was prepared by the Civil Engineering Grants Committee as an input to the 1985 NSERC five-year plan, and consisted largely of a summary of existing research and its relevance to different socio-economic sectors. Efforts to identify areas where 'research is required urgently' (NSERC, 1984, p.10) yielded little more than a list of the main civil engineering subfields, and the inevitable plea for increased resources was not accompanied by any attempt to specify areas with high scientific potential or societal impact. Consequently, those outside the field found the arguments unconvincing.*

The chemistry review was carried out by an *ad hoc* committee with financial support from NSERC and several professional associations. Problems arose when the committee re-wrote its terms of reference without consulting the sponsors, and decided to adopt the priorities set out in the US Pimentel report (see Chapter 5) rather than identifying areas of special promise for Canada as had been requested. The result was that 'the committee slithered away from foresight in science to looking solely at existing finances and pushing for more money',* instead focussing on the organization of Canadian chemistry and recommending that the highest priority for NSERC was to implement a more selective funding policy (Adams *et al.*, 1986).

The draft report met with a strong reaction from chemists, who questioned the basic assumption that the future for the field in Canada would parallel that in the United States. Aside from the different R&D needs of the chemical industries in the two countries, no account was taken of the fact that the structure of research support in the United States is more pluralistic than in Canada, where the level of selectivity was already high. The committee was also attacked for producing a generally poorly researched report, for presenting inaccurate and inadequately interpreted statistics, and for putting little effort into developing consensus on future priorities.* The NSERC Council debated the report but did not endorse its recommendations, and the only policy impact was to reinforce an earlier decision to institute five-year research grants. Agency officials felt that a foresight study was still needed to identify chemistry priorities, and steps were being taken in 1988 to launch another review to assess research areas likely to have an impact over the next twenty years.*

(iii) Strategic Grants Programme

The Strategic Grants Programme (SGP) was established in 1977 following pressure on the National Research Council to relate its work more closely to societal needs, and in 1987 represented just under 10 per cent of NSERC funds. It is distinguished from a parallel University-Industry Programme by its longer-term nature and emphasis on coupling potential utility with scientific novelty (Halliwell, 1985).

The scope for employing foresight to set priorities for SGP is intrinsically greater than for operating grants where scientific merit is the principal if not the sole criterion. However, there has been much debate within NSERC on: (a) the extent to which selection of strategic areas should be left to academics whose expertise enables them to identify novel science-push opportunities; (b) how much emphasis should be given to demand-pull considerations specified by research users; and (c) whether priorities should be set only at a high level of aggregation, leaving researchers free to propose projects within these broad constraints. The scientific community initially resisted the notion of targeting, only acceding because new funds were promised and because the three areas identified by an expert committee were rather general (oceans, energy and environmental toxicology). Invited proposals were then judged on scientific merit and potential contribution to strategic national goals, including the likelihood of results being transferred to identifiable users.

Further areas were selected more systematically in 1981 following an increase in programme funds. Universities were asked to identify fields of Canadian research strength with longer-term potential for yielding socio-economic benefits. However, the views of users were also taken into account by drawing on the results of a 1979 Delphi study in which around 100 respondents (including industrialists) ranked twenty-six research areas in terms of several criteria (e.g. 'potential for technological innovation' — Blais, 1979).[28] The additional areas targeted were food/agriculture and communications/computers, together with an 'open category'. The latter represented a concession to critics arguing against too *dirigiste* a policy, and subsequently proved valuable as 'a sensor, a lightning rod enabling us to adjust rapidly to change'.* For example, an increase in 'open' applications in biotechnology and advanced materials was responsible for these fields later being accorded priority status.

Discussions over the NSERC five-year plan coincided with budgetary constraints in 1985. This resulted in a one-year moratorium on new projects in the oceans, energy and food/agriculture areas, with greater emphasis being given to discipline-based work. At the time, advocates of targeting were arguing that national needs should be given greater weight, and this led to some conflict within the Council and a decision to review SGP. The evaluation report cautioned against too great a stress on targeting, the results of an international survey supporting the majority view within NSERC that researchers should play the primary role in identifying priorities. Moreover, an assessment of projects in different target areas showed that those in the open category obtained high rankings not only in meeting the research goals specified in proposals, but also in the subsequent utility of the results (DPA Group, 1987).

The evaluation report was not universally accepted in NSERC and, despite opposition to employing user needs in determining priorities, it was decided in 1987 to commission SCC to carry out a survey of potential topics for SGP. This aimed to establish a systematic basis for programme selection by identifying a list of thirty broad themes for precompetitive research in the three national priority areas (IT, biotechnology and advanced materials) and in other promising strategic fields (SCC, 1988a, p.10). Previous SCC experience with foresight had shown the need to build policy implementation structures into such exercises from the outset. This was made simpler here given the expressed intention of NSERC officials to use the results in setting priorities for SGP. However, it was clear that acceptance of any recommendations would be contingent upon consultation with all relevant interests, and the first step was therefore to identify the SGP user-community so that a list of interviewees could be agreed.[29] A draft list prepared by SCC was amended with the assistance of NSERC staff and relevant government officials, yielding 350 potential interviewees (over half from companies which included all major R&D performers). Although the emphasis was on future demands for research, academics were also invited to participate to provide a science-push perspective.

After piloting a questionnaire, four consultants conducted the interviews largely by telephone (but in some cases on a face-to-face basis). Judgements were sought on: (a) emerging technologies and the underlying scientific specialties meriting priority support; (b) the primary objectives for SGP programmes in these areas (fundamental research, training of scientists, etc.); (c) why the fields identified were strategically important and which sectors might benefit; (d) mechanisms for organizing research; (e) the optimal time-scale for research (short-, medium- and long-term); (f) the current state of Canadian activities in the field (including the strength of universities and the 'receptor potential' of companies); and (g) the level of investment required by universities to guarantee significant progress. The results were analysed by the consultants and a synthesis report produced by SCC.

A one-day workshop was next held to discuss the report and establish consensus on priorities. Discussions were organized within four 'round tables' covering the three enabling technologies and an 'open' category. A well-designed consultative process was again seen as vital to successful implementation of the resulting recommendations, and the participants in each roundtable were therefore selected to comprise the 'prime-movers and shakers' from: (a) SGP committees; (b) federal agencies and advisory committees; (c) current and potential research users; and (d) experts in relevant fields. In addition, rapporteurs and resource staff were provided by MOSST, SCC and other agencies. Following these discussions, various mechanisms like iterative ballots were employed to agree priorities for each area. This yielded twenty-seven 'first-tier' priorities (e.g. photonics and information-carrier technologies) and twenty-one 'second-tier' ones. More specific lines of research were also identified for each priority topic (e.g. mariculture in the case of marine biotechnology).

Finally, SCC prepared a report synthesizing the agreed recommendations on future policy for SGP. Besides the themes identified during the workshop, several other priorities were proposed such as training initiatives and increasing the average size of projects to achieve the necessary critical research mass. In addition, the SCC Council decided, after considering the report, to recommend to ISTC that programme funding be quadrupled over a five-year period (SCC, 1989, p.9), lending support to those in NSERC arguing for expansion of SGP and a more focussed approach taking into account industrial and societal needs. This ran contrary to recommendations from the DPA evaluation and re-opened the debate on policy for SGP. Some NSERC staff view the report as too impressionistic, but SCC and SGP officials believe the exercise at last succeeded in providing a systematic market-pull dimension to priority-setting, as well as generating significant industrial interest in the planned programme.*

Although it is too early to assess the overall success of the study, the approach was undoubtedly an improvement on that used by SCC in the Emerging Science and Technologies Programme, even though it relied on less sophisticated techniques. NSERC has since adopted a similar procedure in framing a new programme on microelectronics funded by InnovAction, resulting in priorities being agreed for ten themes in design systems (each with a time-horizon of up to a decade before benefits are likely to be realized). This exercise was also judged successful, reinforcing the view of those administering strategic research programmes that industrial involvement in foresight and priority-setting is essential. Within NSERC, however, the debate on targeting is by no means over, even though the traditional attitude that the judgement of academic scientists should be paramount has less credibility in government now that the implications of many areas of university research for future industrial competitiveness have been recognized.

(c) Social Sciences and Humanities Research Council

The SSHRC finances academic research in all social science and humanities disciplines and had a budget of around C$70 million in 1987. Three main support mechanisms are employed: (i) discipline-based project grants; (ii) strategic research programmes; and (iii) studentships and fellowships. Although SSHRC has carried out little systematic foresight, two elements of its planning process have a longer-term perspective. The first involves preparation of the federally-mandated five-year plan, the most recent of which discussed broad expenditure priorities for the period 1985–90 under two budgetary scenarios (SSHRC, 1985). This was drafted under the guidance of the SSHRC Council following consultation with academic researchers, and represents no more than a formal response to an official requirement. It is not a strategic plan and no attempt was made to set overall priorities, for example, through detailed studies of the prospects for different fields.*

Second, despite a strong emphasis on peer-review, SSHRC has not escaped the government policy of targeting research, and strategic programmes focussing on nationally important areas introduced in 1979 now account for 15 per cent of its overall expenditure. Although the Council was initially opposed to targeting, the C$2 million earmarked for the programme proved too tempting to resist. SSHRC did, however, gain government agreement that the areas be chosen by researchers and that methodological and conceptual pluralism should continue. With the exception of the first programme, selection has been based on biennial seminars aided by commissioned reviews of opportunities in specific fields. Programmes have a minimum duration of five years, with peer-review used to assess proposals submitted on an annual basis. Priority areas have included 'women and work' and 'the human context of science and technology' (SSHRC, 1985, pp.95–111). Even though these are fairly general topics, most academics still have a clear preference for undirected research.*

4. STRATEGIC RESEARCH AND BASIC TECHNOLOGY

Although foresight by provincial governments has been discussed only briefly, it is important to note the current emphasis on framing long-term R&D strategies to stimulate the technological renewal of local industries. While Ontario and Quebec have the most developed science-policy infrastructures, all the provinces acknowledge the need for prospective analysis in determining R&D priorities (Dufour, 1985; Davis, 1988). Like the federal government, they tend to rely upon consensus-seeking workshops and committees, but have made some use of systematic techniques (e.g. in a 1982 exercise by the British Columbia Research Council).*

In what follows, we focus on the National Research Council (NRC) which, since its creation in 1916, has had to assume a variety of roles as the federal R&D system has evolved. In the postwar period, the formation of SCC, MOSST and then NABST resulted in a loss of its national advisory function, while universities gradually assumed responsibility for most fundamental research. During a long period of restructuring under severe budgetary constraints in the 1980s, there was widespread concern that NRC had not responded adequately to the more strategic mission set by government.* Despite earlier research successes (e.g. in digital technology), parts of the organization had become moribund, with vested interests in established laboratories and divisions making it difficult to pursue new priorities or to curtail existing programmes (see NRCTG, 1987). NRC also came under criticism for inadequate financial accountability (ibid., pp. 13–14) and for the quality of its basic research relative to universities, with NABST (1988a–e) recommending various measures to improve the planning, management and evaluation of its R&D activities. In response, NRC has been developing a more focussed 'leadership role in the evolution, maturing and selection of new important technologies' (NRCTG, 1987, p.11). It operates a network of laboratories in strategic fields like biotechnology and materials science, and funds a number of programmes promoting the diffusion and exploitation of science and technology. NRC has also retained responsibility for big science. Its 1987 budget was around C$470 million, 70 per cent of which was devoted to intramural work (NRC, 1987a).

A policy of decentralizing responsibility for decision-making to divisions and laboratories meant that there was previously little emphasis on research foresight. The situation began to change around 1980 when NRC was required to produce its first five-year plan. This was prepared by an internal working group following informal consultations with senior laboratory staff.

Besides being used for advocacy purposes, it provided an opportunity to situate future R&D programmes in a wider economic and social context. It also resulted in consideration being given to the potential utility of research forecasting:

> It is clearly not possible to predict what advances will result from basic research activities twenty years from now. But the future course of technology can be predicted with more confidence, for about a decade or so, since many technological innovations that are likely to be available in five years are already in an advanced state of development ... An important role is to evaluate medium- (5–10 years) and long-term (10–20 years) options for scientific and engineering research, and to select those most likely to contribute to long-term economic development. [NRC, 1980, p.15]

One recommendation of the first plan was that a higher proportion of the budget should be allocated to longer-term work on core technologies and related basic sciences, but the emphasis on targeting stopped short of identifying specific priorities. It was also proposed that the time-horizon in future plans be extended to ten years. Because NRC from the outset tried to use the results for strategic purposes (in marked contrast to NSERC), the plans have been updated annually. The first update in 1981, for example, responded to the government decision to fund the lowest of three budgetary options proposed the previous year, emphasizing research priorities meeting identified national needs (e.g. cold-region engineering) and the establishment of formal mechanisms for assessing new programme initiatives. This move towards targeting was reinforced in the next update when the priorities identified included major investments in laboratories for advanced materials and biotechnology (NRC, 1983).

The second long-term plan was framed in the context of government guidelines that greater effort should be devoted to strategic research, technology transfer, and collaboration with industry and universities (NRC, 1985). Among its main proposals was the need for institutional restructuring to provide 'a more suitable framework within which to define priorities, develop strategic objectives and establish goals for organizational units' (NRC, 1985, p.38).[30] The plan attempted to justify arguments for increased funding with data on the socio-economic impact of previous NRC research,[31] and results from a study of technological needs and opportunities up to AD 2000 (NRC, 1984, p.1).

The latter study was organized by the NRC Industry Policy Analysis Group which convened a committee of nine senior managers and in-house researchers. Their first task was to construct a list of potentially important technologies, drawing upon reports produced in Canada and abroad. After consulting several leading NRC scientists, they agreed definitions of each technology area and narrowed the list to sixteen topics (e.g. silviculture). Around 200 NRC technology advisers[32] next interviewed industrial and other research users to obtain views on the economic potential of individual technologies. The results were synthesized by advisers and submitted to the committee together with a smaller number of returns received from NRC laboratory directors and heads of research groups. They were then analysed by the committee using a matrix approach to depict the nature and extent of impact of each technology on different socio-economic sectors (in terms of estimated contribution to GDP). Technologies were classified according to their likely overall impact (ranked on a three-point scale) and their pervasiveness across sectors. In addition, an evaluation was undertaken of areas where threats might arise which called for concentrated research efforts. The final report proposed that resources be targeted on six 'key technologies' showing special promise (e.g. microstructures) and that closer collaboration be developed with other agencies (NRC, 1984).

NRC officials believe the exercise stimulated discussion on generic technologies and the restructuring needed to permit an effective response to emerging opportunities. The six priorities figured prominently in the 1985 long-term plan, while microstructure technology, the importance of which had previously gone unrecognized, soon received increased support (e.g. for work on thin-film architecture and surface phenomena). NRC had also been unaware that its generous support for aviation technology was out of line with the minor contribution of the aerospace sector to GDP. Funding was consequently reduced and researchers encouraged to develop other applications for underlying technologies like aerodynamics.*

As in many attempts to introduce foresight into priority-setting, a number of problems arose which limited the impact of the results. One was that insufficient attempts were made to involve the wider NRC research community in developing consensus on priorities, giving rise to resentment among laboratory staff, especially in areas later subject to budgetary cuts. This was not helped by the fact that the process by which data on likely economic impacts were produced was unclear. Second, the use of pre-defined technology categories to structure the survey exercise implicitly discriminated against emerging interdisciplinary areas, which can generally only be identified through a more interactive procedure. Moreover, the impacts are typically diffuse and difficult to quantify, and are therefore not always amenable to evaluation by the type of econometric matrix-based method employed.[33] Finally, because too little effort was put into diffusing the results, there was no real debate on their broader implications either within or outside NRC.[34] Nor was there any opportunity for other agencies like SCC (which later adopted a similar matrix approach) to learn from the difficulties encountered in the exercise.*

NRC has, however, built upon this experience in extending its foresight activities. Follow-up reports (e.g. on superconducting materials) have been produced exploring opportunities within the six priority areas in greater detail. The long-term benefits likely to arise from the national programme on advanced industrial materials have also been evaluated using a procedure in which individual fields were rated in terms of their 'domestic relevance' and the existence of effective 'technological embodiers' (i.e. internationally competitive companies). The results are shown in Table 8.2, the wide-ranging impact of polymer and composite technology coming as something of a surprise to NRC staff.* Overall, benefits of C$1,400 million were estimated as likely to emerge from the programme, far in excess of its C$90 million cost.[35]

The increased importance of research foresight has been reflected in the decision to establish a Technology Assessment Office within NRC, one task of which will be to carry out longer-term appraisals of new technologies and the underlying fields of science. In addition, overseas monitoring of

R&D Area \ Industrial Sector	Agri-food	Wood	Primary Metals	Non- Metallic Minerals	Petroleum Coal	Chemicals	Transport	Machinery	Electrical Electronic	Other manuf. & services	Sub-totals (points)	OVERALL ECONOMIC IMPACT
Polymers & Composites	1	2	1	1	1	2	3	2	3	2	18	$290M
Membranes	1					2				1	4	$80M
Optical & Electronic Materials	1			2		1	2	2	3		11	$175M
Ceramics			1	1		1	1	3	2	1	10	$160M
Superconductors				1		1	2	2	2	1	9	$140M
Advanced Materials Processing		1	2	2		1		2	2	1	11	$175M
Space & Aeronautical		1	1	1			1	1	1	1	7	$110M
Surface Transportation	1	1	1	1	1			1	1	1	8	$130M
Cold Environment	1	1	1	1	1		1	2	1	1	10	$160M
Sub-totals (points) ECONOMIC IMPACT	5 $80M	6 $95M	7 $110M	10 $160M	3 $48M	8 $130M	10 $160M	15 $240M	15 $240M	9 $140M	88	$1400M

Source: NRC (1987b, p. 4)
Key: Impact is rated on four-point scale, with three being the highest and absence of any rating representing a neglible effect.

Table 8.2 Likely impact of areas within advanced industrial materials programme: ratings of researchers and estimated economic benefits for Canada 'developments in technologies significant to Canadian industry' (NRC, 1985, p.53) is to be expanded in order to pick up emerging trends at the earliest possible opportunity. NRC has explicitly decided against attempting to predict potential breakthroughs. Apart from the considerable uncertainties involved, it is thought unlikely that successful pre-emptive investments in very basic areas would result in significant domestic advantages because of external technology transfer by the foreign multinationals which dominate the Canadian economy.*

5. FORESIGHT IN INDUSTRY

Approaches to foresight within the private sector are probably most developed within Bell Northern Research (BNR). Owned jointly by Bell Canada and Northern Telecom, BNR has grown by 10 per cent per annum in recent years. With 6,000 staff and a budget of around C$600 million in 1987, it is Canada's largest industrial research organization. While R&D is primarily product-oriented, 10 per cent of funds are allocated to the Corporate Development Group which is charged with strategic technology planning and initiating exploratory research in areas with high commercial potential (e.g. gallium-arsenide technology and biomaterials). BNR has relied upon a relatively decentralized approach to selecting new lines of R&D, experience having shown the dangers of top-down planning.[36] The criteria employed are explicitly market-oriented and relate to such questions as: how does the proposed development fit within our technology systems portfolio? Do we need to sell the product, and, if so, is it essential to make it ourselves? In which fields is it necessary to be at the technological forefront and what R&D resources are required? The investment to gain a world lead in a particular technology is considerable and work of this type accounts for most long-term research funds. The company's previously established dominance in certain areas resulted in a tendency to see the technological future as being created largely in-house, and monitoring of competitors was until recently confined mainly to AT&T and IBM. However, concern has surfaced about failing to respond promptly to challenges such as those from Japanese firms, with their much larger investments in basic technologies.*

BNR's planning system operates on three time-scales — short-term (one year), medium-term (three to five years) and long-term (up to fifteen years). One component of the last of these (which is carried out by the Corporate Development Group) is the comprehensive vision of future developments produced every ten years. In 1975, for example, *Digital World* foresaw the move into integrated digital communications earlier than most, and Northern Telecom's investment in the technology enabled it to achieve rapid growth in the late 1970s as successive families of digital systems were brought to the market. The 1986 vision, *Open World*, predicted that the principal trend up to the year 2000 would be increased flexibility and compatibility between telecommunications systems (for text, data, voice and images) within an integrated digital services network. The exercise was carried out primarily through consultation with in-house R&D staff, and included studies of likely trends in technologies central to the company's future (e.g. optoelectronics and software engineering) as well as potential developments in less certain fields (e.g. the use of organic molecular structures for parallel processing). These analyses were set in the context of long-term growth projections for different product sectors and the effects of changing lifestyles and other factors on market demand.

Given BNR's preference for decentralized decision-making, the main impact of long-term plans has been to provide a holistic perspective for strategic technological choices rather than a means of framing detailed priorities. Experience has shown that the systemic nature of technological progress in telecommunications tends to limit the impact of individual scientific breakthroughs, and reasonably reliable predictions for important fields can therefore be made up to ten or even fifteen years ahead. While the quality of researchers recruited is paramount, the company also attempts to identify the generic skills required over the next ten years (these are currently in such fields as real-time software, computer architecture and high-speed optoelectronic signal processing).

BNR has recognized that the rapid progress being made in novel basic technologies (especially in Japan) requires more sophisticated approaches to R&D monitoring, as well as better links with the scientific community. It understands that 'to partake in the global knowledge party, you have to pay for an entry ticket',* and the organization's annual output of 500 scientific publications is therefore seen as the price of invitation to the specialist conferences where breakthroughs are often first reported. In addition, up to ten academic scientists at any one time spend their sabbaticals in the company, and twelve new research professorships have been established in key fields. BNR views its annual expenditure of C$2.5 million on research in leading academic departments as, first and foremost, an investment in keeping open the supply line for trained scientists. Overall, BNR regards itself as having an effective planning structure for linear technology development, but sees the need to strengthen longer-term foresight to cope with increasing uncertainty in various areas of strategic research.*

The second of the organizations considered, the Canadian Advanced Technology Association (CATA), seeks to advance the interests of its 250 members through improved legislation and infrastructural support for R&D. Although this mainly involves providing a focus for industrial lobbying of federal agencies, two recent initiatives have involved research foresight. The first concerns the National Technology Policy Roundtable — a series of workshops attended by 200 leading experts from companies, government and universities held during the 1986 Canadian High Technology Week. The purpose was to discuss research strategies for winning the 'war for national economic survival in which technological leadership is ... the strategic factor in determining national economic performance'. Government and industry were urged to take steps 'quickly and decisively to reorient our priorities: science and technology must become the cornerstones of a national strategy for economic renewal' (CATA, 1986, p.2). In an increasingly knowledge-intensive society, the government needed to strengthen the science and technology infrastructure, in particular to train the researchers likely to be required over the next decade. Although the roundtable discussions were rather general, they contributed to shaping the consensus now emerging within industry on the importance of longer-term R&D, as well as reinforcing the need for federal policy initiatives like InnovAction.*

The second foresight activity has involved CATA in helping coordinate attempts by industry to develop a more strategic approach to R&D in resource-based sectors. Although strong in international terms, research in the area has tended to be rather applied and site-specific. Firms now realize that, in order to remain internationally competitive, they must add greater value to primary products. This in turn depends upon mastering the emerging basic technologies, a task which requires collective effort since most companies are individually too small. CATA therefore participated in an exercise by the Mining Association of Canada to identify high-technology needs for the sector. This was carried out by a committee of R&D directors from major companies which interviewed experts in mineral extraction and processing fields to identify technological gaps between desired and existing production systems. The results revealed surprising consensus on both technological needs and the likely impact of developments in areas like biotechnology and automated materials-handling. There was also agreement that efforts to overcome emerging technological bottlenecks would require inputs from basic research (e.g. in sensor technology and artificial intelligence — MAC, 1985). As a result, a consortium of firms decided to fund a university–industry programme (MITECH) supporting academic centres of excellence working on relevant longer-term research. Such a development would have been almost inconceivable only five years ago, and illustrates the influence of recent debates on the role of R&D in economic regeneration.*

The mining initiative was followed by an evaluation of R&D opportunities and needs in forestry. Undertaken jointly with the British Columbia Science Council and MOSST, it brought together executives from forestry and high technology companies to consider options for developing novel 'Canadian-made technologies'. Officials from federal and provincial government departments, funding agencies and universities were also invited. To provide background information, a consultant was commissioned to prepare a review of emerging technologies in different forest products sectors. This specified a number of innovations likely to be required before the end of the century (e.g. in chemical pulp processing) and pointed to a need for long-term R&D in fields such as control technology, robotics and remote-sensing (CATA/SCBC, 1987). The forum therefore recommended joint action by industry and government to improve the linkages between the forestry industry and knowledge-based sectors like electronics. In particular, it stressed the importance of developing 'a long-term strategy to identify problems and goals, necessary targets for high-technology, and areas for special funding' (ibid., p.3), with the Japanese Fifth Generation Computer approach providing a possible model.[37] The main effect was to reinforce federal and provincial government attempts to coordinate policies for targeting R&D, and to highlight the need for further prospective analysis of the cross-field and intersectoral impacts of emerging technologies.*

6. CONCLUSIONS

Although many of the lessons emerging from our review of foresight in Canada are the result of less than successful experience, progress has undoubtedly been made despite an environment until recently hostile to the framing of longer-term research policy. Some of the lessons are similar to those from other countries, but there are important differences stemming from Canada's federal structure and its unique economic, socio-cultural and geographical situation.

(a) For medium-sized economies like Canada, many aspects of foresight are country specific. The probability of realizing a particular future depends upon the relative international strength of the research community, the existence of established science-based companies capable of exploiting novel results, and on other factors like natural resources and climate. Consequently, Canadian foresight studies have often focussed on the R&D needed to add value to primary-sector production or to capitalize upon opportunities in niche areas like environmental or cold-region engineering.

(b) At the national level, foresight is unlikely to influence policy appreciably unless there is a prior commitment to planning for the longer term.[38] Such a commitment only began to emerge in the 1980s when 'the combination of national reconciliation (with provincial governments), political will regarding science and technology, and consensus building among the country's social partners (industry, labor, academics), meant that the conditions for national science and technology policy were ripe' (Dufour and Gingras, 1988, p.16). This provided a context for the establishment of mechanisms enabling the federal government, provinces and industry to agree that greater priority should be given to promising fields of strategic research. Although some use has been made of systematic techniques like cross-impact matrices and Delphi methods in identifying target areas, the emphasis has been on consensus-seeking approaches employing committees, workshops and informal consultation. This is one consequence of a federal R&D system in which the priority-setting process tends to be highly politicized because of the involvement of numerous powerful stakeholders with interests reflecting divergent regional needs and cultural traditions.

(c) A major factor in the introduction of foresight has been the linkage of longer-term planning with budget-setting. This underlay the increased emphasis on prospective analysis by NRC when it became evident that the federally-mandated five-year corporate plan had real implications for future funding from government. As a result, new initiatives have been developed in emerging fields like microstructures, and activities reduced in established programme areas (e.g. aerospace R&D) which would otherwise have been difficult to cut. In contrast, the distribution of institutional funding for academic research largely on the basis of student numbers has proved inimical to longer-term planning within the university sector, while the lack of any formal integration of advice on national priorities with decisions on the overall federal R&D budget is one reason NABST has so far made little use of systematic macro-level foresight.

(d) More specifically, problems have arisen in several foresight exercises because of a failure to identify the audience for the results and build in implementation structures at a sufficiently early stage. Given the Canadian emphasis on consensus, it is crucial that all relevant interests are consulted during the process of initiating and carrying out studies. The lack of any effective consensus-building mechanism in the National Consultation is now viewed by SCC as a mistake, but one from which valuable lessons could be drawn for its follow-up study to determine priorities for the NSERC Strategic Grants Programme.

(e) The example of SCC also shows how the institutional location of an organization can determine the types of foresight it is best placed to carry out. In particular, SCC's independence from government has enabled it to question the long-term wisdom of federal R&D policies (e.g. in the environmental field). However, that independence also made it difficult to maintain an influence on high-level policy formulation after MOSST assumed a broad coordinating function and NABST was established. Likewise, the decision not to transfer MOSST's responsibilities for federal R&D policy to a new central agency when ISTC was created seems likely to constrain any attempt to launch foresight initiatives aimed at identifying overall government research priorities. This is because of the conflict of interest with other ministries which support R&D. On the other hand, ISTC may find it easier to extend the use of foresight in determining policy for those areas of basic and strategic research it administers.

(f) Another lesson from SCC relates to the difficulty faced by independent agencies in maintaining a balance between contributing critical scientific advice and responding to the longer-term policy problems regarded as important by the paymasters in government. The halving of SCC's budget in 1986 stemmed in part from doubts concerning its political neutrality. In addition, the organization had left itself vulnerable to other types of attack, such as over whether too much emphasis had been given to work highlighting the adverse effects of R&D.

(g) While some of the criticism of SCC by federal agencies seems well-founded, undue stress appears to have been placed on evaluating its foresight activities in terms of their short-term utility for policy formulation. The impact of studies oriented to providing an early warning of emerging societal threats is typically long-term and diffuse — contributing background information to framing policy, generating awareness about emerging needs and opportunities, and helping to catalyse the formation of consensus. Assessments of foresight therefore need to take into account criteria other than immediate influence on policy-making. The recent re-emergence of research on the environment as a central political issue has demonstrated the importance of previous SCC foresight in this area and has resulted in greater prominence being given to its activities within government.

(h) A strength of early foresight initiatives by SCC and NRC was their recognition that a holistic approach is required to accommodate the interdisciplinary nature and pervasive impact of emerging strategic fields. Studies focussing on the opportunities arising from the fusion of technologies such as forestry and advanced electronics have also proved productive, as was demonstrated by the workshop series which CATA helped organize. However, the approaches and time-horizon employed in fields like IT, which are characterized by a systematic development trajectory and national corporate technological strength, are different from those appropriate in biotechnology which is science-driven and less well-established as an industry in Canada. An important issue in such strategic fields has been the demand for specialized researchers, especially by industry, following a growth in confidence that generic skill requirements can be predicted fairly reliably up to ten years ahead.

(i) It is widely recognized in Canada that foresight needs to take into account both 'science and technology push' and end-user or 'receptor' needs, although the relative emphasis given to each will vary with the nature and orientation of different research fields. While there has been disagreement within agencies over exactly where this balance should lie (e.g. in respect of policy for the NSERC Strategic Grants Programme), the earth sciences review demonstrates that academic scientists can accommodate to demands for socio-economic relevance without jeopardizing their intellectual freedom.

(j) The introduction of foresight to aid decision-making in academic funding agencies has, as elsewhere, been resisted by researchers concerned with the threat posed to their autonomy. This was minimized in NSERC, for example, by confining the assessment of prospective options to broad areas only, thereby ensuring scientists continued to retain significant freedom in proposing their own research. The worries of scientists have also been reduced by linking the establishment of selective support mechanisms to the provision of new funds, and using an 'open category' in targeted initiatives to guarantee flexibility in responding to emerging opportunities.

(k) Experience in Canada highlights the importance of involving a 'third party' or 'external broker' in foresight, particularly in relation to achieving acceptance of the results by government, industry and the wider research community. This was evident, for example, in the role of SCC in the priority-setting exercise for the NSERC Strategic Grants Programme, and in the function played by industry in planning exercises by MOSST for longer-term initiatives in biotechnology and information technology where the direct beneficiaries of research funds have mainly been in the public sector. Similarly, the Canadian Plastics Institute functioned as a broker in the study on emerging plastics technologies, enabling SCC to bring together researchers from government, academia and industry to produce an influential report identifying nationally important longer-term research opportunities. In comparison, devolving responsibility for an exercise wholly to the scientists who will be affected by any recommendations tends to lead to reports dominated by existing disciplinary and institutional interests (as typified by most NSERC field surveys) and a reluctance to identify priorities other than for considerably increased overall funding.

(l) Although the design of an appropriate consultative process is an essential precondition for success in foresight, the importance of methodology and technique should not be underestimated. For example, the use of a matrix-based approach combining judgements on pre-defined technology categories with estimated economic impact introduced a conservative bias in the NRC survey, constraining the possibilities for identifying emerging cross-disciplinary areas. This was also a major limitation of the NSERC physics review which employed committees based on traditional subfields. The existence of such problems in several agencies highlights a pressing need that has emerged within Canada for professional competence in science policy, forecasting methodology and R&D indicators.[39] The previous lack of any academic institute specializing in these areas has prevented organizations like ISTC from developing a capacity to explore certain types of foresight approach — for example, the use of information-science techniques to monitor trends in R&D.

Finally, it is important to note that, despite the setbacks experienced, there has been a genuine desire within government agencies and departments to learn from previous foresight exercises. Officials have gradually developed a better appreciation of the role of foresight in policy-making, how to organize it most effectively, and the advantages of different approaches. This is well illustrated by the positive way in which SCC ensured that mistakes in the National Consultation were not repeated in later studies for NSERC. As a result, it is now widely recognized that 'in S&T policy-making, where norms are uncertain and goals are frequently not clearly defined, the very process of policy research is often as important as the product' (Steed, 1988, p.180). Nevertheless, a lengthy learning process still remains, especially in relation to gaining agreement from provinces and other powerful stakeholders to set aside their specific interests so that macro-level foresight can play a part in setting national research priorities. As NABST (1988d, p.4) has stressed, '[success] does not happen automatically. Canada will have to work hard to create the right institutions of consultation to overcome the deeply rooted "zero-sum" mentality of interest groups.'

NOTES

* Refers to material collected during interviews in November 1987, while † denotes information drawn from subsequent correspondence.

1. Canadian science policy is reviewed in various articles included in de la Mothe (1988) and in recent reports by the National Advisory Board on Science and Technology (NABST, 1988a–f). Emerging problems with R&D were also discussed in the Canada Tomorrow Conference (MOSST, 1983a).
2. These figures differ from those reported in other official statistical series (e.g. MOSST, 1987a) because of revisions made in 1987 following improved coverage of R&D in smaller firms.†
3. Although some provincial governments have made considerable use of prospective techniques in research policy-making, it was unfortunately not possible to review these activities in detail. The emphasis in this chapter is very much on federal agencies, which means that our coverage of foresight in Canada should not be regarded as comprehensive.
4. As Mulroney (1987) has stressed: 'We must now all change, to try with every means at our disposal to prepare Canadians and the Canadian economy for the challenges of the 1990's and the new century beyond ... In this new age, we must compete with our brains, our ingenuity and our creativity ... This is why my government put science and technology at the top of the national agenda. We have created the National Advisory Board on Science and Technology [which] is going to become an intellectual laboratory, a continuous brain-storming session to build and improve Canada's scientific and technological goals, policies and programs.'
5. The requirement for many of these new policy structures was identified by the Task Force on Federal Policies and Programmes for Technology Development (MOSST, 1984) and in MOSST (1985).

6. In addition, it is envisaged that the *Annual Overview of Federal R&D* will play a wider role in decision-making, being submitted to the Finance, Privy Council Office and Treasury Board Secretariats for use in fiscal planning and budgetary discussions (MOSST, 1987d, p.16).
7. This was initiated following discussions between provincial and federal science and technology ministers in February 1985 (MOSST, 1985). It was their first meeting since 1978, highlighting the previous problems in coordinating research policy (CSTM, 1987).
8. Apart from the Prime Minister and eight ex-officio representatives (who include the Minister of State for Science and the heads of agencies like the NRC), the majority of NABST members are eminent academics and industrialists drawn from a cross-section of provinces (Office of the Prime Minister, 1987).
9. However, the University Committee stressed that increased targeting should not come at the expense of basic science, and this required a doubling of Canadian expenditure on higher education research over coming years (NABST, 1988b).
10. Although no direct use was made of formal foresight techniques, a variety of prospective studies by the Science Council of Canada and National Research Council helped 'set the scene' for the biotechnology initiative.†
11. Consequently, new initiatives like the *Annual Overview of Federal R&D* are now viewed as being in danger of degenerating into 'just another addition to the internal government paper burden unless it ... [becomes] a dynamic document with real implications for resource allocation' (NABST, 1988e, p.3).
12. This description of SCC foresight activities is drawn in part from Steed (1988) who analyses the types of impact that can be made by science-policy research.
13. The total costs of the EST programme were just under $1 million, of which MOSST provided around a quarter.*
14. See the discussion in Chapter 2 of the three different types of research foresight ('responsive', 'reflective' and 'participatory awareness') identified by Elzinga (1983).
15. Particular difficulties were encountered in relating industrial classification schemes to technology fields.*
16. The sample was selected to ensure satisfactory coverage of major technological and industrial sectors, as well as geographical regions. Further interviews were carried out with forty-one researchers in small information technology firms and in technology centres to obtain more specific views on R&D options and needs in the IT field (Steed and Tiffin, 1986, p.30).
17. For example, an EST was defined as a non-commercialized technology either in the process of emerging from the laboratory or likely to do so within the next five years.
18. Unpublished analyses of the results were, however, also undertaken by MOSST and the Ontario government.†
19. A conceptual overview of the foresight methods and techniques employed in the National Consultation, together with an assessment of the lessons learned, is provided by Elzinga (1985; 1986).
20. SCC had also planned to undertake several sectoral studies, but only the first on forest products (Hayter, 1988) proved feasible after government budget cuts. This demonstrated how the international competitiveness of the industry could be enhanced by improving the R&D infrastructure, and was reportedly well received by policy-makers.†
21. As one government official stressed, 'the major mistake SCC has made is to forget who pays the bills. They respond too much to proposals for studies made by Council members'. As a result, NABST filled some of the gaps in science-policy advice which might well have been within the remit of SCC.*
22. This conflict is part of a wider problem stemming from what one senior official described as 'the chaotic nature of Canada' in respect of determining longer-term

priorities for research. 'In this key area ... there is, at present, no accepted leadership by any key group in the country to establish a coherent national approach for research and development policies and foresight.' Instead, the situation is one of 'MOSST criticizing the Science Council, academics criticizing everybody, and the granting councils setting out on their own'.†

23. According to this viewpoint, the primary reason for the 1986 budget cuts was the lack of a responsible minister able and willing to defend SCC in the face of the growth in MOSST's own policy efforts and the plans for NABST.*
24. As the President of Northern Telecom has stressed, 'Canada must demonstrate its commitment to higher education as a force for competitive renewal and industrial regeneration. Nothing less than our future prosperity is at stake' (Vice, cited in CATA, 1987, p.1).
25. Differing perspectives on these problems are provided by NSERC (1983), Johnson (1985), MOSST (1987b), Davis (1988), CATA (1987), SCC (1988d) and NABST (1988b).
26. The top fifteen of the sixty-seven Canadian degree-granting institutions gained around 80 per cent of research council grants in 1987.*
27. This is despite the fact that the Physics and Astronomy Programme Advisory Committee has a specific mandate 'to make recommendations on the long-term planning of NSERC funding in the area' (NSERC, 1987b, p.2). It was hoped that new guidelines might result from the subatomic physics review.*
28. Although not directly used in selecting priorities, the study helped define the criteria to be employed (Halliwell, 1985, p.3).
29. As an SCC official stressed: 'The question of process was crucial since you build in commitment to the results by getting this right. We didn't want to get criticized for talking to the wrong people. This is why we first had to gain agreement on who should be interviewed.'*
30. Formal 'decision rules' have been established for evaluating proposed R&D programmes, including six criteria for assessing generic technologies (e.g. 'potential for significant impact' and 'scientific and technical merit' — NRC, 1986).
31. Two studies were carried out by consultants. The first employed cost–benefit techniques to demonstrate high payback to industry from NRC R&D, together with considerable benefits in the form of improved safety, health and quality of life. The second surveyed 500 companies and found 80 per cent had made substantial use of NRC research results (NRC, 1985, pp.30–1).
32. Located throughout Canada, these provide advice to mainly smaller companies and monitor emerging R&D needs.
33. The conservative nature of the exercise was perhaps accentuated by the reliance on inputs from technological advisers not actively involved in R&D.*
34. Critics argued that the results should have been made public and that informal circulation to organizations like MOSST of a short mimeographed summary was insufficient.*
35. Technical doubts were, however, expressed about the methodology employed in this assessment, especially concerning the precision possible in estimating the impact of more basic research.*
36. In the case of semiconductor lasers, for example, R&D carried out covertly by researchers in the early 1970s after a management decision to cease support subsequently proved extremely important to BNR.*
37. Further seminars were planned on technological opportunities and needs in specific forest-product sectors.
38. There was undoubtedly previously some truth in the view that 'never was so much written by so few with so little impact as a burgeoning literature on science and technology planning' (Jarrell, 1985, cited in Dufour and Gingras, 1988, p.14).

39. As a result, SSHRC and NSERC have recently agreed to fund jointly postgraduate training awards in the field of science policy.

REFERENCES

Adams, G.H.M. *et al.* (1986), 'Towards Excellence in Chemistry: The Report of a Committee to Study the Health of Research in Chemistry in Canada', *Canadian Chemical News,* vol.38, no.8, Special Supplement.

Anderson, F., Bobkowicz, A. and Gerson, F. (1985), *Emerging Plastics Technologies and the Canadian Plastics Industry,* Ottawa, Science Council of Canada.

Blais, R.A. (1979), *Report of Delphi Study on Areas of National Concern for University Research in Natural Sciences and Engineering,* Montreal, École Polytechnique.

CAP (1981a), *Study of Future Research Opportunities in Physics: Atomic and Molecular Physics,* Ottawa, Canadian Association of Physicists.

—— (1981b), *Future Research Opportunities in Canada in Plasma Physics and Fusion Research,* Ottawa, Canadian Association of Physicists.

—— (1981c), *A Study of Future Research Opportunities in Nuclear Physics: Intermediate Energy Physics and High Energy Physics,* Ottawa, Canadian Association of Physicists.

—— (1981d), *Future Opportunities in Condensed Matter Physics Research in Canada,* Ottawa, Canadian Association of Physicists.

—— (1981e), *Study of Future Research Opportunities in Physics: Optics,* Ottawa, Canadian Association of Physicists.

—— (1982), 'Review Meeting on NSERC Study of Future Research Opportunities in Physics' (mimeo), Ottawa, Canadian Association of Physicists.

CATA (1986), *Report of the National Technology Policy Roundtable,* Ottawa, Canadian Advanced Technology Association.

—— (1987), *The People Prescription: Educating for the Knowledge-Based Economy,* Ottawa, Canadian Advanced Technology Association.

CATA/SCBC (1987), *Advanced Forestry Technology Forum: An Executive Forum on Science and Technology Issues and Developments,* Vancouver, Science Council of British Columbia.

CGC (1986), *Lithoprobe: The Third Dimension — Phase II Proposal,* Ottawa, Geoscience Council of Canada.

CSTM (1987), *The National Science and Technology Policy,* Ottawa, Council of Science and Technology Ministers.

Davis, C.H. (1988), 'Science and Technology Policy in Quebec', *Science and Public Policy,* vol.15, pp.26–34.

de la Mothe, J. (ed.) (1988), 'Science and Technology in Canada', *Science and Public Policy,* vol.15, pp.13–65.

—— (ed.) (1989), 'Science and Technology Policy under Free Trade', *Technology in Society,* May, pp.3–200.

DPA Group (1986), *Evaluation of the NSERC Operating Grants Program,* Ottawa, Natural Sciences and Engineering Research Council.

—— (1987), *Evaluation of the NSERC Strategic Grants Program,* Ottawa, Natural Sciences and Engineering Research Council.

Dufour, P. (1985), 'The Science and Technology Policy Orientations of Provincial Governments' (mimeo), Ottawa, Ministry of State for Science and Technology.

—— and Gingras, Y. (1988), 'Development of Canadian Science and Technology Policy', *Science and Public Policy,* vol.15, pp.13–18.

Elzinga, A. (1983), *Foresight as Anticipatory Intelligence: On the Organization, Structure and Methodology of Foresight Activities for Science and Technology Policy,* Ottawa, Science Council of Canada.

—— (1985), 'Mandated Science in the Universities and Picking the Winners: What are the Implications?', in Becher, T. *et al.* (eds), *Higher Education and Research*

Organizations, Report No.34, Department of Political Science, University of Stockholm.

—— (1986), 'Foresighting Canada's Emerging Science and Technologies', in Brotchie J.F. *et al.* (eds), *Spatial Impact of Technological Change,* London, Croom Helm.

—— and McCutcheon, S. (1985), *Advanced Industrial Materials: Canadian Perspectives and Opportunities,* Ottawa, Science Council of Canada.

Fyfe, W.S. and Rust, B.R. (1981), 'The Next Decade of Earth Science Research in Canadian Universities: Proceedings of the Earth Science Workshop, 1981', *Geoscience Canada,* vol.8, no.3, pp.113–6.

Geoscience Canada (1981), 'Special Issue on Canadian Earth Sciences in the Eighties', vol.8, no.3.

Halliwell, J.E. (1985), 'The Evolution of the Strategic Grants Program' (mimeo), Ottawa, Natural Sciences and Engineering Research Council.

Hayter, R. (1988), *Technology and the Canadian Forest-Product Industries: A Policy Perspective,* Ottawa, Science Council of Canada.

ICPD (1988), *Networks of Centres of Excellence: Competition Announcement,* Ottawa, Inter-Council Program Directorate.

Jarrell, R.A. (1985), 'Government, Technology and Industry: The Emergence of the Canadian Pattern', *Atkinson Review of Canadian Studies,* vol.2, no.2, pp.26–32.

Johnson, A.W. (1985), *Giving Greater Point and Purpose to the Federal Financing of Post-Secondary Education and Research in Canada,* Ottawa, Secretary of State.

Katz, S. (1987), *Emerging Sensor Technology: The Promise and the Threat,* Ottawa, Science Council of Canada.

MAC (1985), *High Tech Development Needs in the Canadian Mining Industry,* Ottawa, Mining Association of Canada.

Mascolo, D., Wright, P.M. and Slemon, G.R. (1985), *Engineering Education in Canada,* Ottawa, Science Council of Canada.

Mitchell, G. and McBean, E. (1985), *Water Resources Research in Canada: Issues and Opportunities,* Ottawa, Canadian Ministry of Environment.

MOSST (1981), *Biotechnology: A Development Plan for Canada,* Ottawa, Ministry of State for Science and Technology.

—— (1982), *Annual Report 1981–1982,* Ottawa, Ministry of State for Science and Technology.

—— (1983a), *The Canada Tomorrow Conference — Proceedings,* Ottawa, Ministry of State for Science and Technology.

—— (1983b), *Towards 1990: Technology Development for Canada,* Ottawa, Ministry of State for Science and Technology.

—— (1984), *Report on the Task Force on Federal Policies and Programs for Technology Development,* Ottawa, Ministry of State for Science and Technology.

—— (1985), *Science, Technology and Economic Development,* Ottawa, Ministry of State for Science and Technology.

—— (1986a), *National Biotechnology Advisory Committee, Annual Report 1985–1986,* Ottawa, Ministry of State for Science and Technology.

—— (1986b), *Advanced Materials Technologies: Underpinnings of Industrial Competitiveness,* Interdepartmental Working Group on Advanced Industrial Materials, Ottawa, Ministry of State for Science and Technology.

—— (1987a), *Science and Technology Resource Allocation Statistics 1987/88,* Ottawa, Ministry of State for Science and Technology.

—— (1987b), *University Research in Canada: A Background Paper for the National Forum on Post-Secondary Education,* Ottawa, Ministry of State for Science and Technology.

—— (1987c), *InnovAction: The Canadian Strategy for Science and Technology,* Ottawa, Ministry of State for Science and Technology.

—— (1987d), *A Decision Framework for Science and Technology,* Ottawa, Ministry of State for Science and Technology.

—— (1987e), *Annual Report 1985–86,* Ottawa, Ministry of State for Science and Technology.
—— (1987f), *Ministry of State for Science and Technology: 1987–88 Estimates — Part III Expenditure Plan,* Ottawa, Canadian Government Publishing Centre.
—— (1987g), *University Research in Canada: A Background Paper for the National Forum on Post-Secondary Education,* Ottawa, Ministry of State for Science and Technology.
—— (1987h), *InnovAction: The Federal Microelectronics Strategy,* Ottawa, Ministry of State for Science and Technology.
—— (1987i), *Advanced Industrial Materials: Report on Consultations,* Ottawa, Ministry of State for Science and Technology.
—— (1988), 'Government Announces New Initiatives to Increase Funding for University Research' (mimeo), Ottawa, Ministry of State for Science and Technology.
Mulroney, B. (1987), 'Notes for an Address on Research and Development at the University of Waterloo, March 4th' (mimeo), Ottawa, Prime Minister's Office.
NABST (1988a), *Industry Committee Report,* Ottawa, National Advisory Board on Science and Technology.
—— (1988b), *University Committee Report,* Ottawa, National Advisory Board on Science and Technology.
—— (1988c), *Department of Industry, Science and Technology Report,* Ottawa, National Advisory Board on Science and Technology.
—— (1988d), *Government Committee Report on InnovAction,* Ottawa, National Advisory Board on Science and Technology.
—— (1988e), *Government Committee Report,* Ottawa, National Advisory Board on Science and Technology.
—— (1988f), *Economic Summit Proposal Committee Report,* Ottawa, National Advisory Board on Science and Technology.
NRC (1980), *The Urgent Investment: A Long-Range Plan for the National Research Council of Canada,* Ottawa, National Research Council.
—— (1983), *The Urgent Investment: Year Three,* Ottawa, National Research Council.
—— (1984), *Technological Areas of National Economic Importance,* Industry Policy Analysis Discussion Paper No.45, Ottawa, National Research Council.
—— (1985), *A Practical Perspective: The NRC Plan 1986-1990,* Ottawa, National Research Council.
—— (1986), *A Practical Perspective: First Update 1986,* Ottawa, National Research Council.
—— (1987a), *Annual Report 1986/87,* Ottawa, National Research Council.
—— (1987b), *Estimated Economic Impacts of Proposed Advanced Industrial Materials Program,* Industry Policy Analysis Discussion Paper No.81, Ottawa, National Research Council.
NRCTG (1987), *Report of the National Research Council Task Group,* Ottawa, Ministry of State for Science and Technology.
NSERC (1983), *Brief to the Royal Commission on the Economic Union and Development Prospects for Canada,* Ottawa, Natural Sciences and Engineering Research Council.
—— (1984), *Engineering for Life: Civil Engineering Research Needs 1985–1990,* Ottawa, Natural Sciences and Engineering Research Council.
—— (1985), *Completing the Bridge to the 1990s: NSERCs Second Five-Year Plan,* Ottawa, Natural Sciences and Engineering Research Council.
—— (1987a), *Report of the President 1986-87,* Ottawa, Natural Sciences and Engineering Research Council.
—— (1987b), *Awards Guide,* Ottawa, Natural Sciences and Engineering Research Council.

—— (1987c), *Planning in Sub-Atomic Physics,* Ottawa, Natural Sciences and Engineering Research Council.
Office of the Prime Minister (1987), 'Inaugural Meeting of National Advisory Board on Science and Technology' (mimeo), 16 February, Ottawa, Office of the Prime Minister.
OPC (1986), *Competing in the Global Economy,* Toronto, Ontario Premier's Council.
—— (1987), *Centers of Excellence: The Ontario Experience,* Toronto, Ontario Premier's Council.
RAGHO (1987), *Hydrogen: National Mission for Canada,* Ottawa, Ministry of State for Science and Technology and Ministry of Energy, Mines and Resources.
SCC (1972), *Policy Objectives for Basic Research in Canada,* Ottawa, Science Council of Canada.
—— (1979), *Forging the Links: A Technology Policy for Canada,* Ottawa, Science Council of Canada.
—— (1984), *The Canadian Science Counsellors,* Ottawa, Science Council of Canada.
—— (1985a), *Seeds of Renewal: Biotechnology and Canada's Resource Industries,* Ottawa, Science Council of Canada.
—— (1985b), *Aquaculture: An Opportunity for Canadians,* Ottawa, Science Council of Canada.
—— (1985c), *Social Science Research in Canada: Stagnation or Regeneration,* Ottawa, Science Council of Canada.
—— (1986a), *The National Science and Technology Policy Forum,* Ottawa, Science Council of Canada.
—— (1986b), *A Growing Concern: Soil Degradation in Canada,* Ottawa, Science Council of Canada.
—— (1986c), *Annual Report 1985–86,* Ottawa, Science Council of Canada.
—— (1987a), *Annual Report 1986-87,* Ottawa, Science Council of Canada.
—— (1987b), *A National Workshop on the Role of Epidemiology in the Risk Assessment Process in Canada,* Ottawa, Science Council of Canada.
—— (1987c), *A Sectoral Approach to Innovation: The Case of the Forest-Product Industries,* Ottawa, Science Council of Canada.
—— (1988a), *Priority Topics for Strategic Grants Research,* Report to NSERC, Ottawa, Science Council of Canada.
—— (1988b), *Water 2020: Sustainable Use for Water in the 21st Century,* Ottawa, Science Council of Canada.
—— (1988c), *Environmental Peacekeepers: Science, Technology, and Sustainable Development in Canada,* Ottawa, Science Council of Canada.
—— (1988d), *Winning in a World Economy: University-Industry Interaction and Economic Renewal in Canada,* Ottawa, Science Council of Canada.
—— (1989), *Enabling Technologies: Springboard for a Competitive Future,* Ottawa, Science Council of Canada.
SSHRC (1985), *Five-Year Plan for Financing Research in the Social Sciences and Humanities, 1985–1990,* Ottawa, Social Sciences and Humanities Research Council.
—— (1987), *Annual Report 1986–1987,* Ottawa, Social Sciences and Humanities Research Council.
Statistics Canada (1988a), *Science Statistics,* vol.12, no.4, Ottawa, Statistics Canada.
—— (1988b), *Science Statistics,* vol.12, no.6, Ottawa, Statistics Canada.
Steed, G. (1988), 'Alerting Canadians: The Scope and Use of Policy Research by the Science Council of Canada', *Technology in Society,* vol.10, pp.165–83.
—— and Tiffin, S. (1986), *A National Consultation on Emerging Technology,* Ottawa, Science Council of Canada.
Wheeler, J.O. (1981), 'Canadian Earth Sciences in the Eighties', *Geoscience Canada,* vol.8, no.3, p.105.

9 Research foresight in Sweden

1. INTRODUCTION

The structure of the Swedish R&D system is broadly similar to that of most other European nations, as can be seen from Figure 9.1.[1] Infrastructural support to universities is provided by government on the advice of the National Board of Universities and Colleges (UHÄ), while project-based finance comes from four research councils and the National Board for Technical Development (STU). A particularly distinctive characteristic is the virtual absence of a government laboratory sector, except in defence where clear national needs have to be met. As a result, universities necessarily undertake a certain amount of mission-oriented research funded by numerous sectoral and regional agencies. Private-sector research is carried out principally by companies, but there is also a network of 'cooperative institutes' financed jointly by industry and government.

After a long period in which government stressed the 'relevance' of R&D as its main funding criterion, the emphasis has switched to basic research. Industry has encouraged this shift, recognizing that technical innovation will increasingly depend upon the ability rapidly to master new generations of science-based technologies. Another feature of Swedish research policy has been increasing central coordination, with overall priorities and funding levels now agreed triennially in a Government Research Policy Bill. The extensive consultative process involved has provided an opportunity for a growing coalition of public and private sector interests to argue for additional research funding. Government expenditure is projected to grow in real terms by around 1.5 per cent per annum over 1987–90, much of the increase going to the universities for programmes in priority areas. Industrial research has been rising even more rapidly, particularly in specialized engineering sectors and pharmaceuticals, bringing Sweden closer to the national goal of 3 per cent of GDP devoted to R&D.

In what follows, we look first at the relatively limited foresight activities undertaken by UHÄ and two research councils — the Natural Science Research Council (NFR) and Council for Planning and Coordination of Research (FRN). More methodologically oriented work by academic science-policy departments is also considered. Next, the role of foresight in strategic and applied research is examined, several initiatives of STU, the Royal Academy of Engineering Sciences and the Defence Research Institute being analysed. Finally, we describe research foresight in two large companies before moving on to identify the general lessons that can be drawn from recent Swedish experience.

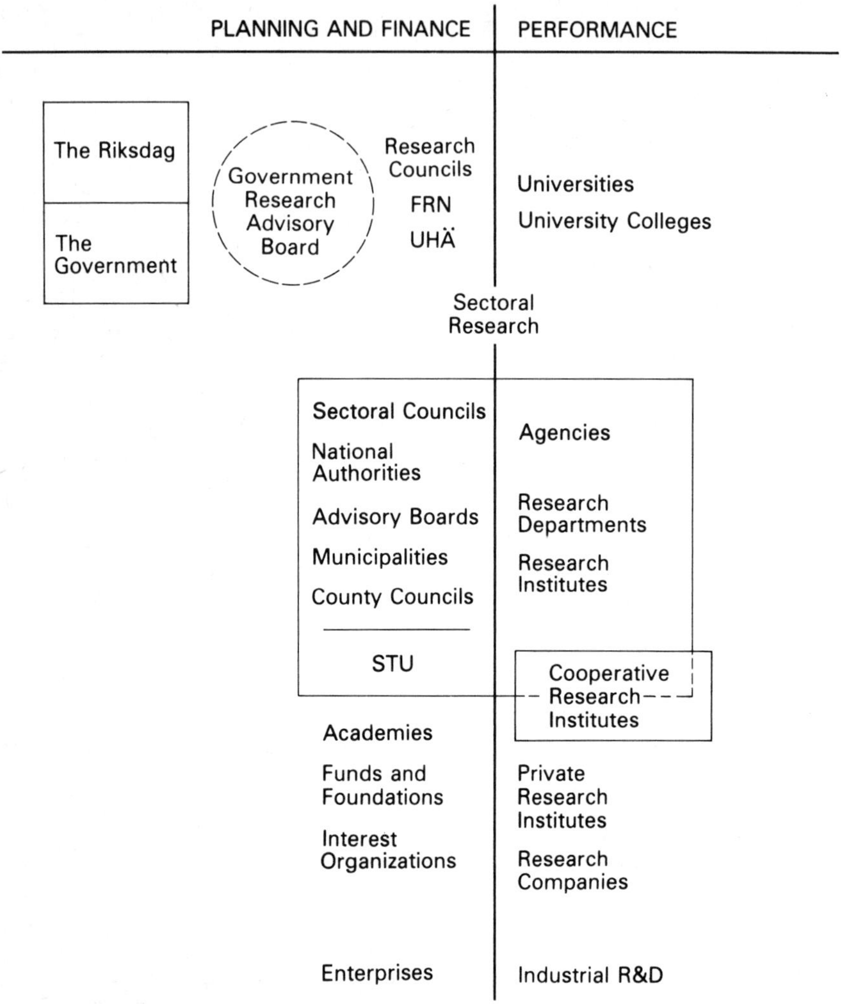

Source: Dyring (1985, p. 11) and, in modified form, in OECD (1987, p. 33). Reproduced by kind permission of the Swedish Institute.

Figure 9.1 Outline of the Swedish R&D system

2. NATIONAL R&D STRATEGY

The government emphasis during the 1970s on 'relevance' resulted in sectoral agencies assuming increased responsibility for distributing public R&D funds. Priorities were set in a relatively decentralized manner in line with agency missions (for example, environmental protection) and an industrial policy stressing general infrastructural support rather than targeting of key sectors. The Conservative administration began to change this policy in the early 1980s following a period of national debate over the causes of Sweden's apparent decline in international competitiveness. One result was the 1982 Research Policy Bill which proposed a variety of measures to improve the coordination and planning of R&D, with overall responsibility vested in the Cabinet Office. The Bill also identified a number of broad priority fields, including several set out in an earlier Parliamentary act on industrial policy (e.g. biotechnology and new materials) and two concerning socio-cultural research (see OECD, 1987, p.15). Most importantly, it proposed that a five-year planning system be introduced, requiring agencies to formulate their R&D plans within a triennial budgetary framework, and thereby enabling government to become more involved in coordinating overall priorities.[2]

The Socialist administration in power in 1984 followed these guidelines in framing a new Research Policy Bill but placed more stress on basic science, and in particular on strengthening the universities. A subsequent 1987 Bill reinforced this policy, expanding the overall level of R&D funding and channelling 'these increases ... primarily into higher education, with the emphasis on recruiting new generations of researchers and work in certain priority areas of research' (Cabinet Office, 1987, p.7). Much of the additional funding has gone to enhancing capabilities in basic technologies like biotechnology (especially cell and molecular biology) and information technology (systems engineering, semiconductor research and IT applications). The increasing stress on earmarked programmes is interpreted by many as reflecting a trend towards centralized control of research although this is nowhere formally stated to be government policy.*

The move to a quinquennial planning horizon encouraged agencies to develop new corporate planning systems which employ research foresight to varying extents. The responses of several agencies to this change are discussed below, the National Board for Technical Development having been particularly successful in introducing systematic approaches for identifying longer-term research priorities. In most cases, however, the triennial budget request remains little more than a document listing existing and projected activities. Consequently, in preparing the 1990 Research Policy Bill, the Cabinet Office will place greater emphasis on direct consultation with researchers in universities, government agencies and industry.

More recently, the Prime Minister, Ingvar Carlsson, decided to strengthen the Government Research Advisory Board. This high-level body previously had little impact — meetings dealt with too many issues, were unstructured, and the membership was large and reflected an unmanageably diverse range of political interests.* The members now include only the heads of the four principal research-oriented ministries, the Government Chief Scientific Adviser, the Director of Research Planning in the Cabinet Office, and six scientists chosen for their personal qualities and experience. The Prime Minister chairs the four main meetings held each year.

In addition to business meetings, periodic seminars are arranged with invited researchers on issues of future national importance — for example, on promising developments in emerging cross-disciplinary fields like biotechnology. These are again attended by the Prime Minister and have a foresight function in that they provide a direct mechanism for scientists to outline to government officials potentially significant research opportunities. The Board is thus seen as 'an informal forum for contact, information and debate. It is acting in a sense as a 'sounding board' for different proposals and ways of solving problems. [It] is a link between the political and scientific communities, identifying difficulties and contributing ideas and suggestions' (Fredga, 1985, p.5).

One result of these meetings has been confidence-building between politicians and the scientific community, probably contributing to the current policy of increasing R&D spending (albeit modestly) while other government expenditures are falling. In addition, they have served to generate commitment to new research initiatives — according to the Chief Scientific Adviser, the seminars played an important part in determining the priorities set out in the 1987 Bill.*

3. FORESIGHT IN BASIC RESEARCH

Basic research is largely carried out within the university sector which comes under the aegis of the Ministry of Education and Cultural Affairs. The Ministry is rather small and most policy functions are devolved to the agencies it administers — the National Board of Universities and Colleges (UHÄ) and four research councils (the Medical Research Council, Natural Science Research Council, Council for Research in the Humanities and Social Sciences, and Council for Planning and Coordination of Research).

(a) National Board of Universities and Colleges

The principal task of UHÄ is to advise government on the distribution of core funding to Sweden's forty or so universities and colleges. It is governed by a Board of Directors assisted by five planning committees. These cover the main disciplinary areas (e.g. technical sciences) and have an advisory role in relation to both teaching and research. Members are appointed by government and are drawn from a diverse range of socio-economic sectors (UHÄ, 1984).[3] Among the decisions with longer-term implications are those concerning the establishment of new university chairs and the renewal of existing positions on the retirement of incumbent professors. These are made centrally by government on the recommendation of UHÄ, although there has been pressure for this responsibility to be devolved to the universities and regions.[4]

While UHÄ makes little use of foresight, it did previously operate a small Unit for Long-Term Planning. Established in 1980, this worked on three studies: (i) a review of methods for research planning in Swedish and foreign agencies; (ii) an evaluation of undergraduate and doctoral programmes (focussing on measurement of skill requirements and student output); and (iii) an assessment of the long-term recruitment needs of university faculties (which predicted problems were likely to emerge in mathematics). According to the unit's ex-head, the first study was especially instructive in highlighting the lesson that 'if there is no interested receiver, you cannot do longer-term planning'.* This was because of the reliance at the time on an annual budgetary system, decentralized decision-making, and the existence of an institutional ideology generally hostile to strategic planning.[5] The unit was closed in 1983, ostensibly because of financial cuts but also as a result of conflict between its remit and the 'ordinary planning' required by the new triennial budgetary procedure instituted in the Research Policy Bill.

The only substantive foresight activity undertaken by UHÄ to date has concerned preparation of an outline five-year research programme for the universities, including identification of priorities, requested by the Ministry of Education and Cultural Affairs as an input to its task of drafting the 1984 Bill. The resultant submission took the form of a multi-volume compilation of current and projected 'faculty programmes' in the main academic fields like medicine or the social sciences. This was essentially a detailed listing of the information on research activities obtained from individual universities,[6] and little effort was devoted to synthesizing an overview of future trends or to recommending priorities. Consequently, although UHÄ invested considerable effort in producing similar faculty programmes for the following 1987 Bill (for example, see UHÄ, 1986), the drafting committee in the Cabinet Office declined to use them.

Despite its obvious limitations, UHÄ officials felt the 1984 exercise had nevertheless been useful in providing the first comprehensive summary of future university research intentions. In addition, it helped pinpoint structural problems in the supply of researchers in several areas, prompting an increase in government funds for doctoral and postdoctoral positions. Most importantly, comparison of the pattern of UHÄ finance of universities with that of research councils and sectoral agencies identified certain mismatches. The lengthy discussions which ensued resulted in sectoral agencies agreeing to take a longer-term perspective in their academic research funding, and to support programmes and doctoral studentships as well as individual projects.

The tendency towards more centralized policy coordination since 1984 has, together with the move to set R&D budgets triennially, undoubtedly led to a greater awareness of the need for longer-term thinking. One consequence was a UHÄ initiative in 1985 to consult companies about their future personnel requirements, which involved visits to twelve major firms as well as a national conference organized jointly with other ministries and government agencies. This resulted in a strong plea from industry that universities should concentrate on building high quality competence in fundamental research, but with certain broad areas like engineering and computer science being identified as especially important over the next decade (Ministries of Education and Industry, 1986).

(b) Natural Science Research Council

The Natural Science Research Council (NFR) is the largest of the funding agencies under the Ministry of Education and Cultural Affairs. Its budget of around 400 million Skr in 1987 supported research in a wide range of physical, biological and environmental fields (see NFR, n.d.). The criteria for determining research priorities relate mainly to scientific merit (although there are exceptions like high-temperature superconductivity). The view is that other agencies (in particular, the National Board for Technical Development) take care of strategic and applied research, and NFR's mission should therefore be to maintain a strong fundamental effort in the universities. Overall policy is set by the NFR Council, several members of which are chosen by an electoral college nominated by the scientific community. Since the Council also appoints an active researcher as its Secretary-General, NFR is in effect run largely by academic scientists and 'external' considerations like socio-economic potential have so far had little impact on resource-allocation decisions.

Besides the NFR Council, several discipline-based programme committees (consisting of ten to fifteen scientists) are responsible for decisions to initiate work in new areas, and for assessing proposals and evaluating completed research. There is also an *ad hoc* committee, chaired by the Secretary-General, which among other things deals with applications for interdisciplinary research grants. Although only a small budget is involved (around 1 per cent of the total), this provides a mechanism for NFR to fund work in fields interfacing with the responsibilities of other agencies and with goals over and above those of creating scientific knowledge (Lindqvist, 1985).[7]

The dominance of academic scientists in NFR means that the organization tends to operate in a responsive mode, only initiating new lines of research after significant proposal pressure. Consequently, little systematic foresight is undertaken by Council staff, the main exception concerning future instrumentation needs in big science.[8] Instead, NFR places greatest stress on ensuring that the quality of research undertaken is at least equal to that in other industrial nations, and that programme areas are congruent with what the international scientific community sees as especially promising for the future. This is achieved by regular reviews of disciplines and subfields (e.g. structural chemistry or condensed matter physics — NFR, 1986; 1987) which are generally carried out by committees of foreign scientists chaired by eminent Swedish researchers. As well as evaluating the international standing of recent research, committees assess individual groups and grant-holders and make recommendations on future NFR funding policy.

Since 1979, some fifty evaluations have been carried out. NFR regards such exercises as extremely valuable since the involvement of foreign experts is seen as the only way of overcoming the inherent deficiencies of peer-review in a small country (NFR, 1981).[9] They also help identify emerging research opportunities and contribute an important foresight input to policy-making in that the recommendations of review committees are generally influential in setting future priorities (Luukkonen-Gronow and Ståhle, 1989). NFR's traditional approach to resource allocation has, however, come under strain as a result of the growing emphasis on financing national priority programmes. This began with the energy, microelectronics and information technology programmes, with others like biotechnology currently under way. Discussions on establishing programmes take place during the inter-agency consultations held prior to drafting the Research Policy Bill, and NFR involvement means that external criteria are *de facto* being incorporated in the

Council's decision-making process (Van Rossum, 1985, p.77). The policy has been to minimize the impact by contributing relatively basic research (for example, on semiconductor physics) and to participate only if the resources made available are additional to its core programme for non-targeted work (Lindqvist, 1985).

Moreover, the move to triennial budget-setting means that NFR is now required to identify medium-term priorities. These are determined by each programme committee on an informal basis and are rather broad — for example, priorities for physics in 1987–8 were the basic structure of matter, rapid dynamic processes in atomic systems, the interaction between plasma physics and astrophysics, the interface between mathematics and computer science, and high-temperature superconductivity. Whether these have much effect on the research performed is doubtful since it is possible to reclassify existing activities in line with such priorities. Until recently at least, NFR has regarded priority-setting more as an administrative necessity than an intrinsic requirement for effective support of high quality science.*

(c) Council for Planning and Coordination of Research

The Council for Planning and Coordination of Research (FRN) was established in 1977 to initiate and support interdisciplinary research addressing societal needs. This includes work in such areas as the ageing society, regional development, chemical health hazards and natural resources. FRN also promotes the public dissemination of research results and administers a national fund for the purchase of larger items of scientific equipment. Its expenditure in 1986–7 was 131 million Skr, two-thirds of which was earmarked for the equipment programme.

The diverse activities of FRN are reflected in its Board which contains representatives from a wide range of economic and social sectors (political parties, trade unions, industry, municipal and local authorities, and the other research councils). This helps it both to fulfil a networking function, and to identify areas where research competence is needed but which are inadequately supported by other agencies. The Board is often proactive in proposing lines of future research, although it also responds to requests from government and elsewhere.

Once an area has been identified as a candidate for support, an *ad hoc* committee is usually established to identify research needs and opportunities and to propose a programme. Committees are typically cross-disciplinary and include both research 'producers' and 'consumers'. They rely principally on internal discussion coupled with consultation of researchers and other interested parties. While conferences and workshops are sometimes used, emphasis is rarely placed on systematic foresight techniques. Although most committee reports are oriented to prospective developments over the medium or long term, they tend to be driven by demand-side considerations — for example, emerging ecological problems (FRN, 1982) or difficulties faced by the handicapped (FRN, 1985). Only in a few cases is the focus on capitalizing upon new scientific opportunities — as, for example, in a recent study on phytochemistry (FRN, 1987). On completion of a report, the normal practice is to initiate a research programme of up to ten years duration.

The previous prominence of FRN in setting national research priorities and in futures work more generally has perhaps declined over recent years. The Council was, for example, charged by the previous Conservative government with defining the broad priorities set out in the 1982 Research Policy Bill. This was undertaken by an internal working group which engaged in extensive consultations with other agencies and held hearings with the scientific community (see FRN, 1981). However, the fact that FRN reports to the Ministry of Education and Cultural Affairs apparently made it difficult for this role to be maintained when priority-setting became tied to the task of determining overall national R&D budgets. Consequently, the Cabinet Office and Government Research Advisory Board have now taken over some of these functions.*

FRN was also responsible for futures studies in Sweden over the period 1980–7 through the Secretariat for Futures Studies (SFS). Originally situated within the Cabinet Office, SFS had a wide mandate to carry out future-oriented research relevant to emerging policy problems. It was difficult, however, to reconcile such an institutional location with studies which sometimes questioned government activities (for example, criticism of the country's energy policy by the well-known *Solar vs Nuclear* report), and FRN was consequently given responsibility in 1980. Parallel to this, FRN has also administered the Committee for Future Oriented Research (SALFO) which continues to have an innovative potential, at least within the realm of cross-disciplinary research. However, SFS for its part became progressively less concerned with futures research *per se*, and more with advocating action-oriented programmes to create alternative social futures. This led to growing criticism of its activities within government, industry and academia (although not within alternative social movements), which culminated in an evaluation undertaken in 1984. The resultant report criticized the methodological and technical weaknesses of SFS and recommended that it either be closed or radically restructured (Wittrock *et al.*, 1985).

After a further enquiry by a Parliamentary committee (Framtidsstudie-kommittén, 1986), it was decided to launch a new Institute of Futures Studies independent of FRN and any other government agency.[10] Established in 1987 with an annual budget of 10 million Skr, the Institute intends to focus more on research foresight issues and is beginning with a study on Biosis. This aims to assess the prospects and problems (especially ethical and moral concerns) associated with the rapid development of biomedical science — for example, in respect of cloning. There will also be a strong interdisciplinary orientation, with inputs from natural and social scientists as well as engineers. In addition, it is intended to create 'quasi-customers' for all institutionally funded projects to ensure that the results of its work are relevant and adequately disseminated. Certain doubts exist, however, as to whether an institute independent of both government and the universities will be able to achieve significantly greater impact on public R&D policy than SFS.*

(d) Academic research on foresight

Two academic groups have undertaken studies aiming to develop improved approaches to research foresight.[11] The Research Policy Institute at Lund University has concentrated on strategic research commissioned by STU (considered below), while the Department of Theory of Science at Gothenburg University has been engaged in a novel programme on the arts and humanities.[12] The first of two studies under way in 1988 is concerned with 'knowledge and vision in the humanities'. Financed by UHÄ, the intention is to contribute to the debate in Sweden over whether it is possible to formulate a research policy for the arts and humanities.

The study began with an extensive literature review exploring different perspectives on the role of arts and humanities in society: should universities, for example, concentrate on developing basic knowledge in traditional disciplines? Should they instead stress the pragmatic role of knowledge (as is often advocated by those in cognitive science, linguistics or philosophy)? Or should the emancipationist function of arts and humanities in creating new social identities and societal structures be emphasized? A series of seven seminars was then held which brought together academics and practitioners from various disciplines (including history, music, literary criticism and theory of art) in an attempt to arrive at an understanding of: (i) current thinking on the 'philosophy of science' in each discipline; and (ii) likely directions in the discipline's development and the possible impacts on society. This is being followed by interviews with a sample of practitioners to determine perceptions about the roles of different disciplines in the year 2000, especially in relation to the development of high-technology society. These will be complemented by interviews with policy-makers in government agencies.

It is hoped that the results will stimulate thinking about how to set in place a distinctive research policy for the arts and humanities. While cultural sciences constitute one of the targeted fields identified in the 1987 Bill, the relatively low funding priority is seen as stemming from an unfair comparison with natural sciences and engineering using criteria specifically constructed for determining science as opposed to cultural policy. Another intended output is an anthology in which prominent personalities will outline how they envisage their fields developing over the next fifteen years. This will be prefaced by an article attempting to provide an overall perspective on policy for the arts and humanities.*

The second project on 'humanities research and industry' is being financed by a private foundation. Aiming to develop a better understanding of the potential role of humanities research, it is being carried out primarily by means of a questionnaire survey. A list of 300 industrialists having an appreciation of the humanities was first compiled with the assistance of STU, the Royal Academy of Engineering Sciences, and several collective research institutes undertaking industrially-oriented work. Respondents are being asked to make judgements on: (i) the future need for humanities expertise in industry; (ii) what disciplinary skills, if any, have been utilized by their company (e.g. ethnology or archaeology); (iii) whether the firm is likely to make greater use of humanities and arts, and, if so, of what fields and types of knowledge; (iv) the merits of doctoral training in the humanities from the viewpoint of industry; and (v) general images of academic researchers in the humanities. It is planned to publish the results of the survey in *Business Weekly*, a magazine widely read within Swedish companies. A further stage in the project will involve in-depth interviews with up to thirty experts after which the results will be written up in greater detail. The lack of previous prospective research in the arts and humanities means that the Gothenburg work seems likely to make a pioneering contribution to the field.

4. STRATEGIC RESEARCH AND BASIC TECHNOLOGY

(a) *Ministry of Industry*

While primary responsibility for technologically-oriented research rests with the Ministry of Industry, most policy and administrative functions are devolved to the National Board for Technical Development (STU) and other essentially autonomous agencies. Longer-term policy is set out in triennial government bills, the most recent stressing the need for industrial diversification and continued structural change oriented towards research-intensive products and internationalization of the economy (Ministry of Industry, 1987). The 1987 Bill also argued that the trend towards increased support of R&D by government and industry should continue, with priority given to longer-term generic research feeding into Sweden's traditional areas of competitive strength — transportation equipment, industrial automation and energy equipment — as well as into newer sectors like pharmaceuticals. Ministry of Industry funding for R&D amounted to around 1,300 million Skr[13] in 1987–8 and covered: (i) promotion of R&D for industrial renewal (subsidies for STU, the Industrial Development Fund and the Board for Space Activities); (ii) finance for national technology-development programmes in high-priority areas (such as satellite telecommunications); and (iii) expenditure on the technological infrastructure (including the Swedish Foundation for Technological Attachés and Royal Academy of Engineering Sciences).

Given its small size, little systematic foresight is undertaken within the Ministry.[14] Future socio-economic and technological potential are nevertheless key criteria in the triennial budget-setting process. Five-year plans are proposed by each of the agencies for which the Ministry is responsible, with informal consultation being the main mechanism for deciding priorities. Although overall budgetary guidelines are set every three years, minor adjustments are made annually drawing upon a contingency fund (for example, to respond to unexpected developments like high-temperature superconductivity) and these form an input to discussions on the Research Policy Bill (Cabinet Office, 1987).

The Ministry sees research planning as relatively straightforward since the country is sufficiently small for the scientific and technological elites to interact frequently. Hence, when advice is sought 'a well-discussed opinion is (normally) obtained' and consensus is easy to achieve.* However, it has become clear that 'the tradition of reaching workable compromises has unhealthy side-effects in that it can result in resources being split between too many projects, and lead to an emphasis on minimizing risk rather than maximizing research opportunities with a consequent loss of dynamism in the research system'.* Thus, while the process has enabled agreement to be reached on broad longer-term research objectives, the Ministry sees the need to develop improved mechanisms for identifying more specific priorities.

Although cross-national studies of emerging technologies have been commissioned from consultants and think-tanks, these have rarely contributed significantly to policy formulation. Rather greater reliance is placed on the extensive monitoring of foreign R&D undertaken by companies, researchers and technical attachés.[15] Given that Sweden conducts less than 2 per cent of the world's R&D, systematic intelligence-gathering is seen as essential for informing public agencies and industry on emerging threats and opportunities.[16]* Ministry staff regard the lack of in-house strategic thinking as a barrier impeding effective use of such information, and have been considering how to strengthen this function. Following concern that industrialized nations are all targeting similar research fields, consideration has been given to whether the Ministry should fund more specialized R&D areas with longer-term implications for particular niche markets. An exploratory study was commissioned from a STU official on Innovation Clusters for the Renewal of Swedish Industry (Stenberg, 1987), which reviewed the characteristics of research areas historically important for Swedish industry, as well as attempting to identify priorities for the next five to fifteen years. Although the findings came too late to be used as an input to the 1987 Research Policy Bill as originally intended, they are nevertheless seen as shedding light on the factors affecting previous Swedish success in technological innovation and providing insights as to how longer-term priorities might be set in future.*

(b) *National Board for Technical Development*

Established in 1968, STU is the principal agency financing technical and industrial research. It is responsible for supporting and facilitating: (i) the creation of knowledge and research competence in areas which are strategic to future economic competitiveness; (ii) the development of new technology and its introduction into established industrial sectors; and (iii) the formation of technology-based companies in high-growth areas of the economy (STU, 1986f; 1987a). Responsibility for research oriented to more specific needs of government departments rests with around 100 sectoral agencies such as the Defence Research Institute (discussed below) and the Council for Building Research (see Ministries of Education and Industry, 1986).

Over the last ten years, STU has devoted increasing resources to longer-term strategic research and generic technologies, primarily through five-year 'framework programmes' in selected fields (STU, n.d.). In particular, STU support for microelectronics and information technology has been combined with that of other agencies in an integrated national programme (STU, 1984), followed recently by an initiative on materials (including ceramics and powder metallurgy). The focus on basic technologies has resulted in a growing proportion of STU expenditure going to the universities — over 30 per cent of its 1987 budget of around 800 million SKr. Originally, academic funding was limited to direct project support, but conflict with the Ministry of Education concerning overheads and UHÄ priorities led the Board to begin limited financing of tenured posts.[17] Since 1984, it has also operated the Technical Research Council (STUF) which provides small grants to academics for research of high scientific merit. Such grants are primarily awarded to the technical universities, where STU finance now constitutes up to 60 per cent of external research income. There has historically been a deliberate policy against creating specialized laboratories which are seen as building inflexibility into the R&D system, although STU has recently supported the establishment of two cooperative institutes in computer science and information technology.

(i) Corporate planning system

The corporate planning process has changed considerably over time, reflecting the agency's developing mission and previous experience with priority-setting. Up to 1975, the principal mechanism for agreeing priorities was a system of ten 'requirements boards' (e.g. for food technology). However, these came to be dominated by industrialists who stressed short-term factors. Not only was it difficult to initiate longer-term research but it was almost impossible to change the allocation of resources to each board in line with new priorities. Consequently, STU decided to abandon this mechanism and instead to evolve a more proactive planning system in which industry still retains an important voice. This is based on triennial Forward Looks in which priorities for the following five years are set out together with the resource requirements (e.g. STU, 1987a). As well as constituting the organization's budget request to government, it provides an input to the process of establishing national priorities in the Research Policy Bill. The Forward Look procedure has been in place since 1981 (STU, 1982).[18]

At the heart of the system are *ad hoc* planning groups (e.g. on carbohydrate chemistry), which are formed when the need for a research initiative becomes apparent (different procedures are employed for more applied technological or product-support activities). Each has eight to ten members drawn from industry, university and government to balance science-push and demand-pull perspectives. Their task is to propose a general programme for research in the field, identifying priorities and budgetary requirements over the next five to ten years. This typically involves extensive consultation with the scientific community, with STU programme officers acting in a coordinating role and providing relevant inputs (for example, information on foreign initiatives obtained from technical attachés).[19] The group is disbanded after the planning phase. If the programme is funded, a steering group is set up (including one or two planning group members) to solicit and assess research proposals. It is also subsequently responsible for mid-term and final assessment of projects. Foreign experts are generally brought in to evaluate the overall programme and they also help identify future research needs (e.g. see STU, 1987b; 1987c).

A difficult part of the Forward Look procedure is to establish an agreed set of overall priorities. This is the responsibility of a central group in the Director-General's Office. Submissions are made by 'project departments' on their planned five-year programmes, which integrate existing commitments with proposed initiatives (including those recommended by planning groups). Since the budgetary requests invariably total more than the funds likely to be made available by government, modifications must be made to the programme submissions. Criteria used include the opportunities associated with emerging scientific and technological trends, the competence available within universities and research institutes, and the long-range options for Swedish industry. A draft budget proposal specifying the programmes to be supported under alternative funding levels is next considered by the STU Industrial and Scientific Councils. After approval by the Governing Board, the final proposal is submitted to the Ministry of Industry. It is then subject to often rather lengthy negotiations with the Ministry of Finance, the Cabinet Office and to a lesser extent other ministries.

The STU officials interviewed felt the Forward Look procedure to be working fairly well, and that an acceptable balance now exists between bottom-up and top-down mechanisms following concern over too great a level of central direction in earlier exercises (STU, 1979). The current approach is regarded as a relatively flexible means of generating research priorities, with some 1,000 experts from planning groups, steering groups and other committees making direct inputs. The function of the corporate planning process is thus seen by STU as being to: 'organize a platform for an ongoing dialogue between researchers, developers and consumers of technology ... This broad input into planning is of subsequent importance in getting into practice the results of research and development, and it enables possible blocks to future introduction to be identified early' (STU 1987a, p.25).

(ii) Identification of emerging interdisciplinary areas

Although STU is reasonably satisfied with the Forward Look process for identifying opportunities in *existing* research fields, there have been worries about it yielding conservative decisions and failing to pick up promising areas emerging through the fusion of several specialties. As a result, an experimental study was conducted as part of the 1986 Forward Look to explore a more structured approach to priority-setting. This was coordinated by the STU Long-Term Planning Group in cooperation with several programme officers, and was supervised by a high-level management committee.

Since many new technology areas are strongly research based, it was decided to concentrate initially on science-push factors and to draw upon the views of younger scientists less wedded to existing disciplinary paradigms. A questionnaire was sent in 1984 to all Swedish postdoctoral researchers in technological and natural science fields asking them: (i) to outline the main research trends in their field; (ii) to describe the changing international scientific frontiers and predict the principal breakthroughs up to 1990; (iii) to assess the likely consequences for future work by Swedish university groups; and (iv) to identify research opportunities in the field which could at the same time constitute a threat to Swedish industry.

The 600 responses received were analysed by the STU Engineering Sciences Advisory Council, which consists mainly of eminent academics but also includes industrialists. Their involvement served both to draw in valuable expertise and to increase the level of commitment to the results within the agency and externally. A report summarizing the findings was then presented at the STU University Day (a biennial event where universities participate in discussions concerning the agency's research strategy). Around sixty of the survey respondents were among those attending, who were invited to identify interdisciplinary trends and assess which emerging fusion areas were likely to become significant. One conclusion was that the links between existing disciplines will become increasingly important not only for the exploitability of research but also for scientific progress itself (STU, 1986a, p.2).

The next step was to integrate demand-pull perspectives into the exercise. In cooperation with the Royal Academy of Engineering Sciences (IVA), experts from different technical fields and industrial sectors were brought together in three large meetings convened to discuss developments in: (i) physics, electronics and computer sciences; (ii) chemistry and biotechnology; and (iii) production engineering and process technology. The aim was to check whether any promising research opportunities had been neglected, and to obtain informed judgements on the potential impact of each interdisciplinary area on Swedish industry. Some sixty firms participated, often represented by directors of planning or corporate R&D.

As a result of the survey and workshop discussions, three 'tentative interdisciplinary fields or competence limiting areas' were identified as essential for technical development in Sweden during the 1990s: (i) 'micronics' — a term coined to describe the synthesis of surface physics and chemistry, electronics, micro and molecular biology, materials chemistry and micromechanics; (ii) expert-based control systems — an integrated research area covering control technology, computer technology and applied mathematics/artificial intelligence; and (iii) algorithms and processors — an intersecting field between mathematics, information processing and computer architecture.

The three areas were then subject to further discussion by invited experts, who realized that there was likely to be appreciable synergy with certain humanities and social science fields. Consequently, a seminar was arranged which identified a fourth interdisciplinary field on 'humanities as a resource for technical development' (covering areas like perceptual psychology, pedagogy, linguistics and phonetics). To ensure wide dissemination of the findings, a science journalist was commissioned to prepare reports summarizing the opportunities in each field (STU, 1986a; 1986c; 1986d).

In addition, it was decided to corroberate the results with foreign peers. A joint seminar was arranged with the Engineering Board of the National Research Council which had been conducting a survey of US engineering research. This concluded that the three technological areas were likely to be significant, and that Sweden could take a world lead if a prompt decision was made to fund the programmes. US experts judged the micronics area to be somewhat unconventional but highly promising. A final discussion was then held at which industrialists were unanimous in endorsing the identified priority fields as important for Swedish industry. Micronics, in particular, was seen as vital in maintaining the country's traditional strengths in specialized engineering.

According to officials interviewed, the exercise was successful in several respects. First, it achieved an immediate policy impact, with a two-year programme being established to develop competence in micronics in thirteen research groups. Moreover, a commitment was made to increase funding substantially in the future, as well as to initiate work in the other three areas. Second, the exercise brought together universities and industry to identify priorities that were both intellectually challenging and technologically important. The consensus thus created, together with the endorsement by the National Research Council, enabled STU and the Ministry of Industry to make a strong case to government that new funds should be allocated to the programme. Third, the approach employed enabled various field-specific interest groups and their industrial constituencies to cooperate in defining common future needs, thus overcoming the intrinsic conservatism of the normal STU planning system. Finally, the exercise succeeded in identifying promising fusion fields, something which NRC failed to achieve when using a panel-based procedure in its study (and which, in the opinion of Swedish interviewees, consequently yielded rather conservative results).*

While this success undoubtedly reflects the contributions made by the research community, a key role was played by the head of the STU Long-Term Planning Group in promoting a more formal approach to priority-setting. Also important was the existence of considerable accumulated research foresight expertise within the group. As a result, in coordinating the exercise care was taken from the outset to ensure widespread participation of relevant groups, and to generate the high-level organizational and industrial commitment necessary to implement the ensuing recommendation for a new interdisciplinary programme.

(iii) Infrastructure for foresight activities

STU's move to support basic technological research has increased the need for world-wide monitoring in order to identify developing threats and opportunities. The Long-Term Planning Group has also attempted to improve the infrastructure for foresight to assist project departments and programme officers in preparing the STU Forward Look and in planning more generally. Of particular interest is a study comparing activities in various science-based technologies across Sweden, West Germany and Japan. This was carried out by the Research Policy Institute at Lund University and aimed: (a) to test the validity of patent and bibliometric indicators for assessing national strengths and weaknesses in strategic research (e.g. in lasers and fibre-optics); and (b) to develop an understanding of the dynamics of emerging technologies with a view to identifying future growth-points.

The findings of the study (Granberg, 1985; 1986) provided an input to the 1986 Forward Look (STU, 1986e), and raised a number of important questions for STU research policy. First, in some key technological areas it is difficult to adopt a 'watching brief' since developments are taking place so quickly that the country runs the risk of being left behind. Prompt decisions therefore need to be made on whether to support research in such areas or to pull out entirely. Second, a larger fraction of Swedish research on emerging technologies is undertaken in universities than in West Germany and Japan where government institutes and companies are more prominent, highlighting the importance of ensuring that the university–industry interface is working effectively. Third, the costs of developing research programmes across a broad range of strategic technologies mean that Sweden must necessarily engage in more international cooperation (a policy now adopted by STU). Finally, novel institutional arrangements are being developed overseas to support research on basic technology (e.g. the Japanese ERATO Programme) and STU should monitor such initiatives to determine what lessons can be drawn.*

The Research Policy Institute was subsequently commissioned to undertake a further project on Technology Policy Watch and Analysis for Countries, Organizations and Important Research Areas. Covering the main industrial countries, this aims to monitor: (a) changes in research infrastructure; (b) trends in the organization of collaborative R&D; and (c) government support for basic technological research. Particular stress is being placed on studying the success of different policy instruments with a view to their possible adoption in Sweden.[20]

Finally, the Long-Term Planning Group has been evaluating the strengths and weaknesses of different foresight methods, and monitoring experience with their use in priority-setting exercises in other countries. STU analysts had identified a shift away from predictive forecasting techniques (like trend extrapolation and cross-impact analysis) towards consensus-generating approaches, coupled in recent years with an emphasis on early-warning systems often incorporating bibliometric analysis. As a result, a study was commissioned from the US Center for Research Planning exploring the merits of employing co-citation techniques and patent citations to basic science in research foresight (Franklin *et al.*, 1986). According to STU officials, the co-citation approach proved not to be particularly appropriate for policy purposes, but work will continue on evaluating the strengths and weaknesses of more traditional bibliometric techniques of the type used by the Research Policy Institute.*

(c) Royal Academy of Engineering Sciences

Founded in 1919, the Royal Academy of Engineering Sciences (IVA) is an independent institution financed in part through grant-in-aid from the Ministry of Industry. Like other national academies, its role is to promote research and foster national and international scientific links. Its 300 elected members come from industry, government laboratories and academia.

As early as the mid-1960s, the Academy recognized how crucial it was for Sweden to adopt a policy for strategic research that was both long-term and selective if future socio-economic needs were to be met effectively. IVA stressed the necessity of monitoring overseas research since it is difficult for a small country to do more than maintain a watching brief in many areas. It was thus instrumental in setting up a network of overseas technical attachés to track R&D developments globally. The attaché system is well regarded by industry and government agencies, although responsibility has since passed to STATT (a foundation also funded partly by the Ministry of Industry).

Another IVA initiative at about the same time was to establish a Study Group for Futures Research to undertake longer-term forecasting. This was a unique development in Sweden at the time and was strongly supported by industry which was trying to get to grips with the new futurological techniques then in vogue in the United States. A decision was subsequently taken in 1968 to propose to the Ministry of Industry that an Institute of Future Studies be created with two main tasks: (i) to develop a global early-warning system on research developments from which industry might benefit; and (ii) to study the likely long-term impact of science and technology on the nation's economic and social structure. The proposal was rejected, however, by the Social Democratic government on the grounds that it was too technocratic and reflected the sectional interests of industry and the research community.[21] Instead, it was decided to set up a broadly-focussed futurological unit attached to the Cabinet Office, an initiative which, as we have seen, is now regarded as something of a failure.

IVA's response to this rejection was to establish two in-house organizations. The first, the Association for Futures Studies, collected and disseminated information on forecasting techniques. Its membership grew rapidly to 100 companies and government agencies, as well as 500 individuals. The second, the Clearing House for Futures Studies, carried out commissioned projects on a multi-client basis for some seventy corporate members. Both organizations were relatively successful until the late 1970s when consultancy companies like SRI began to provide a wider range of professional services, and firms lost faith in forecasting and disbanded their futures groups. As one IVA official remarked: 'like others, we have picked up the strategy of not trying to foresee the future, but rather magnifying and analysing "weak signals" in time to do something about changes to come.'†

The move to integrate prospective analysis into the planning function instead of treating it as an activity in its own right began in the late 1970s. At the time, IVA was pressing the government to think more about longer-term research priorities, and it launched a major study on Technical Capability and Industrial Competence (IVA, 1979a). This was clearly in the new 'foresight' mould and was carried out over an eighteen-month period during 1978–9 at a cost of 3 million Skr (part of which came from government).

From the outset, the study was accorded great importance. The committee overseeing the work consisted of eleven senior IVA members drawn from government, industry and academia, while the wide-ranging consultative procedure employed enabled most parts of the Academy to contribute in some way. Three issues were addressed: (i) what changes in the industrial environment since 1945 have affected the international competitiveness of Swedish companies? (ii) What future adaptation do these changes require in the structure and performance of Swedish industry? (iii) What is the role of technology in industrial competitiveness, and in what fields is further strengthening required?

In executing the study, four subcommittees were appointed to deal with: (i) the role of technology in postwar industrial development; (ii) cross-national comparisons of technical and industrial competence; (iii) world trends in technological development; and (iv) the future prospects for Swedish industry in the international marketplace (IVA; 1979b; 1979c; 1979d; 1979e; 1979f). Most relevant here is the work of the third subcommittee, which attempted to provide the basis for the main committee's recommendations concerning governmental measures on research and education. It also indicated promising areas requiring increased attention from industry (IVA, 1979d, p.13).

The first step was a literature review to identify potentially important R&D fields, which drew in particular upon the results of a cross-national comparison of technological development trends and social needs (IVA, 1979e). A provisional list of 150 research areas, together with background information on their likely future development, was circulated to experts for comment. Around 100 areas were then selected for inclusion in the study, covering agriculture, electronics, transportation, materials, biotechnology, natural resources and production technology.

This was followed by a Delphi exercise in which some 250 Academy members (selected to take into account both science or technology push and demand-pull) first assessed the likelihood and importance of possible developments in each field. Three main questions were asked: (i) is the

Figure 9.2 *Results of IVA enquiry into technological areas of future importance to Sweden*

Note: Items in bold type: market opportunities exist in the short term and product development should be stressed. Remaining items: efforts should mainly be focussed on long-term scientific research.

Type of effort needed	Degree of importance for Sweden to engage development resources in the area		
	'Strong': 'decisive'	'Moderate': 'selective'	'Limited': 'monitor the field'
Technical/scientific knowledge which is important for industrial needs	Composite materials **Powder metallurgy** Polymers with new properties Surface sciences **Biochemistry for medical purposes** Growth processes in agriculture and forestry **Fermentation technology** Enzyme technology	Picture processing and remote sensing Ceramic materials Refractory materials **Aluminium** **Cement and concrete** Electrochemistry Chemistry of natural products Neurotechnology Industrial applications of genetic engineering **Geotechnology**	Fusion energy Digital technology for audio purposes Holography Super-conductors Amorphous metals and semiconductors Improved biological nitrogen fixation Photosynthesis Microbial methods for metal extraction
Broad development of advanced technology	**Process control equipment** New methods for joining, cutting and surface treatment Substitution plastics and composites **Development of forestry machinery** **Adaptive manufacturing processes** Production with reduced manpower **Automated materials handling systems** **Reliability technology and systems know-how** New materials and manufacturing methods in mechanical engineering	Methods for improved computer programme production Chemistry for agriculture New chemical technologies for mechanical engineering 'Risk-free' production technologies	**Navigation, steering and safety at sea** Controlled environments in agriculture Soil-less cultivation Fishing Production technology for outer space conditions Building information systems

Concentration on advanced technology for components, equipment and systems	**Nuclear hot water reactors** **Heat pumps, solar heating and cooling systems, energy storage** **Telephone systems with end-to-end digital transmission** **Personal and freight-carrying vehicles** **Marine transport, specialised ships** Fine and speciality chemicals **Floating/mobile factories and plant** **Underground and mining equipment**	**Very large scale integrated circuits** Fibre optic technology for use in telecommunications Broadband communications **Chemicals from the forest industry**	Breeder reactors High temperature gas-cooled reactors Electricity from solar energy Widespread wireless telephony **Direct satellite television** Large coastal projects Large self-supporting structures
Methods and systems for increased efficiency in use of resources and for environmentally advantageous production forms	**Efforts to increase energy efficiency in industry** **Husbanding of materials** **Environment protection** **Energy saving technologies for forest industries** **Environment protection in forest industries** Energy saving as a factor in building technology	**Mining and use of coal** **Wind energy** Biomass production Direct and melting reduction of iron-ore Ethanol methanol **Waste water treatment technology** **Materials recovery** Organic waste collection and utilization	**Geothermal energy**
Development of technologies and know-how for important social systems	Health care Teatment of social diseases **Medical care systems and technologies** **Help for the aged and in rehabilitation**	**Decentralized data processing capability** **Integrated supply systems for the retail trade** **Development of computer technology for office work** **Systems for movement of people** **Goods distribution systems** **Security and alarm systems** Improvement of community systems in developing countries Improved functioning of urban conglomerates	**Cheap inter-continental telecommunications** Information systems for the general public **Inter-regional train systems** Artificial mobile habitats

Source: IVA (1979a, p. 30). Reproduced with the kind permission of the Royal Academy of Engineering Sciences.

development probable, and, if so, when are commercially promising breakthroughs likely to occur? (ii) What opportunities or threats will arise for Swedish industry? (iii) What is the nation's current competence in the field, and are additional resources required to strengthen that competence? A summary of the results (which noted marked divergencies in opinion) was then provided to respondents so they could modify their answers in the light of others' views. In each field, judgements were also requested concerning timing (up to twenty years in the future), the level of existing national competence, and recommended priority for future funding (i.e. whether resources should be increased, cut, used more selectively or cover only the costs of a watching brief).

The final step was to analyse the results of the survey (a 90 per cent response rate was achieved) and make recommendations on the scope and need for action to strengthen research in each field. As can be seen from Figure 9.2, the overall priorities took into account the nature of the R&D (whether it was concerned, for example, with scientific/technical knowledge, generic technologies, or component and system technologies), as well as the extent to which the main need was for longer-term research or more applications-oriented activities.

Interviewees identified several lessons which can be drawn from the study. The first concerns the value of a key individual combining a broad technological understanding with expertise in forecasting and science policy, and who also has the contacts to secure active participation by both the academic and industrial communities (thus helping to ensure science-push factors are integrated with future socio-economic needs). Without the professional and organizational skills of the individual coordinating the study, it would reportedly have been difficult even to launch the initiative.

Second, the literature-based procedure for selecting the fields to be studied resulted in a tendency to emphasize existing activities — for example, although surface science was chosen as a priority no radically new fields like micronics were distinguished.[22] STU took this into account in its 1986 exercise to identify emerging cross-disciplinary areas, together with a third lesson concerning the integration of foresight into decision-making. At the time, IVA had been in dispute with STU over government policy for the agency to take over many of its previous administrative functions. Although STU was perhaps the principal audience, the conflict meant that it was difficult for the IVA results to be integrated into the Board's decisions on future priorities. Consequently, the study's main impact related to its general recommendations — for example, to establish a stronger Government Research Advisory Board.

A final lesson concerns the extent to which the implementation of foresight findings is limited by wider political factors. The IVA stress on a coordinated national strategy for supporting priority areas of science and technology was not in accord with the prevailing government industrial policy which was oriented to creating an environment conducive to innovation, leaving strategic choices to the actors involved.* The situation has since changed and a similar study today would probably have more impact. This said, the IVA study undoubtedly contributed to reorienting Swedish research policy towards greater selectivity and longer-term planning.

During the 1980s, IVA has continued to play a general role in promoting foresight activities. In particular, it organizes regular symposia bringing together researchers and industry to discuss emerging developments in technical science and engineering (e.g. IVA, 1983). A review of national needs for trained engineers over the next fifteen years has also been undertaken (IVA, 1986).

(d) Defence Research Institute

The Defence Research Institute (FOA) is Sweden's largest government laboratory, with around 1,250 staff and a budget of 300 million Skr in 1988. Although its primary mission is to meet the needs of defence agencies, considerable basic research is also undertaken. FOA activities are planned in collaboration with both administrative and operational arms of the Ministry, with research foresight playing a role in: (i) the regular five-year corporate planning process; and (ii) periodic 'technical prognosis' studies undertaken in conjunction with defence-policy reviews.

(i) FOA corporate planning

Corporate planning within FOA has changed markedly since 1984. Prior to this, rolling five-year plans were prepared annually on the basis of discussions with operational defence agencies (e.g. the navy and air force). However, this led to a concentration on short-term equipment-development requirements (tasks generally better handled by companies) to the neglect of longer-term research. Consequently, it was decided to adopt a more proactive approach to planning, but with safeguards to prevent FOA interests becoming too dominant.

Two further factors influenced the introduction of a new planning process. First, a period of declining budgets led FOA to conclude that greater selectivity was needed to ensure effective deployment of R&D resources. Second, it was recognized that acceptance by the Ministry of Defence and the government of a longer-term plan would diminish the possibility of future cuts in funding as a result of short-term political or economic exigencies. This could be achieved by placing consensus-building at the centre of the strategic planning process, while enabling institute staff to play a larger part in identifying research needs and opportunities.

The current planning system is based upon a relatively formal process in which major studies are undertaken every few years to inform decisions on medium-term priorities, with minor adjustments made annually within a quinquennial budgetary framework. The first such study was carried out in 1984 by an FOA project group under the direction of senior management. Officials from the various defence agencies were invited to participate both to aid the consultative process with the user community, and to assist in generating consensus on the institute's future research programme. Two main steps were involved, the first being to develop a broad menu of R&D options together with the associated funding requirements. A questionnaire was sent to staff in FOA and other defence agencies requesting judgements on: (a) significant emerging research areas; (b) trends in existing R&D; and (c) likely defence needs. The project group then analysed the results and prepared a costed list of research programmes judged important for defence needs over the next ten to twenty years. The main findings (FOA, 1984) were circulated widely for comment.

The second step was to define overall R&D priorities under two budgetary scenarios — a 9 per cent reduction in real terms over the following five years and a 4 per cent increase. The priorities identified under each scenario were divided into: (i) potentially important applications areas (such as protection against non-conventional weapons); and (ii) the underlying research required to meet those applications needs (FOA, 1985). Extensive consultation of both FOA staff and external research users was again employed in the selection process, with considerable compromise necessary before agreement was reached.

FOA interviewees felt the corporate planning process represents a significant improvement on previous practice, enabling policy-makers to see clearly the implications of budgetary decisions. However, the fact that research foresight is closely constrained by resource-allocation considerations means that it is inevitably politicized, with sectional interests playing an important role. This is less so with the technical prognosis studies undertaken for defence-planning purposes.*

(ii) Technical prognosis studies

Comprehensive foresight exercises have been carried out every ten to fifteen years since the early 1950s on longer-term R&D needs. These are updated every five years to provide an input to the quinquennial defence-policy reviews. The previous technical prognosis (TP) study begun in 1974 aimed to inform the 1977 defence review, and revisions were made in 1978 and 1982. The TP-87 study was still under way in 1988 with the 1992 defence decisions in mind.

The approach employed is comprehensive and methodologically sophisticated, drawing on a wide range of information and taking into account science and technology push as well as user needs. One aim is to generate consensus on research priorities among FOA, FMV (the Defence Materials Administration) and the operational branches of the military. The technical prognosis exercise hence enables strategic decisions on future directions for R&D to be integrated with those on defence policy. It also represents an opportunity to involve industry in discussions on planned programmes (thus facilitating the later task of technology transfer) and to demonstrate the need for military research to politicians. The fact that Sweden is neutral means that NATO technological forecasting exercises cannot be used as a basis for defence-policy decisions.

Although FMV is formally responsible for the technical prognosis studies, they are carried out by the FOA defence analysis group. A preliminary step in TP-87 (phase 0) was to develop a list of 500 keywords representing significant developments in science, technology or defence systems. These were obtained from discussions with in-house and external experts, by reviewing intelligence material and recent priority-setting reports produced by foreign agencies, and by scanning literature abstracts. The keywords were then synthesized to yield twenty broad areas likely to be important in the future (e.g. applied chemistry, propulsion, new materials and small-scale energy systems). Experts from FMV and FOA also completed a brief questionnaire on emerging research areas showing promise for meeting projected defence technology needs, resulting in a further ten areas being identified (e.g. micro-machinery and micro-technology).

The next step (phase 1) involved three tasks. First, thirty panels of researchers[23] (drawn primarily from FOA and FMV, but also from industry and universities) described the principal scientific and technological trends during 1974–87 within each area. They also constructed 'maps' relating developments in their area to those in the other twenty-nine. Second, each group attempted to forecast over the short term (1987–90),[24] medium term (1990–2000), and longer term (2000–10):[25] (a) likely research advances and breakthroughs; (b) novel development programme opportunities; and (c) possible operational uses of new defence technologies or systems. A three-by-three matrix was then constructed for each research area describing predicted developments. Finally, the panels compared forecasts for Sweden and the West with those for the Eastern bloc to identify comparative technological strengths and weaknesses. Information on Soviet and East European R&D was drawn from both open and intelligence sources, with additional inputs made by an FOA group with specialist expertise on the subject.

Phase 2 of TP-87 aimed to integrate 'market' and 'user requirements' perspectives into the forecast. First, industry was consulted on potentially important R&D programmes and whether they would collaborate if funding could be secured. Here, companies expressed a preference to participate in the more basic technological research. Second, representatives of operational military units joined the thirty panels in a modified Delphi exercise. This involved: (a) specifying the defence functions likely to be necessary in the future; (b) assessing how these could best be met with new or improved weapons systems; and (c) considering the contributions which might be made by the research advances forecast in phase 1. Significant developments were thus identified for eight general system areas (e.g. electronic warfare) and a much larger number of sub-areas.

The final step was to articulate within an overall analytical framework the two forecasts of technical possibilities and of defence needs. One element of this entailed constructing a large matrix of possibilities versus needs. Another was to assess the forecasts to identify potential threats and opportunities (such as high-power laser weaponry) with strategic implications for national defence policy. Classified reports had been completed by early 1988 on the two main phases of TP-87 (FMV, 1986; 1987) and a final report was being drafted (together with detailed reviews of the thirty generic technology and eight systems areas).*

Effort was in particular being put into preparing an overview of the principal findings in an attempt to avoid the problems encountered in the TP-74 exercise when few senior officials were willing to read the lengthy final report produced. It had also been recognized that wider diffusion and debate of the findings is necessary within defence agencies, while external presentations have been made to influential bodies like the Government Research Advisory Board. Another lesson taken into account in TP-87 was the crucial role of consensus formation in ensuring subsequent policy implementation. Greater numbers of staff from FMV and operational defence units were therefore involved in panels, and a high-level advisory committee was convened to oversee the exercise. Although the more consultative process was one reason the exercise took around three years, it was still judged to have been successful. This is in part due to the fact that the FOA staff members who organized the TP-87 forecast possessed a wide range of experience in research policy, the official in charge having previously been involved in strategic planning at STU.

5. INDUSTRIAL RESEARCH FORESIGHT

After the period of economic stagnation in the 1970s, industry recognized that the key to future growth lay in concentrating on core areas of Swedish technological strength, together with diversification into emerging areas linked to traditional R&D capabilities and benefiting from receptive domestic markets. This has meant devoting greater emphasis to R&D — hence the rapid increases in industrial expenditures over recent years noted above. Of the two companies studied, Ericsson is a leading multinational in the well-established communications field, while Pharmacia is a smaller firm specializing in the pharmaceuticals and health products sector which is widely regarded as a major growth area for Swedish industry.

At Ericsson, expenditure on R&D rose to 9 per cent of turnover by 1987 from 8 per cent four years previously. Basic research accounts for around 10 per cent of the total, with generic programmes financed mainly by operating divisions but partly through a corporate overhead. Although more basic research used to be undertaken within central laboratories, all R&D is now performed within business areas. Like most major Swedish firms, Ericsson has recently reassessed its overall strategy and introduced a 'core business concept' based upon a three-level model of corporate activity (Ericsson, 1987). At the front-end, 'business areas' are market-oriented and focus on meeting customer needs. The driving force behind new business activity is a 'base-systems' structure which is constituted by the company's specialized technical capabilities and systematized know-how (e.g. in mobile cellular telephone systems). Finally, 'basic technologies' such as optical switching and silicon technology provide the knowledge-base for developing the key technical areas like CAD, VLSI and processors which underpin the base-systems. While willing to engage in joint ventures to develop business areas, and even to undertake collaborative research in basic technology fields, Ericsson refuses to cooperate with other firms in activities which might compromise its base-systems capacity.

Research foresight in Ericsson has primarily concerned basic technologies where the main task is to monitor emerging developments on a global basis. Links with universities and STU are important here, as are the results of longer-term R&D undertaken by a number of sectoral laboratories (such as the Institute of Microwave Technology)[26] which are financed collectively by government and industry. Ericsson also engages in considerable background scanning of the scientific and technical literature, in particular US and Japanese data-bases, as well as occasionally employing consultants to review developments in relevant fields.

Priorities for longer-term research are set by a Scientific Council comprising the technical directors of the fifteen operating groups and chaired by the Corporate Technical Director. Its principal task is to decide upon generic programmes aiming to develop an in-house capacity in more basic fields cross-cutting several business sectors (like computer science). If necessary, *ad hoc* groups are established to identify priorities for specific areas (e.g. artificial intelligence) and submit proposals for new initiatives to the Scientific Council. Although foresight plays a part in these decisions, the tendency has been to rely on the expertise of corporate technical staff coupled with analyses of emerging research needs derived from market-oriented strategic planning. The Technical Director stressed that such an approach has until now been appropriate given the largely systems-oriented nature of Ericsson's business. It has been rare for scientific breakthroughs to occur which have impacted significantly on commercial products in the short or medium term, and this has meant that it has not been worth investing heavily in monitoring basic research. An exception is high-temperature superconductivity where the potential impact was thought likely to be more immediate, and the Scientific Council therefore decided within a few months of the IBM breakthrough in 1986 to initiate a modest programme to develop a basic technical capacity.[27]

The increasing pace of technological change in communications has, however, resulted in Ericsson recently recognizing the need to mount more systematic foresight. One result[28] was a study of semiconductor technology which assessed likely trends in problem areas (such as process development, design, cooling and encapsulation) in the period up to 1995. This was carried out by an expert panel drawn from within the company and from Swedish institutes and universities, but also including an overseas researcher. A Delphi approach was employed, the first stage of which involved experts setting down their views on prospective developments over the next decade. A summary of the results was circulated for comment, and a one-day meeting held with Ericsson staff after which panel members revised their judgements. These were then analysed and the implications for longer-term corporate R&D strategy drawn in a report submitted to the Scientific Council. An important effect of the exercise was to reinforce the judgement of company staff that a revolution was under way in semiconductor technology, but that the impact would only be significant in the medium to long term. Consequently, optical lithography would remain the dominant technology to the middle of the 1990s, while gallium arsenide would only slowly displace silicon (in which competitors were continuing to make major investments). The company's capability in high-density silicon technology therefore needed to be strengthened. Overall, Ericsson regards the more structured approach to foresight employed in this study as relatively successful, and plans to use a similar method to develop a long-range strategy for optical communications research. As is common in rapidly growing fields, divergent views exist within the company on the prospects for specific technologies like optical switching and the main aim will be to achieve greater consensus on future R&D policy.*

The second company, Pharmacia, has also recognized that longer-term research is becoming more central in determining its future competitiveness. This is not necessarily seen as having to be undertaken in-house, and basic research still amounted to less than 10 per cent of total R&D in 1987 (which represented 15 per cent of turnover). In pharmaceuticals, new products often stem directly from scientific breakthroughs so research foresight is intrinsically less certain than is the case in the systems-based technology sectors in which Ericsson operates.

Pharmacia was in 1985 faced with declining economic prospects and undertook a review of R&D policy as part of a wider reappraisal of corporate strategy. This led to the company becoming aware that around 75 per cent of sales stemmed from breakthroughs made at the University of Uppsala and the Karolinska Institute (Pharmacia, 1986). While generally unpredictable, such breakthroughs usually arise from high-quality research in fields linked to the company's core activities. Consequently, Pharmacia decided that the best strategy to strengthen its science base was to widen the existing interface with academia in both Sweden and abroad. As a result, over 200 million Skr was awarded to universities in the form of grants and fellowships in 1987, with preference given to institutions located near operating groups on the basis that experience has shown geographical proximity is vital to effective technology transfer. At the same time, it recognized that too high a level of intramural basic research can give rise to internal vested interests which then impede relationships with universities.*

The 1985 review also involved an audit of the company's technological position relative to competitors. This revealed strengths in biotechnology (protein separation and analysis), opthalmic surgery, allergy and basic pharmaceuticals, with weaknesses in health-care and medical instrumentation. To develop a capacity in weaker areas and strengthen those where a comparative advantage already existed, Pharmacia decided to embark on an extensive acquisition programme, taking over a number of US firms judged to have a technological lead in their fields. However, certain integration problems have subsequently arisen.

As regards longer-term planning, Pharmacia has to date made little use of formal foresight and relies primarily on expert judgement, encouraging all staff to contribute to identifying prospective options. Overall responsibility for building new core technologies rests with the Corporate R&D Director. In addition, a Strategic Planning Group, consisting of corporate management, leading researchers and representatives of operating divisions and subsidiaries, meets twice a year with R&D policy figuring centrally on its agenda. Among the core technologies regarded as important for the future is recombinant DNA. After the failure of a costly joint-venture in Sweden, Pharmacia decided the quickest way to create the R&D capability required was to establish a Californian subsidiary. Although this upset the firm's existing biotechnology group, not to mention the government and universities, the initiative has proved highly successful in capitalizing upon US applied research in the field. Another core technology is biosensors where Pharmacia has set in place an eight-year 'workplan' to explore the commercial potential of research advances made at Linköping University. The first stage involved 'raiding' staff from the university, the defence research institute and one of the company's own laboratories. Although a somewhat traumatic process, it succeeded in bringing together experienced scientists from relevant basic physics and biology disciplines.*

Pharmacia has previously experimented with using systematic foresight in selecting core technologies. In particular, future options in oncology were assessed during 1982 in terms of: (a) scientific developments; (b) internal research competence; and (c) likely medical needs. A prominent immunobiologist helped conduct the review, which led ultimately to Pharmacia taking over a company established to exploit research at Lund University. However, the general experience with structured foresight exercises is that the expert panels employed concentrate on possible advances *within* existing specialties and therefore fail to foresee emerging interdisciplinary developments.*

It should be noted, finally, that Pharmacia devotes considerable effort to monitoring competitors through patent analysis and systematic scanning of the trade press. Citation analysis has also been used to examine the cross-disciplinary impact of particular research areas as well as to monitor competitors, although, like other quantitative approaches, it is regarded as yielding information that is mostly too dated to be employed in strategic planning. Instead, attendance at scientific meetings is seen as providing the earliest opportunity to spot new research efforts by rival firms, in line with the company's preference for informal foresight mechanisms.

6. CONCLUSIONS

Swedish experience with research foresight has overall been relatively positive compared to other European nations. Contributing to this is the commitment to consensus decision-making shared with Japan, the stress on planning to achieve long-term socio-economic stability, and an outward-looking emphasis on monitoring R&D trends and opportunities on a global and continuing basis. While these factors are partly culturally specific, several lessons may nevertheless be drawn:

(a) The acceptance that foresight has a legitimate role to play in policy-making was accelerated by the early recognition that research *forecasting* is beset with conceptual and technical problems. Scientists know only too well that progress in research rarely conforms to the predictions of professional forecasters. However, they are rather less sceptical about *foresight* oriented to picking up, as early as possible, weak signals on emerging developments with scientific or socio-economic promise. Such approaches to foresight have in Sweden been characterized by:

(1) An emphasis on consultation with all relevant actors in order to choose among the available policy options and hence facilitate the creation of one 'future' rather than another.

(2) A stress on global monitoring to ensure that signals are picked up quickly on emerging R&D trends and opportunities. This is the reason STATT built up its comprehensive network of foreign technical attachés, and similarly why agencies like STU and FRN place great store on the recommendations of foreign peers when establishing new lines of research.

(3) A growing use of early-warning indicators to complement traditional expert-based methods for identifying research trends. STU in particular has explored a variety of approaches — scanning the scientific literature and abstracts, citation analysis, surveys of patent applications, patent–patent linkages, and patent citations to basic research. Timeliness is crucial, with Pharmacia stressing that patent-based indicators are normally too historic for research-planning purposes in most science-based areas. Analysis of papers presented at international specialist conferences may be more valuable — hence STU's recent experiment with bibliometric indicators.

(b) A key factor underlying successful foresight has been the recognition that links need to exist with decision-making processes, and that the rationale for the exercise and the audience for the results must be identified at an early stage. Foresight has been integrated with priority-setting in the two agencies having the most effective procedures for research planning — STU and FOA. The triennial Research Policy Bill has provided a conducive climate for foresight, even though it has done little more than stimulate discussion on strategic planning in other agencies. In contrast, the lack of integration between foresight and decision-making was largely responsible for the limited impact of the 1979 IVA initiative and the failure of the UHÄ Longer-Term Planning Unit.

(c) Closely related to this is the requirement to ensure that implementation structures are built into foresight initiatives. Early commitment to the results must be gained from those with the institutional power to influence policy. This can be achieved through various mechanisms — formal consultation at a personal level or through public workshops, involvement in a high-level advisory committee, or more direct participation in an 'authorizing committee' for an exercise. The central problem is how to maximize the likelihood of implementation while keeping powerful interest groups sufficiently at arms-length to avoid politicizing the foresight process. A feature of Swedish experience has been the close interaction between government and researchers through regular seminars on R&D opportunities convened by the Government Research Advisory Board. The involvement of the Prime Minister and Cabinet colleagues has facilitated implementation of decisions to launch new research initiatives, as well as contributing to determining the priority areas set out in the Research Policy Bill.

(d) As in other countries, foresight is most developed in scientific and engineering fields with economic or technological potential. Consequently, greater emphasis has generally been placed on foresight by IVA, STU and FOA than by agencies supporting curiosity-oriented research. This said, Sweden has (together with Norway) been an exception in targeting social and cultural research as a national priority. A longer-term perspective is consequently being taken on the needs and opportunities for such research, and this has led to the unique work at Gothenburg University on foresight in the arts and humanities which aims to explore the possibilities for developing 'cultural policy' akin to science and technology policy.

(e) Besides reinforcing many basic points about research foresight noted in previous chapters, several more specific lessons arise from Swedish experience:

   (1) While demand-pull considerations are an essential element in foresight, they should not be allowed to dominate otherwise conservative and short-term policy perspectives are likely to result (as when requirements boards were employed in STU and FOA).
   (2) The time-horizon in foresight must be sufficiently long-term to take into account the outcome of currently emerging trends, but not be so remote that the uncertainties involved make it impossible to see the relevance to the choices that have to be made today. This was a major

limitation of the scenario-based exercises conducted by FOA in the 1970s. Time-horizons vary across research fields and industrial sectors but fifteen to twenty years is normally the limit.

(3) Individuals have played a prominent role in all the major prospective studies undertaken by IVA, FOA and STU. They have invariably been sufficiently senior to act as a 'product champion' for foresight, while also possessing a range of skills — broad technical knowledge, an understanding of the dynamics of scientific and technological change, good contacts with companies (and other research users), an appreciation of the complexities of industrial innovation, and experience in forecasting and science-policy research.

(4) Failure to give sufficient attention to diffusing foresight results can severely limit their impact. Although a problem with earlier technical prognosis studies carried out by FOA, this has now been remedied by circulating summary reports targeted at specific sections of the defence research community. STU has also stressed dissemination, employing a science journalist to transform the committee reports from its interdisciplinary foresight exercise into glossy, well-written brochures with immediate public appeal.

(5) Considerable benefits appear to follow from integrating foresight with *post hoc* evaluation of research. NFR and STU have achieved this through inviting the international panels established to review fields to pay particular attention to future developments in their policy recommendations.

(f) Probably the most important lesson from Sweden relates to the 1984–5 STU exercise concerning emerging interdisciplinary fields. By initially relying on young postdoctoral researchers to identify promising areas, the problems of conservatism typically associated with committee-based procedures and traditional field categories were largely avoided. The consultative workshops with industrialists also enabled demand-pull factors to be adequately integrated. Furthermore, the process by which the exercise was carried out: (i) had effective implementation structures built in (primarily through the STU Engineering Sciences Advisory Council); (ii) was clearly linked to priority-setting; (iii) enabled the scientific community to reach consensus on fields characterized both by intellectual interest and technological relevance, and to demonstrate that consensus to government; and (iv) ensured that the recommendations were publicly endorsed by the US National Research Council and by Swedish industry. Such a structured consultative approach to foresight might usefully be explored by agencies in other countries.

(g) Another feature of Swedish research foresight has been the importance attached to theoretical and methodological issues. The contrasting experiences of Pharmacia and Ericsson show that the dynamics of scientific and technological development differ markedly across the pharmaceutical and communications sectors. In the former, unpredictable scientific breakthroughs can in themselves lead to new generations of products, while in the latter the process is more systemic and predictable. The implications for foresight have been recognized by STU, which has begun to explore the merits of bibliometric and patent indicators for mapping the growth dynamics of strategic fields. Such studies have shown that developments are occurring so rapidly in some areas that a selective watching brief approach is often unrealistic, and that an active research capacity must be maintained in fields judged to have particular long-term promise. Academic researchers have made a major contribution to developing the capability to use information-science techniques for analysing R&D indicator data.

(h) Finally, Swedish experience with research foresight provides a reminder that success should not be judged solely in terms of impact on decision-making. Foresight can also result in an early warning of emerging economic, technological and environmental problems, and these may well pose a challenge to conventional wisdom or threaten the interests of established institutions. The example of the Future Studies Secretariat demonstrates the tensions which can arise when critical prospective studies are undertaken, and points to the importance of ensuring an appropriate institutional location. It also highlights the dangers of researchers becoming too politically involved with realizing the goals identified in foresight. Achieving a balance between disinterested analysis and taking steps to ensure foresight results are implemented is undoubtedly difficult. In the view of the government at least, the action-oriented approach of the Future Studies Secretariat is a case where this mark was clearly seen as having been overstepped.

NOTES

* Refers to material collected during interviews with Swedish officials in early 1988, while † denotes information drawn from subsequent correspondence.

1. Overviews of the Swedish R&D system and the principal policy issues are provided in Dyring (1985), OECD (1987) and Ministries of Education and Industry (1986). A review of earlier developments in R&D policy is given in Elzinga (1980).
2. The triennial cycle of Research Policy Bills is so timed that preparatory work is concentrated in the year following national elections. In this way, it is hoped to promote continuity in long-term planning.
3. The UHÄ planning system was being evaluated in 1988 and changes were anticipated.†
4. The procedure for appointing professors was also under review in 1988 following a request by the Minister of Education.
5. This was contrasted with the foresight approaches employed by the National Defence Research Institute, whose mission makes strategic planning essential.
6. They were compared unfavourably by UHÄ officials with the Forward Look documents produced by the National Board for Technical Development.
7. Another important development in 1984 was the establishment by STU of the Technical Research Council (STUF). This funds projects spanning the responsibilities of STU and NFR which previously might have gone unsupported.
8. Reviews are also periodically undertaken of needs for other research instrumentation, especially where equipment supports scientists from several fields (e.g. see NFR, 1983).
9. Luukkonen-Gronow and Ståhle (1989, p.20) argue that such evaluations are not always so neutral and objective as is claimed, and the involvement of the research community in selecting panel members often results in the choice of 'experts with

whom they had had frequent previous contacts and who were more leniently disposed potentially towards them or likely to promote the options they favoured ... Thus, the pitfalls of user review are not avoided by using international experts'.

10. The history of future studies in Sweden is outlined in Glimmel and Laestadius (1987) and Framtidsstudiekommittén (1986). While FRN interviewees disagreed with the results of the evaluation by Wittrock *et al.* (1985), other government officials thought the criticism justified. A positive assessment of SFS achievements is given in Leyten (1987, pp.20–6).
11. A third group on Technology and Social Change at the University of Linköping has been more concerned with technology assessment than with policy-oriented work on research foresight (Leyten, 1987, p.36).
12. One reason for conducting these studies is the lack of future-oriented work in the arts and humanities. The head of the Gothenburg group, Aant Elzinga, was central in the foresight exercise undertaken by the Science Council of Canada (discussed in Chapter 8) and was therefore able to bring considerable expertise to both projects.
13. This excludes R&D carried out in state-owned manufacturing companies.
14. The work of the Ministry's Structural Secretariat is primarily concerned with industrial policy rather than R&D-related issues.
15. The Swedish Foundation for Technical Attachés has offices in all the main industrialized nations, including three in the United States.
16. International exchanges and projects are also regarded as valuable mechanisms for tapping into promising R&D developments, with the United States, West Germany and Japan having been targeted as particularly important.
17. During the early 1980s, the technical universities were critical of what they saw as the tendency of the Ministry of Education and Cultural Affairs to favour initiatives in the humanities and social sciences at the expense of engineering. This was regarded as a symptom of the anti-technocratic ideology prevalent within sectors of the Swedish political and governmental elite. They therefore requested that responsibility for their administration be taken over by the Ministry of Industry. Although the situation has changed to some extent, one interviewee stressed that 'technology has had to be forced on the Ministry of Education — they are dominated by very basic considerations and more by social science'.*
18. The first attempt at corporate planning began in 1978 and resulted in *STU Perspectives 1979*. However, this was produced by a 'planning department' attached to the directorate and with no direct links to programme management divisions. Consequently, its impact was limited.*
19. Inputs may also be made by studies undertaken by the STU Longer-Term Planning Group or by commissioned consultancy reports — for example, a survey of Japanese research in optoelectronics by Scantech helped define a recent initiative on photonics (STU, 1988). In addition, overseas visits by planning groups often play an important role.
20. For example, the UK initiatives targeted for study are the Advisory Council on Science and Technology, the Centre for Exploitation of Science and Technology, the Alvey Information Technology Programme, the DTI Link Programme, and public-sector efforts to attract foreign science-based companies.*
21. This is indicative of a wider ideological dispute which previously existed between industrially-oriented organizations like IVA and sections of the Social Democratic Party.*
22. However, several priorities identified by IVA concern the application of research within a particular field to apparently unrelated industrial sectors — for example, 'new chemical technologies for mechanical engineering' (see Figure 9.2).
23. On average, groups had ten members so overall some 300 researchers were involved in TP-87.

24. Neither FOA nor FMV were particularly interested in the results of the short-term forecast. However, experience with TP-74 had shown that groups insist on making such predictions, whether asked or not, and it is better to make this analytical distinction beforehand to ensure short-term and medium-term judgements are not conflated.
25. Up to the early 1970s, FOA experimented with even longer-term forecasting using scenario techniques. The tendency is now to rely more on trend analysis because defence options are regarded as being clearer. In addition, normative forecasting was found to yield relatively conservative predictions which were, in any case, of limited utility in planning because the time-horizon was too distant.*
26. Several collective research institutes are specifically charged with maintaining a global watch on new technologies and identifying potential opportunities for Swedish industry. For example, the Institute for Manufacturing Technology monitors progress in fields like automation and robotics, corrosion and surface chemistry.*
27. This involved recruiting a leading academic to develop an in-house capacity in the field. Additional resources were allocated for cooperative research with STU and other companies, and to support research in two universities (to ensure a future supply of trained staff). One of these universities, Gothenburg, had been working on superconductivity since 1962, and Ericsson cited this as an example of the value of funding academic research in areas not currently seen as strategically important. Such research must, however, be of high quality, and the performance of Swedish universities was generally judged to be rather uneven.
28. Prior to this, the last systematic foresight study was undertaken in the 1970s and concerned traffic theory (e.g. stochastic processes and queuing theory), a field in which Sweden held a leading position. This was a core technology for Ericsson when communication was based on electro-mechanical systems.*

REFERENCES

Cabinet Office (1987), *1987 Government Bill on Research: Main Issues,* Stockholm, Cabinet Office.

Dyring, A. (1985), *Swedish Research: Policy, Issues, Organization,* Stockholm, Swedish Institute.

Elzinga, A. (1980), 'Science Policy in Sweden: Sectorization and Adjustment to Crisis', *Research Policy,* vol.9, pp.116–46.

Ericsson (1987), *Ericsson Core Business Concept,* Stockholm, Ericsson.

FMV (1986), *Technological Prognosis Part A: Technological Areas,* Stockholm, Defence Materials Administration, Ministry of Defence (Classified).

—— (1987), *Technological Prognosis Part B: Systems Areas,* Stockholm, Defence Materials Administration, Ministry of Defence (Classified).

FOA (1984), *Defence Research for the Future,* Stockholm, Defence Research Institute.

—— (1985), *Defence Research Towards the 1990s,* Stockholm, Defence Research Institute.

—— (1986), *FOA in a Nutshell,* Stockholm, Defence Research Institute.

Framtidsstudiekommittén (1987), *To Choose a Future,* Stockholm, Statens Offentliga Ultredningar.

Franklin, J.J., Coward, H.R. and Simon, L. (1986), *Identifying Areas of Swedish Research Strength: A Comparison of Bibliometric Models and Peer Review Evaluations in Two Fields of Science,* report to Swedish National Board for Technical Development, Philadelphia, Center for Research Planning.

Fredga, K. (1985), 'The Swedish Government Research Advisory Board', paper presented at Conference on Science Policy Organizations in Small Industrial Countries, Amsterdam, April.

FRN (1981), *Swedish Research 1982-86: Ways, Values and Choices,* Stockholm, Council for Planning and Coordination of Research.

—— (1982), *Renewable Natural Resources: A Swedish Research Programme for the Future,* Stockholm, Council for Planning and Coordination of Research.

—— (1985), *Research on Handicap,* Stockholm, Council for Planning and Coordination of Research.

—— (1987), *Phytochemistry,* Stockholm, Council for Planning and Coordination of Research.

Glimell, H. and Laestadius, S. (1987), 'Swedish Future Studies in Transition', *Futures,* December, pp.635–50.

Granberg, R. (1985), *A Bibliometric Survey of Fiber-Optics Research in Sweden, West Germany and Japan,* Research Policy Institute, University of Lund.

—— (1986), *A Bibliometric Survey of Laser Research in Sweden, West Germany and Japan,* Research Policy Institute, University of Lund.

IVA (1979a), *Technical Capability and Industrial Competence,* Stockholm, Royal Academy of Engineering Sciences.

—— (1979b), *Technology and Industrial Structure: The Economic Crisis of the 1970s in International Perspective,* Stockholm, Royal Academy of Engineering Sciences.

—— (1979c), *Knowledge as a Basis for Industrial Development,* Stockholm, Royal Academy of Engineering Sciences.

—— (1979d), *Technological Imperatives for the 1980s,* Stockholm, Royal Academy of Engineering Sciences.

—— (1979e), *Technological Development Trends and Societal Needs — An International Perspective,* Stockholm, Royal Academy of Engineering Sciences.

—— (1979f), *The Long-Term Potential for Development of Swedish Industry in an International Context,* Stockholm, Royal Academy of Engineering Sciences.

—— (1983), *Important Technological Trends: Proceedings of the Fourth Convocation of Engineering Academies,* Stockholm, Royal Academy of Engineering Sciences.

—— (1986), *Engineers for the Future,* Stockholm, Royal Academy of Engineering Sciences.

Leyten, J. (1987), 'Aspects of the Integration of Science and Technology in Swedish Society', in Ministry of Education and Science, *Technology Assessment: An Opportunity for Europe,* The Hague, Government Printing Office.

Lindqvist, I. (1985), 'Priorities of the Society and Maintenance of Scientific Quality', in European Science Foundation, *Rise and Fall of a Priority Field,* Strasbourg, European Science Foundation, pp.163–5.

Luukkonen-Gronow, T. and Ståhle, B. (1989), 'Quality Evaluations in the Management of Basic and Applied Research', paper presented at ECE Seminar on Evaluation in the Management of R&D, Madrid, April.

Ministries of Education and Industry (1986), *Science and Technology Policies in Sweden,* Stockholm, Norstedts Tryckeri.

Ministry of Industry (1987), *Swedish Industry Facing the 90s: Main Issues,* Government Bill 1986/87:74, Stockholm, Ministry of Industry.

NFR (n.d.), *The Swedish Natural Science Research Council,* Stockholm, Natural Science Research Council.

—— (1981), *International Evaluations of Research Projects Supported by the Swedish Natural Science Research Council,* Stockholm, Natural Science Research Council.

—— (1983), *A Supercomputer for Swedish Science,* Stockholm, Natural Science Research Council.

—— (1986), *International Evaluation of Condensed Matter Physics,* Stockholm, Natural Science Research Council.

—— (1987), *International Evaluation of Structural Chemistry with Diffraction Methods,* Stockholm, Natural Science Research Council.

OECD (1987), *Reviews of National Science and Technology Policy: Sweden,* Paris, Organization for Economic Cooperation and Development.

Pharmacia (1986), *Nittonhundraelva Nittonhundraattiosex Pharmacia Sjuttiofem Ar*, Uppsala, Pharmacia.

Stankiewicz, R. (1982), 'The Place of Basic Technologies in the R&D Policies of the Small Industrialized Countries', in P.H. Kristensen and R. Stankiewicz, *Technology Policy and Industrial Development in Scandinavia*, Research Policy Institute, University of Lund, pp.88–101.

Stenberg, L. (1987), *Development Areas for the Renewal of Swedish Industry*, Stockholm, Ministry of Industry.

STU (n.d.), *Long-Term Development of Competence in Engineering Sciences: Our Guarantee of Progress*, Stockholm, National Board for Technical Development.

—— (1979), *Technology for the Future: STU Perspectives 1979*, Stockholm, National Board for Technical Development.

—— (1982), *The 1981 STU Forward Look: International Dependence and Technical Renewal of Swedish Industry*, Stockholm, National Board for Technical Development.

—— (1984), *National Microelectronics Programme*, Stockholm, National Board for Technical Development.

—— (1986a), *Interdisciplinary Research: Research on Micronics*, Stockholm, National Board for Technical Development.

—— (1986b), *Interdisciplinary Research: Research on Algorithms and Processors*, Stockholm, National Board for Technical Development.

—— (1986c), *Interdisciplinary Research: Research on Expert-Based Control Systems*, Stockholm, National Board for Technical Development.

—— (1986d), *Interdisciplinary Research: Research on Humanities as a Resource for Technical Development*, Stockholm, National Board for Technical Development.

—— (1986e), *Longer-Term Technological Research in Sweden, West Germany and Japan*, Stockholm, National Board for Technical Development.

—— (1986f), *The Swedish National Board for Technical Development*, Stockholm, National Board for Technical Development.

—— (1987a), *The 1986 STU Forward Look: Executive Summary*, Stockholm, National Board for Technical Development.

—— (1987b), *Evaluation of Swedish Research in Adaptive Controlled Industrial Robots*, Stockholm, National Board for Technical Development.

—— (1987c), *Report of the International Evaluation Group on Carbohydrate Chemistry*, Stockholm, National Board for Technical Development.

—— (1988), *Photonics: A Possibility for Industrial Renewal that Sweden Should Not Miss*, Stockholm, National Board for Technical Development.

UHÄ (1986), *Fakultetsprogram: Teknisk och Matematisk-Naturvetenskaplig Forskning*, Stockholm, National Board of Universities and Colleges.

Van Rossum, W. (1985), 'Research Councils and Priority Setting: The Cases of the German DFG, the Swedish NFR and the Swiss NF', in European Science Foundation, *Rise and Fall of a Priority Field*, Strasbourg, European Science Foundation, pp.65–96.

Wittrock, B., deLeon, P. and Nowotny, H. (1985), *Choosing Futures: Evaluating the Secretariat for Futures Studies*, Stockholm, Council for Planning and Coordination of Research.

10 Research foresight in Norway

1. INTRODUCTION

The last country case-study concerns Norway, which is in many respects unique with its small population, large coastline and northerly location. Its highly specialized economy is oriented around natural resources (fisheries and forestry), oil and petrochemicals, basic metals and electronics. The cyclical nature of the oil sector from which a large part of national income is derived has caused structural adjustment problems, and it is recognized that manufacturing industry needs to be strengthened if the country is to have a secure future (OECD, 1985). One result has been a rapid rise in research funding in recent years. Industrial R&D expenditure increased from 1,900 million Nkr in 1983 to an estimated 5,000 million in 1987, while government research spending on the university and institute sector grew from 2,800 million Nkr to approximately 4,100 million over the same period (FSU, 1987, p.2).

The rise in government funding has come primarily in the form of targeted priority programmes, the achievement of consensus on these areas being an important reason for the commitment of additional resources at a time of austerity. It has also been accompanied by increasing central coordination of research policy and renewed emphasis on longer-term basic programmes. Let us consider first the role of foresight in establishing national research priorities, in particular the work of the Science Policy Council. We next discuss foresight in basic science agencies, focussing on the Research Council for Science and the Humanities. This is followed by a description of research foresight in the Ministry of Industries, the Royal Norwegian Council for Scientific and Industrial Research and the Defence Research Establishment, which are all primarily concerned with strategic and applied R&D. Finally, we briefly discuss foresight activities in a Norwegian multinational company and a more general forecasting exercise on Scenarios for Norway in the Year 2000. As we shall see, little systematic research foresight has been carried out in Norway. However, the experience with setting national research priorities is of some interest insofar as it led to consensus among the scientific community and strengthened the case made to government for budgetary increases.

2. FORESIGHT IN NATIONAL PRIORITY-SETTING

Until recently, the Norwegian research system was highly decentralized, with government adopting an approach of 'science in all policies, rather than one

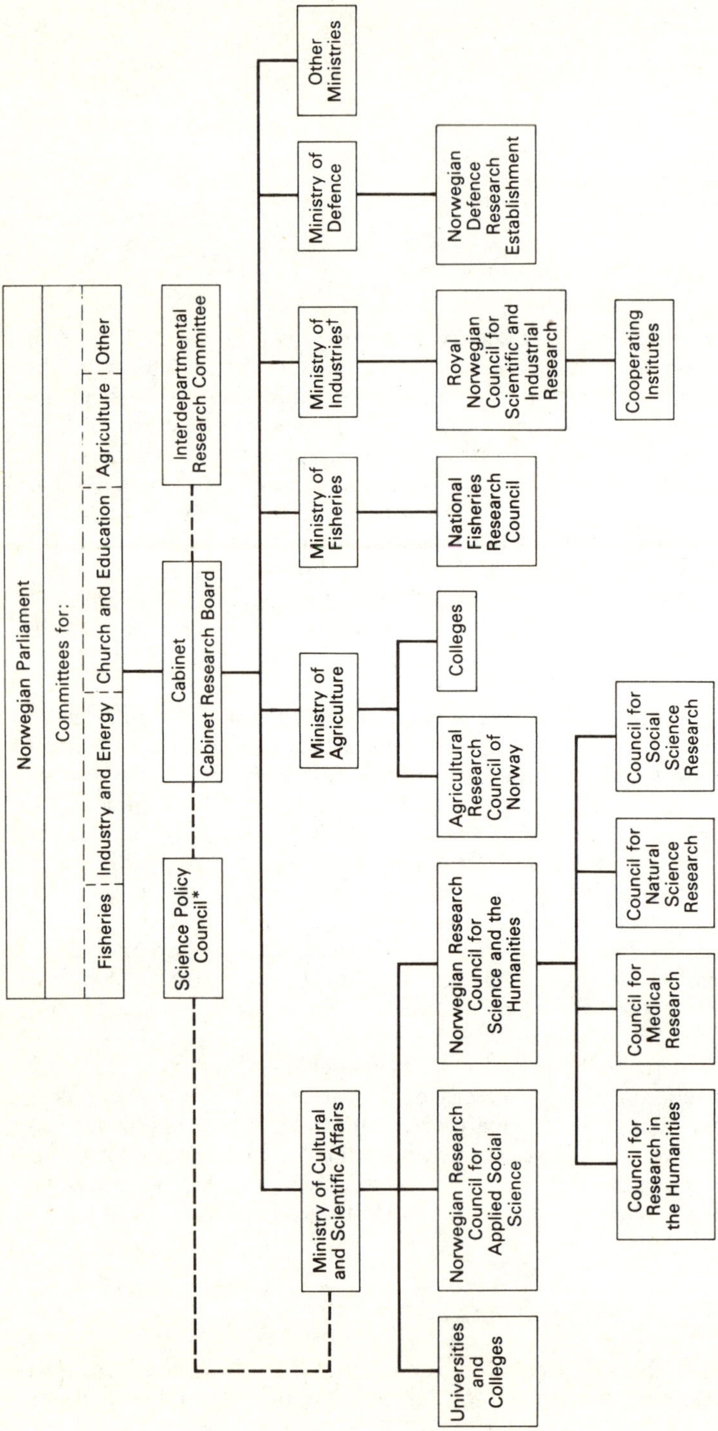

Source: NAVF Institute for Studies in Research and Higher Education.

Notes: * The Science Policy Council was abolished in early 1988 and its functions will be taken over by a new organization whose remit has yet to be decided.
† The name was changed from the Ministry of Industry in early 1988.

Figure 10.1 Outline of the Norwegian R&D system

science policy' (Westgaard, 1985, p.127). Competition among ministries made it almost impossible to agree on research activities spanning the university, governmental and industrial sectors, or to set overall R&D priorities. Since 1985, however, the situation has to some extent changed with a white paper, *On Scientific Research in Norway,* laying the basis for a longer-term, comprehensive research policy. This argued that government must strengthen the basic research infrastructure in order to meet future national needs, and identified a set of priority areas (MCSA, 1985a).

The Conservative government then in power had recognized the need for increased selectivity, and decided in 1983 that the quadrennial white papers on research policy should be used as a mechanism for achieving greater coordination among ministries and funding agencies. The Ministry of Cultural and Scientific Affairs delegated the task of drafting the 1985 white paper to the Director of the Science Policy Council (SPC), Tore Olsen. The SPC had only recently been established (replacing the Central Committee for Norwegian Research) following criticism of the lack of an effective institutional mechanism for framing national R&D policy (e.g. OECD, 1985). Until its abolition in 1988, the SPC reported both to the Cabinet Research Board (which brings together ministers with research interests) and to government via the Ministry of Cultural and Scientific Affairs. This is shown in Figure 10.1, which outlines the structure of the Norwegian research system.

The process by which the white-paper priorities were established employed a significant though methodologically simple foresight exercise.[1] According to Olsen, the key problem was to integrate this with a consultative procedure involving relevant ministries and funding agencies, while ensuring that inputs could be made by the research community.* An *ad hoc* steering committee of seven ministerial advisers acted as a 'sounding board', helping ensure inter-departmental participation and commitment to the final results. This represented a break from tradition in that previous white papers had been prepared by individuals.[2]

The main task in the exercise was to undertake a survey of future research needs and opportunities. A letter was sent by the Minister for Cultural and Scientific Affairs to the five research councils, the Norwegian Employers Federation, seventeen ministries (which in turn circulated it to their laboratories and agencies) and the four universities. This contained the following request:

> In the national budget for 1984, the government has expressed its wish to prioritize research that can stimulate technical development and economic growth, long-term competence-building and basic science. Taking into account these guidelines and the needs of different sectors, we are asking ministries and research councils to answer two questions: (i) In what disciplines and areas is the lack of knowledge a 'bottleneck' to meet future challenges and to further the solution of future societal needs? (ii) What major R&D programmes should be established and prioritized to develop this knowledge? Answers should be given in a form as detailed as is normal in white papers. Furthermore, we hope that you will do this within the framework of a relatively long-term time-horizon — for example, two four-year periods. Finally, all suggestions should be prioritized. [MCSA, 1985a, p.97, unofficial translation]

Responses varied, ranging from individual project proposals to a comprehensive identification of priority fields, with universities stressing the need to build up technical skills to deal with an uncertain future. Analysis showed appreciable consensus among respondents, although the areas proposed were generally similar to priorities identified in a survey of research agencies in France, Sweden and the United Kingdom. Proposed fields were divided into two groups: (i) 'science and technology push' areas of potential importance to Norwegian industry (e.g. biotechnology and information technology); and (ii) areas meeting identifiable future socio-economic needs (e.g. offshore technology R&D). Applied social science was mentioned by several ministries as important, especially for developing improved procedures for policy formulation, analysis and implementation.[3]

The next step was to narrow down the number of possible areas and arrive at a provisional list of priorities. One input was the set of target R&D fields established by the Ministry of Industry in its plans for the 1985 budget (IT, aquaculture, materials research and offshore technology — see Department of Industry, 1984). The provisional priorities were discussed by the steering committee during which 'cultural sciences' was added to the list (largely through the intervention of the Chairman of the Science Policy Council, Francis Sejersted). Research agencies and companies were also consulted. A draft of the white paper was then prepared and, after further discussion by the committee, submitted to the Minister for Cultural and Scientific Affairs.

After the draft had been considered at a Cabinet meeting, ministries reassessed the implications for their future activities, as a result of which members of the steering committee requested further modifications prior to its submission to Parliament. This led to disagreement among the Ministries of Agriculture, Fisheries and Industry over responsibility for aquaculture, which was first renamed 'aquaculture-biotechnology', then 'biotechnology-aquaculture' and ultimately 'biotechnology'. Consequently, the final list of priority areas was: information technology/telematics; biotechnology; research relating to oil and gas; organization, management and administrative systems; and research on tradition and culture (MCSA, 1986, p.25).

The subsequent parliamentary debate resulted in 'health, welfare and the environment' being accorded priority status after lobbying by medical pressure groups. Intervention by the Ministry of Industry also led to 'materials research' being *de facto* classified as a priority field. While the omission of the health area is now recognized by Olsen as a mistake, he still feels that the original decision not to target materials research was correct since the field had been generously supported over a long period.* This view led at the time to conflict with the Ministry of Industry.

By all accounts, the exercise was reasonably successful.* The procedure for obtaining inputs was more open than previously, encouraging the scientific community to participate. Also important was the decision to target broad fields only, leaving agencies to determine their own sub-priorities (using peer-review in the case of research councils). In the previous two years, there had been much debate among researchers over the 'picking winners' approach being adopted in other countries, and it had been agreed that this would be acceptable in Norway only if confined to the identification of general priorities.[4]

While the high-level steering committee constrained the available choices and introduced inter-departmental politics into the exercise, it did help ensure that the recommendations were implemented (which is by no means automatic in Norway where white papers act primarily as parliamentary discussion documents). Also important were Sejersted's personal links with the Conservative government and Cabinet, which gave the Council the authority needed to secure the full participation of ministries.[5]

Most importantly, the white paper provided an opportunity for the scientific community to demonstrate to government that it could play a part in the task of national technological regeneration. As a research council official remarked, 'target areas make science policy more concrete in the minds of politicians, and for this reason they have been more willing to increase the funds for research'.* The fact that most fields, even humanities, had a target area with which they could associate, contributed to preserving consensus among researchers. For Olsen, the crucial task was thus to:

> take the R&D community with [us]. It was a simple psychological point of trying to concentrate on the positive aspects of identifying research needs and opportunities. Consequently, we deliberately steered away from discussing cuts in funding for existing areas or possible controversial restructuring of research councils.*

As regards implementation, national programmes on IT and materials were agreed in the 1986 budgetary discussions. Only when the Socialist government took office, however, were significant funding increases forthcoming, with the national programmes being accorded high priority in the period up to 1990. Larger programmes such as biotechnology are being coordinated by national steering committees which for the first time have brought ministries and agencies together to cooperate in longer-term interdisciplinary research. Stress is also being placed on developing competence in targeted fields, and additional resources have been allocated for foreign visits by scientists as well as for participation in international collaborative research.[6]

Overall, the exercise represents 'a step in the development of a more long-ranging and holistic research policy' (Westgaard, 1985, p.127). In a traditionally decentralized research system, this is a considerable achievement. It was made possible by Olsen's wide science-policy experience, the contacts in government and industry that he and Sejersted were able to draw upon in agreeing the compromises inevitably necessary, the authority of the Science Policy Council at the time, and an environment conducive to research targeting following a successful priority-setting exercise by the Ministry of Industry. One result has been a growing recognition that a more centralized research policy has certain advantages, although at the same time the long-standing conflict over responsibility for this function has been revived. The demise of the SPC was partly hastened by the desire of the Ministry of Cultural and Scientific Affairs to assume greater responsibility for negotiating inter-departmental research priorities. However, other ministries have pressed for this task to be allocated instead to a new organization (dealing with both science and technology) attached to the Prime Minister's Office so as to limit the possible influence which can be exercised by individual departmental interests. If they are successful, the remit of the replacement body for SPC planned by the Ministry of Cultural and Scientific Affairs will be confined to university research.*

3. FORESIGHT IN BASIC SCIENCE

(a) *Ministry of Cultural and Scientific Affairs*

Around 75 per cent of basic research is undertaken within the academic sector. The Ministry of Cultural and Scientific Affairs is primarily responsible for such research and uses two funding mechanisms: (i) general infrastructural support for universities and colleges; and (ii) separately budgeted finance disbursed through the Research Council for Science and the Humanities. In addition, agencies under other ministries, especially the Royal Norwegian Council for Scientific and Industrial Research, provide project-based finance in strategic and applied areas. Most academic research support comes in the form of institutional core-funding (MCSA, 1986, p.46).

To date, little use has been made of foresight by the Ministry of Cultural and Scientific Affairs in determining policy for higher education research. However, the 1985 white-paper exercise showed the benefits of strategic planning, and in late 1987 the Ministry established a committee to consider the aims, organization and priorities for higher education and research up to the year 2000 (MCSA, 1987). Norwegian universities have traditionally enjoyed considerable autonomy, making it rather difficult to concentrate resources in areas of strength or promise. The Ministry aims to introduce more coordinated planning into the higher education system, and a period of restructuring seems likely. The new study is also examining the long-term needs of Norwegian society for trained scientists and engineers, and the findings will form an input to the 1989 white paper on research policy.

(b) *Research Council for Science and the Humanities*

The Research Council for Science and the Humanities (NAVF) is an umbrella agency responsible for four sub-councils — for Humanities Research, Medical Research, Natural Science Research and Social Science Research respectively. The NAVF budget for 1987 was 350 million Nkr, with the Council for Natural Science Research accounting for over a third of total spending.

The 1985 white paper had two effects on NAVF. First, the Council for Societal Planning was removed from its control and reconstituted as a separate Research Council for Applied Social Science. Second, the NAVF Board was made independent of the four sub-councils so that strategic planning and priority-setting might be carried out more effectively, one of NAVF's new roles being to 'prepare long-term plans, based on analyses and studies of the need for research' (NAVF, n.d., p.2). This led to discussions on corporate strategy and in 1987 a long-term planning process was set in place. Previously, NAVF relied upon rolling five-year financial plans, prepared by combining sub-councils' submissions and then engaging in 'horse-trading' until the overall programme fitted the budget available. As a result, it was difficult to alter appreciably the proportions of funds going to each sub-council.

The aim of the new planning procedure is to inject longer-term thinking into the budgetary process. For the first time, the Board can take a comprehensive overview of NAVF activities and make strategic choices. A joint secretariat from the four sub-councils was in 1988 preparing the draft plan, and officials in each agency responsible for major programme areas (e.g. economics) had been requested to identify promising research directions and priorities. The NAVF Board has final responsibility for the overview document. According to interviewees, acceptance of this approach has been assisted by the recognition that politicians are more likely to increase research funding if presented with a well-argued case. The resultant corporate plan, which is being modelled in part on those of UK research councils, is also intended to inform decision-making within the Ministry of Cultural and Scientific Affairs.

Aside from this, individual sub-councils have undertaken various research-planning exercises, but have made little use of systematic foresight techniques, relying instead on expert committees coupled with proposal-generated suggestions for future lines of work. The Council for Natural Science Research (RNF) has perhaps been most active in testing novel planning approaches, experimenting since 1984 with the Swedish method of field evaluations. In these, panels of foreign experts assess selected fields in terms of: (i) Norway's relative scientific strength; (ii) the past record of Norwegian research groups; and (iii) strategic priorities for future funding. Recent evaluations have included 'structural chemistry and physics' and 'condensed matter physics' (NAVF, 1986a; 1986b). Quantitative techniques such as citation analysis have occasionally been used, but generally the approach combines site-visits with committee discussions. While the quality of reports has been mixed, the main problem has been RNF's inability to act upon many of the recommendations — for example, to increase substantially the funding for a particular subfield or to restructure research around more successful groups. Although RNF generally implements the less expensive recommendations, panel reports have too often created unduly high expectations among researchers which have little chance of being met.* Consequently, the question of whether such evaluations should continue was under consideration in 1988, although the new NAVF planning system may well provide greater scope for implementing panel recommendations.

Ad hoc studies of specific fields are also sometimes undertaken — for example, to consider Norwegian participation in international space research or high-energy physics initiatives. These tend, however, to be carried out 'partly to justify formally a decision that has already been made informally'.*

Recently RNF has introduced a mechanism for selective research support similar to the *Schwerpunktprogramme* of the Deutsche Forschungsgemeinschaft. Priorities in 1987 included polar research, arctic marine biology and high-temperature superconductivity. These were not identified in any formal way, but arose from pressure within the scientific community. In framing a new programme, the normal procedure is to establish an expert committee to identify longer-term priorities within the area. Seminars in which scientists participate are also held. Other sub-councils under NAVF likewise tend to rely upon committee-based procedures and 'hearings' with the research community. These have increasingly adopted a longer-term perspective — for example, the Council for Social Science Research (RSF) carried out field surveys in 1986 for its main disciplines (e.g. economics and education). Hearings were held in all four universities to identify priorities over the next ten years (e.g. the economics survey report recommended increased emphasis on population studies). RSF was in 1988 considering carrying out an overall review of future priorities along the lines of the British report on *Horizons and Opportunities in the Social Sciences* (ESRC, 1987) which, together with the field surveys, would provide an input to the 1989 NAVF Corporate Plan.

All four sub-councils have faced management problems since funding for the national programmes began in 1986. This produced an unprecedented expansion in a few fields,[7] compared with the level budgets available to most non-targeted areas. NAVF has primary responsibility for the programmes on tradition and culture and health and the environment, and less problems have reportedly arisen than in those like biotechnology run by inter-agency steering committees.* Five-year programmes have been developed by NAVF committees in consultation with the research community; in the case of health and welfare, this included organizing three national conferences (e.g. RSF, 1987) as well as involving an element of foresight in that the programme committee tried to identify research areas likely to become important over the next fifteen years (e.g. the ageing population) if existing social trends continue (NAVF, 1987). Similar trends in other countries were also assessed (e.g. Rimpelä, 1987).

Overall, NAVF officials feel that the national programmes have stimulated a necessary discussion on longer-term strategy for basic science — something which would have been difficult without the increase in funding for such initiatives. Their main worries concern how and when priority areas will be terminated, and the possible consequences (e.g. an inability to attract new students) for important fields like chemistry which have not been accorded targeted support.[8] In an attempt to maintain scientific quality, research proposals in priority fields are still subject to normal peer-review, but there are doubts over whether the overall quality of applications is as high as for normal programmes.*

4. FORESIGHT IN STRATEGIC RESEARCH

As Figure 10.1 shows, responsibility for strategic and applied research is distributed among several departments. However, we shall focus primarily on the Ministry of Industries which oversees the Royal Norwegian Council for Scientific and Industrial Research (NTNF), historically the most influential funding agency and supporting much of the nation's strategic research. We also consider the foresight activities of the Defence Research Establishment, which has an international reputation in key fields like optoelectronics and information technology (Irvine *et al.*, 1981).

(a) Ministry of Industries

The Ministry of Industries undertakes no research foresight itself, leaving this to NTNF which is responsible for most planning and priority-setting tasks. Instead, the Ministry concentrates on establishing overall guidelines for strategic and applied industrial research. These are agreed with NTNF during: (i) the long-term planning process; (ii) negotiations over the annual budget submission to government; and (iii) discussions concerning national priorities for the quadrennial white paper on research policy.

As in Sweden, industrial research policy has changed considerably over time, with targeted support becoming more widely accepted since around 1983. This was partly in response to policy developments abroad, and partly to the Thulin Report which recommended greater stress on basic research and clearer priorities for longer-term R&D (Department of Industry, 1981). Prior to this, the emphasis was on general infrastructural support.

Another impetus to target priority areas, however, came with the realization by Ministry officials that longer-term generic research might prove popular to government and industry. In 1984, NTNF had identified six priority fields (deep-water processes, IT, aquaculture, materials science, CAD/CAM and geological mapping/prospecting) during the preparation of its long-term plan (considered below). Following discussions with the Ministry, these were narrowed down to four characterized by both national competitive advantages (e.g. health and medical services in IT for which a sizeable home market exists) and strong domestic technological competence. In seeking government support, it was crucial that there was consensus on the future socio-economic impact of target areas, and the Ministry played a key role in 'encouraging the research community to speak with one voice'.* For example, efforts were made to ensure that the two reports on IT research (NTNF, 1985b; STRAPIT Committee, 1985) 'came to the same conclusions so that it would be easy for government to make its decision on funding'.* One result is that longer-term industrial research was given high priority, the budget rising in real terms by 24 per cent between 1986 and 1988.

(b) Royal Norwegian Council for Scientific and Industrial Research

NTNF is by far the largest Norwegian research council with a budget of 757 million Nkr in 1987. A major beneficiary of the increase in government support for R&D, its spending is set to exceed 1,000 million Nkr by the end of the decade (NTNF, 1987a). In 1987, 55 per cent of its effort was targeted on aquaculture, biotechnology, IT, offshore technology and new materials, and this proportion will rise to 60 per cent by 1990. NTNF is central in supporting technical and industrial research in universities, public-sector institutes and companies. Over time, its responsibilities have changed, becoming closer to those of STU in Sweden. The emphasis on basic research and the university sector is increasing, and NTNF now finances studentships in technical science and engineering.

Following the Thulin Commission report, NTNF was substantially restructured, with its research institutes being granted greater autonomy so that the Council now operates primarily as a funding agency. The aim was to reduce previous conflicts of interest between intramural work in institutes and external research support, while at the same time enabling NTNF to take a more strategic role in determining longer-term policy. The Thulin Commission also recommended a reduction in the number and influence of NTNF committees, but this has not yet been fully achieved, with established interests (especially the large research institutes) retaining some of their political power.

(i) Long-term planning

Executive responsibility in NTNF is held by a Board of Directors which includes two representatives nominated by the sixteen 'cooperating institutes'. Officially the NTNF Council is charged with formulating long-term strategy, although in practice priorities continue to be influenced by the committee system which contains representatives from the main research performers. The long-term planning system is based on four-year forward looks which are updated biennially.[9] These are prepared by the Department for Planning (part of the NTNF Directorate) in consultation with the Board, and set out priorities and projected funding levels (NTNF, 1987b). During the year following each long-term plan, committees prepare their own plans as an input for the next forward look (e.g. NTNF, 1986a). Committee plans depend heavily on the 600 proposals submitted each year and have, until recently at least, represented reactive financial projections rather than proactive strategic decision-making.

An attempt was made in 1985 to introduce significantly greater centralized control of long-term planning. Rather than relying on inputs from committees, the official responsible for preparing the plan translated the broad directives of the newly established NTNF Council into a discussion document on alternative policy options (NTNF, 1985a). Two innovations were introduced: first, differing scenarios for NTNF research were proposed assuming budgetary increases of 2 per cent and 10 per cent respectively. In both scenarios, funding for priority areas would rise significantly, but in the former growth would come at the expense of existing activities. Second, the question of initiating new priority areas and terminating older ones was explicitly raised by earmarking over 5 per cent of the 1988 budget for a yet to be specified programme. This, it was hoped, would focus attention on strategic planning and introduce more flexibility into the system. The changes were, however, too radical and the effect was rather to cause an adverse reaction on the part of NTNF committees. The 1987 long-term plan consequently reverted to the traditional approach, dropping the notions of an undesignated priority area and alternative budget scenarios (NTNF, 1987b). Senior officials recognized that this reduced NTNF's ability to develop a longer-term strategy, and therefore decided to modify further the committee system and introduce centralized programme budgeting in 1989. The intention is that strategic planning should be more than merely a synthesis of committee plans.*

It is also hoped to expand the level of foresight in NTNF. In recent years, several future-oriented studies have been carried out to aid longer-term thinking. In particular, assessments of Norwegian industry's technical needs in the 1990s were commissioned from the NTNF Resource Policy Group and Stanford Research Institute (SRI, 1987; Nilsson, 1987; Resource Policy Group, 1986a). However, the findings of these reports were judged to be too general, and the impact on NTNF policy was marginal.* Another study by Arthur D. Little on Strategies for the Maritime Sector was equally disappointing, concentrating on management and organizational problems in the shipping industry rather than future R&D needs (A.D. Little, 1987). In general, foreign consultants have found it difficult getting to grips with the specific problems of Norwegian research, tending to apply inappropriate perspectives developed during previous analyses of R&D trends and opportunities in larger, less specialized economies.*

A study on Petroleum Technologies Towards the 1990s conducted at the same time had rather more impact (NTNF, 1986b).[10] In part, this reflects the fact that it was produced for a clear purpose (to plan research collaboration between Norwegian institutes and foreign multinationals required under the 'technology agreements' introduced in the early 1980s)[11] and had a defined user (the Ministry of Petroleum and Energy). However, the study was also technically more sophisticated. Building upon an earlier fifteen-year perspective for offshore research, it involved wide-ranging consultations with large numbers of researchers in institutes and companies, thereby helping to build consensus as well as ensuring that both technology-push and user-needs were considered.

Probably the most influential recent study has been a report on future needs for technical personnel (NTNF, 1986c). This was carried out by a committee established by NTNF to investigate problems identified in a previous white paper (MCSA, 1985b). Chaired by a leading industrialist, it used economic projections prepared by the Ministry of Finance to estimate skill requirements in different sectors over the period to 1995. The report concluded there would be an eventual shortfall of 1,500 graduate engineers, and recommended that student numbers be increased from 960 per annum in 1986 to 1,320 in 1995, with foreign universities being used for training where necessary.

NTNF intends to expand the Department for Planning to play a greater role in undertaking research foresight, and has increased the seniority of the planning officer. The key task remains the development of an acceptable institutional framework for longer-term planning which provides sufficient opportunities for participation by NTNF divisions and committees while still allowing strategic decisions to be taken. Whether such a balance is in practice possible remains to be seen since any further centralization of strategic planning is likely to be resisted by the committees.*

(ii) Priority research programmes

As noted earlier, changing political attitudes towards selectivity led in 1984 to the first attempt by NTNF and the Ministry of Industry to identify priorities. Although fairly informal, this did open up the possibility for subsequent more systematic foresight, in particular during the formulation of the national plan for information technology in 1985.

While opinions differ,[12] the IT plan can be seen as stemming from two parallel initiatives, each responding to widespread concern within companies, the Ministry of Industry and NTNF that Norway needed to follow other nations in developing a concentrated research effort in the field. The first involved the STRAPIT Committee, established by the Norwegian Association of Electronics Industries to propose a long-term IT strategy (STRAPIT, 1985). The Committee set up an expert working group (with representatives from companies, universities and research institutes) to assess the R&D needs of industry over the next fifteen years under different growth scenarios for the IT sector. It found that, even with modest growth, bottlenecks in the supply of scientists and engineers were likely across all major disciplines. Consequently, the Committee's main recommendations were that the government should at least double the output of graduates, and strengthen longer-term research within universities and institutes in fields like automation and process-control, sensor technology, microelectronic component technology and systems architecture. The need for increased government support of industrial R&D was also stressed (in particular, through public procurement programmes), as was the desirability of improved links with foreign researchers.

In the second initiative, an NTNF committee (established three months after STRAPIT) was given the task of developing priorities for a costed R&D programme to be submitted to the Ministry of Industry in time for the 1986 budget discussions. This came to broadly similar conclusions as the STRAPIT report,[13] giving greatest priority, at least in the initial stages, to graduate training and the promotion of basic research (NTNF, 1985b). One principle guiding its recommendations was that Norway has little choice other than to be a 'technological follower'. It should therefore concentrate on building the broad technical capacity needed to pick up emerging research developments abroad and exploit them in niche areas like the application of IT in medical services (where public procurement can guarantee a strong domestic market). The report also recognized that international contacts and collaborative research need to be encouraged if an effective 'watching brief' on new basic technologies is to be maintained.

Cross-membership of the two committees, together with some coordination by the Ministry of Industry, ensured that their recommendations did not differ markedly, the STRAPIT report demonstrating to government that industry judged the planned NTNF programme to be in the long-term national interest. After delays caused by changes in the Minister for Industry and then in the government itself (as well as a widening of the programme to bring in other agencies like the Ministry of Telecommunications), funding was finally authorized in 1987.

The growing stress on priority programmes in NTNF has further weakened the power of committees since target areas tend to cross-cut traditional industrial sectors and technical fields. This has resulted in conflict between committees and the secretariats of national programmes, the latter now having considerable influence on NTNF policy while retaining appreciable autonomy.* This conflict is one reason the national IT programme was placed under the direct control of the Ministry of Industries in 1988, although the main aim was to ensure that greater attention is paid by the research community to education and regional development. However, NTNF still provides the secretariat for the programme and thus retains considerable influence over its direction. In addition, the Council services the inter-agency steering committees for the national programmes on offshore technology and materials research (which also act as advisory committees for NTNF's own activities in the two fields).

(iii) Resource Policy Group

The Resource Policy Group (GRS) was set up by NTNF to provide an internal research policy capability as well as to analyse longer-term issues relating to resources and the economy. Its first head was a member of the MIT Limits to Growth forecasting team in the early 1970s, and he was given the task of building up a specialist research group for foresight studies and modelling. Although continuing to undertake work for NTNF (e.g. on Norway's industrial future), GRS is now more autonomous and has carried out future-oriented studies for other organizations like the Science Policy Council (e.g. on mechanisms for establishing research priorities — Resource Policy Group, 1984; 1986a).

The most recent GRS project with a major foresight component entailed organizing three seminars on Research and the Future during 1986 and 1987. The aim was to assist the Science Policy Council in identifying priority issues to be addressed in its background submission to the 1989 white paper on research (SPC, 1988b). The first seminar focussed on future challenges arising from the changing structure of industrial society, with lectures being given by two well-known foreign social scientists on lifestyles, leisure-time and the post-industrial society. The second examined trends in Norwegian society, papers being presented on visions of the future within the spheres of culture, the family, political institutions and the production system. The final seminar explored challenges that these trends pose for future research in the natural sciences, humanities, economics, medicine and biology (Resource Policy Group, 1986b; 1987a; 1987b).

Although several issues identified in the seminars were taken up in the SPC report (such as the links between higher education and research, and the special needs of smaller countries), the overall results of the exercise were regarded as somewhat disappointing.* It had been hoped to attract politicians and industrialists, but only members of the Science Policy Council and academics attended. Even so, the number of participants was much larger than planned with the result that it was impossible to keep the debate structured. Matters were not helped by the wide diversity of the lectures, which added to the difficulties faced by GRS in drawing coherent conclusions. Perhaps the main problem, however, was that too little effort was put into synthesizing the papers and discussions.*

Overall, the Resource Policy Group has yet to find an effective means of relating to national research-policy discussions. In part, this is because of the essentially political process through which priorities are set, where background research is not necessarily a very important determinant in decision-making. However, it may also be related to a view among some government officials that the group is not equipped with a sufficiently broad range of science-policy skills, its staff being predominantly social scientists and economists.

(c) *Norwegian Defence Research Establishment*

The Defence Research Establishment (FFI) is widely recognized as the most innovative laboratory in the government sector. Around 10 per cent of its R&D programme is devoted to strategic research in areas like integrated electro-optics and image-processing, where the time-horizon for application in new military technology is up to fifteen years in the future. One of its strengths has been the emphasis placed by management on strategic planning and foresight. Much effort is devoted to monitoring R&D developments world-wide, primarily through informal links with defence laboratories in other NATO countries and with US high-technology companies, but also through attendance at specialized international conferences.[14] FFI has a long-established policy of encouraging younger staff to spend a year or so as visiting researchers in foreign defence laboratories and companies, thus providing a mechanism for monitoring and accessing information on emerging areas as well as building up the network of overseas contacts vital for a scientifically peripheral country like Norway. This policy was largely responsible, for example, for FFI's success with imported digital-processing technology in the early 1960s, leading to the establishment by former staff of Norsk Data (Irvine *et al.*, 1981).[15] It has also been a key factor in recent work on digital communications, which has reportedly resulted in Norway becoming the first NATO country to operationalize a fully-integrated military communications system.*

In identifying new research opportunities, an attempt is made to analyse the dynamics of important emerging technologies. This is undertaken by senior staff with a broad overview of different research fields and an understanding of Norway's very specific military requirements. The aim is to develop basic technological concepts to guide future R&D rather than to identify specific incremental developments which industry and larger foreign laboratories are generally better placed to undertake.

Although longer-term perspectives figure centrally in research planning, little use is made by FFI of systematic forecasting techniques. There is, however, a formal system of annually updated five-year plans. These include: (i) an analysis of developments in science and technology with military potential; (ii) an examination of technology-related military problems stemming from Norway's strategic situation (produced by the FFI Systems Analysis Group); and (iii) an assessment of available resources. The aim is to set out a detailed programme for the five-year period, taking into account defence requirements and the laboratory's technical capabilities. As an input, experts undertake surveys of research opportunities and likely trends over the next fifteen years in core areas like image processing. The plan is then discussed and revised by the Defence Research Policy Board (whose members include senior military staff, laboratory officials and researchers). Meetings are also held with military personnel in operational branches to discuss technical details of possible projects before the plan is finally submitted to the Defence Minister.

The Director of FFI stressed that the role of the formal planning system is more to communicate with military users, and gain their agreement, than to select priorities. Decision-making is in practice relatively informal, with senior staff (who are in regular contact with the armed forces and so are aware of user-needs) having a good understanding of the technologies likely to impact on military activities over the next decade. An important mechanism for generating internal consensus on laboratory priorities is an annual two-day meeting within each FFI division to speculate on future technical developments. This is seen as useful in developing longer-term thinking among researchers.*

Senior FFI staff have in the past been especially vocal advocates of greater concentration of government research expenditure on priority areas and increased integration of R&D policy. Finn Lied, former director of the laboratory, was instrumental in initiating discussions on targeted national programmes,[16] while Karl Holberg, the ex-research director, was involved in the STRAPIT Committee. However, despite the undoubted progress in recent years, FFI management feel that too much Norwegian research remains characterized by mediocrity due to an inability to concentrate funds on priority areas and the strongest groups. The consensus approach to decision-making inevitably involves political and institutional compromise, which limits the possibilities for implementing the results of research foresight.* As a recent OECD report on Norway observed:

> The mutual 'consultation' and feedback procedure — especially that of calling for comments (høring) from all government authorities and all bodies concerned — no doubt takes the edge off controversy over policy. But at the same time it dampens any inclination to innovate or put forward original proposals to recast the system. [OECD, 1985, p.20]

5. RESEARCH FORESIGHT IN INDUSTRY

Preliminary discussions with Norwegian firms established that little longer-term research foresight is undertaken and only one company, Norsk Hydro, was therefore visited. It specializes in light metals, fertilizers, oil and petrochemicals, and had a turnover of 55,000 million Nkr in 1986. The company previously faced severe problems as a result of the downturn in the oil sector, and has been diversifying into science-based areas (such as aquaculture and marine biochemicals) linked to existing technological strengths. Although still only spending just over 1 per cent of turnover on R&D, the company has been expanding its activities and now operates Norway's largest industrial research laboratory.

Strategic research in Norsk Hydro is principally funded from corporate sources rather than through divisions, and in 1987 accounted for 15 per cent of total R&D. Formal techniques have occasionally been employed to aid decision-making on longer-term research investments — for example, to move into biotechnology in 1970. Consultants identified market opportunities for biotechnology products and areas where the company might profitably establish a presence. In this and other foresight exercises, far greater stress has generally been placed on market forecasts than on technology- push factors. Quantitative techniques developed by US consultants are regularly used to assess possible R&D projects in terms of (i) cost, (ii) potential production savings or product income, and (iii) probability of technical success. The estimated 'percentage likelihood of technical success' is then plotted against 'economic potential divided by cost', the aim being to identify projects which combine a high probability of technical success with significant economic returns. It is claimed that the technique helps to preserve a balance between 'bread and butter' research and more speculative but risky work.*

In 1985, a high-level 'innovation committee' was established to oversee an expansion of strategic research, and this led to moves to strengthen longer-term planning which was regarded as having previously been too conservative. A set of core research themes has since been identified which relate to areas of current and future diversification such as new materials (e.g. light metal composites). More attention is also being given to monitoring and analysing the opportunities arising from research elsewhere. Given its increasing reliance on basic research, the company supports recent government initiatives to emphasize priority fields. However, significant restructuring of universities and government-funded institutes is felt to be necessary if the public research system is to be upgraded to meet the future needs of industry.*

6. SCENARIOS FOR NORWAY IN THE YEAR 2000

The final foresight exercise to be considered was conducted by a multidisciplinary team at the Norwegian School of Management (under the direction of Terje Osmundsen, a policy adviser to the former Conservative Prime Minister). Its origins stem from a belief that the tradition of consensus in Norway has politicized long-term planning, preventing the strategic decisions necessary to revitalize the economy (Osmundsen, 1986). Work has concentrated on mapping the socio-economic, political and technological trends that will influence Norway's future development. This has involved the use of scenarios to explore the potential longer-term effects of differing sets of structural changes. Here, 'the basic idea [is] thus not to forecast or plan for a given future, but rather to contribute to the understanding of those forces, uncertainties and opportunities that shape the future' (BI Institute, 1987b, p.1). The project's budget of 8 million Nkr for the period 1986–8 was provided by a consortium of ministries, companies and NTNF.

The study is interesting in that it represents one of the few recent attempts to use the global scenario-based techniques which went out of fashion in the late 1970s. Three scenarios were constructed to characterize possible directions of Norwegian development: (i) 'suicide and comfort' — the continued availability of oil revenues, coupled with the increasing political and economic power of women, results in greater expenditure on the 'caring society' to the detriment of investment on industry and research; (ii) 'slowly developing economic crisis and erosion of political legitimacy' — the country is squeezed by over-dependence on oil and commodities, with successive cyclical downturns in the international economy placing greater strains on collective decision-making mechanisms, resulting in individuals and organizations choosing an 'exit solution' to escape from their own problems at the expense of the community; (iii) 'pragmatic adjustment to change' — the country's traditional ability to adapt gradually to essential institutional changes reasserts itself (for example, tax reforms are carried out and EEC membership is sought). In developing these scenarios, varying assumptions about the role of technology in social change were made, with detailed studies being undertaken of the future impact of new materials research, IT and biotechnology (BI Institute, 1987a; 1987b).

According to Osmundsen, the results of the study have affected strategic thinking in both government and industry, contributing, for example, to the recent decision to slow down exploration of the continental shelf in order to ensure that Norway does not become too dependent on oil. Osmundsen believes this impact has much to do with the processes by which the study was carried out (with a steering committee of sponsors involved from the outset) and the results were disseminated (through regular seminars and briefings — for example, to the Ministries of Industries and Finance). As far as research foresight is concerned, the impact is likely to be limited to stimulating longer-term thinking in companies and government agencies as to future technological needs, a strategic activity Osmundsen believes has previously been lacking. It is, however, too early to assess what effect the study will have on research-policy decisions in government, funding agencies and companies.

7. CONCLUSIONS

The overall level of Norwegian research foresight is low compared to that evident in most of the countries considered previously, perhaps not surprisingly given the nation's small population and its dependence on oil and petrochemicals, primary products and a few specialized industrial areas which have not been especially science-intensive. Until now, the aim of government has been to ensure that a broad research infrastructure exists so that R&D opportunities emerging overseas can be exploited domestically. The rapid progress in strategic fields like biotechnology (with its implications for aquaculture and wood products) and information technology means that there has, nevertheless, been a move towards greater targeting in recent years. Consequently, the main conclusions that can be drawn concern the process of identifying national priorities employed in preparing the 1985 white paper on research policy.

First, despite the politicized nature of decision-making in Norway and the history of competition between ministries with R&D responsibilities, it was still possible to achieve consensus on longer-term priorities. This was facilitated by a foresight process which: (i) encouraged a wide variety of inputs from the scientific community through a more open consultative framework than hitherto employed; (ii) focussed discussion on priorities around both science-push and demand-pull concerns over the next ten years; (iii) limited the scope for conflict by establishing broad target areas only; (iv) ensured that most research interests were satisfied by targeting fields spanning the entire spectrum from cultural studies to engineering; (v) maximized the likelihood of implementation by involving an advisory committee of high-level ministry officials; and (vi) concentrated only on making the case for additional expenditure on new lines of research, steering clear of questions concerning restructuring and identification of low priority areas which might have derailed the exercise. Although such a model probably has limited general applicability, it may be relevant to smaller countries with socio-political structures which make it difficult to introduce selective R&D support policies.

Second, while the time was ripe for such an initiative (with priorities already having been established by the Department of Industry and NTNF), a key role still needed to be played by individuals such as Olsen and Sejersted in achieving consensus among ministries, funding agencies and researchers, in ensuring institutional legitimacy, in encouraging inter-departmental participation and in providing a direct channel for recommendations to sympathetic ministers. As in other countries, the existence of a strategically placed 'product champion' was crucial to the success of such a foresight exercise aiming to set national priorities for the first time.

Third and finally, Norwegian experience shows that targeting of promising research need not be limited to natural science and engineering disciplines with technological or industrial relevance. In the survey of research needs and opportunities over the next decade, applied social science was widely judged to have an important role in improving the quality and effectiveness of decision-making, while a programme on Norwegian culture and tradition was also accorded priority support. Among the most valuable lessons for other countries is the recognition that social and cultural research is not only important to the future of society, but may also make significant indirect contributions to national economic and technological regeneration. As such, decisions on long-term support for these areas should not be relegated to the margins of research policy.

NOTES

* Refers to material collected during interviews with Norwegian officials in late 1987, while † denotes information drawn from subsequent correspondence.

1. A background document on 'research for industrial renewal' was also commissioned from the NTNF Resource Policy Group with the aim of identifying mechanisms for establishing priorities (Resource Policy Group, 1984). This produced only very general recommendations and apparently had limited utility.*
2. The Science Policy Council played an important role during preparation of the report, providing comments and inputs to drafts. Much of the white paper's content was derived from an earlier green paper published by the Council's predecessor in November 1982.†
3. In part, this reflected criticism by ministries of what they saw as the highly academic and ideological nature of much social science.*
4. This debate was initiated by Finn Lied (ex-director of the Defence Research Institute) and Johannes Moe (director of SINTEF), two influential figures in the research community. Subsequently, several institute directors wrote to the Prime Minister outlining the case for a coordinated national effort to target research areas important for future competitiveness. The Thulin Commission (Department of Industry, 1981) also helped place priority-setting on the political agenda. It was, however, the Ministry of Industry which first established strategic priority areas in 1984 (in association with NTNF), and this provided a model for the wider exercise described above.
5. One result was that the personal adviser to the Prime Minister, Terje Osmudsen, became interested in the initiative and gave the Prime Minister's Office a role to play in securing the active participation of relevant ministries.†
6. In particular, Norway became a member of the European Molecular Biology Laboratory in 1986 and of the European Space Agency in 1987.
7. For example, funds provided to NAVF for biotechnology rose appreciably in 1986 and 1987 (Furseth, 1988, p.43).
8. A study of the impact of priority programmes undertaken by the NAVF Institute for Studies in Research and Higher Education (Furseth, 1988) found they had developed in more applied directions than originally intended, and that a shortage of skilled researchers was evident in certain areas (e.g. biotechnology). There was also evidence that the mechanisms for inter-agency coordination of some programmes were not working as well as expected, organizational rivalries making it difficult to develop a coherent plan. Finally, concern was expressed about whether the scientific community had promised economic and technological benefits that stood little chance of being delivered within the time-horizon of programmes.* Also relevant here is a recent review of NAVF which emphasized 'the need clearly to stand for enhancing the quality of Norwegian science. For officials taking this view, the essential danger which NAVF must face up to is less irrelevancy than mediocrity' (SPC, 1988a, p.95).
9. Prior to 1986, long-term plans were updated annually, but this placed undue pressure on the NTNF administrative system.*
10. This was carried out by the NTNF Petroleum Technology Secretariat on behalf of the Ministry of Petroleum and Energy.
11. The 'technology agreements' were primarily intended to ensure that technology transfer was built into the North Sea activities of multinational oil companies.

12. There have been two studies on the origins of the national information technology programme. The first, funded by the Science Policy Council and carried out by a university group at Trondheim, stressed the primacy of Olsen's white paper (Tøssebro and Buland, 1987). The second, undertaken by the Resource Policy Group, accorded major responsibility to the STRAPIT Committee (Borg *et al.*, 1987). Interviewees felt that, while both studies provided interesting historical descriptions, they contributed little to improved R&D management.*
13. One difference was that STRAPIT stressed hardware technology, while the NTNF committee regarded the development of software expertise as equally important.
14. Staff are encouraged to publish widely and to develop a presence at the forefront of their field. The relatively open publication policy is also seen as important for motivating scientists, especially those just embarking on research careers.
15. There are other examples where longer-term basic research by FFI laid the basis for the establishment of successful science-based companies — for example, Simrad Optronics grew out of the laboratory's work on electro-optics (Irvine *et al.*, 1981).
16. See note 4 above.

REFERENCES

A.D. Little (1987), *Strategy for the Maritime Sector,* report to NTNF, Oslo, Royal Norwegian Council for Scientific and Industrial Research.

BI Institute (1987a), *Søkelys på Fremtiden — Noen Byggeklosser,* Oslo, Norwegian School of Management.

—— (1987b), *S-2000: Ståstedet 1987 — Spenninger, Trender og Utfordringer,* Oslo, Norwegian School of Management.

Borg, A. Lotherington, A.T. and Stenseth, Å. (1987), *Framveksten av Satsingsområder: Prosessen som Forte til det Nasjonale Handlingsprogrammet for Informasjonsteknologi,* Oslo, Resource Policy Group.

Department of Industry (1981), *Forskning, Teknisk Utvikling og Industriell Innovasjon,* NOU: 30A and 30B, Oslo, Universitetsforlaget.

—— (1984), 'Recommendations from a Working Group Evaluating Growth Stimulation Measures in Connection with the 1985 Budget' (mimeo), 5 May, Oslo, Department of Industry.

ESRC (1987), *Horizons and Opportunities in the Social Sciences,* London, Economic and Social Research Council.

FSU (1985), *National Plan for Biotechnology,* Oslo, Joint Committee of the Research Councils.

—— (1987), *R&D Statistics 1985–87,* Oslo, Committee for R&D Statistics of the Norwegian Research Councils.

Furseth, I. (1988), *Hovedinnsatsområdene i Norsk Forskning,* Oslo, NAVFs Utredningsinstitutt.

Irvine, J., Martin, B.R. and Schwarz, M. (1981), *Government Support for Industrial Research in Norway — A SPRU Report,* Appendix 4 of Department of Industry (1981), op.cit.

MCSA (1985a), *Om Forskningen i Norge — St Meld Nr.60 (1984-85),* Oslo, Ministry of Cultural and Scientific Affairs.

—— (1985b), *On Higher Education, White Paper No.66 (1984-85),* Oslo, Ministry of Cultural and Scientific Affairs.

—— (1986), *On Scientific Research in Norway — White Paper No.60 (1984-85)* (English version), Oslo, Ministry of Cultural and Scientific Affairs.

—— (1987), *Norwegian R&D Policy,* Statement Presented at OECD Ministerial Conference, October, Oslo, Ministry of Cultural and Scientific Affairs.

NAVF (n.d.), *Norwegian Research Council for Science and the Humanities,* Oslo, Research Council for Science and the Humanities.
—— (1986a), *International Evaluation: Norwegian Structural Chemistry and Physics — Research Using Diffraction Techniques,* Oslo, Research Council for Science and the Humanities.
—— (1986b), *International Evaluation: Research on Condensed Matter Physics in Norway,* Oslo, Research Council for Science and the Humanities.
—— (1987), *HEMIL Nasjonal Handlingsplan 1987-1992,* Oslo, Research Council for Science and the Humanities.
Nilsson, J.E. (1987), 'A Perspective on Norway's Industrial Future', *Futures,*vol.19, pp.455–63.
Norsk Hydro (n.d.), *Norsk Hydro,* Oslo, Norsk Hydro.
Norsk Hydro (1986), *Hydro,* no.2 (English edition), Oslo, Norsk Hydro.
NTNF (1985a), *Langtidsplan 1986–1990,* Oslo, Royal Norwegian Council for Scientific and Industrial Research.
—— (1985b), *Et Forslag til et Satsingsprogram pa Informasjonsteknologi (IT) i Norge,* Oslo, Royal Norwegian Council for Scientific and Industrial Research.
—— (1986a), *Komite for Bygg- og Anleggsteknisk Forskning (Committee for Building and Construction) — Langtidsplan 1987–1990,* Oslo, Royal Norwegian Council for Scientific and Industrial Research.
—— (1986b), *Research and Development for Norwegian Industry — Petroleum Technologies Towards the 1990s,* Oslo, Royal Norwegian Council for Scientific and Industrial Research.
—— (1986c), *Future Needs of Highly Qualified Technological Personnel,* Oslo, Royal Norwegian Council for Scientific and Industrial Research.
—— (1987a), *Strategy 1988–1992,* Oslo, Royal Norwegian Council for Scientific and Industrial Research.
—— (1987b), *Langtidsplan 1988–1992,* Oslo, Royal Norwegian Council for Scientific and Industrial Research.
OECD (1985), *Reviews of National Science Policy — Norway,* Paris, Organization for Economic Cooperation and Development.
Osmundsen, T. (1986), 'Scenarios for Norway Year 2000', *Futures,* vol.18, pp.549–52.
Resource Policy Group (1984), *Research and the Future: Research Renewal,* vol.3, no.2, Oslo, Resource Policy Group.
—— (1986a), *Norway's Industrial Future,* Oslo, Royal Norwegian Council for Scientific and Industrial Research.
—— (1986b), *Research and the Future: A New Industrial Model,* vol.5, no.2, Oslo, Resource Policy Group.
—— (1987a), *Research and the Future: Trends in Norway,* vol.6, no.1, Oslo, Resource Policy Group.
—— (1987b), *Long-Term Societal Changes and Research Policy Priorities,* Oslo, Resource Policy Group.
Rimpelä, M. (1987), *Trends and Prospects for Lifestyle and Health Promotion Research,* Oslo, Council for Social Science Research.
RSF (1987), *Health, Environment and Living Conditions: Preparatory Report No.3 for National Programme on Health and Environment,* Oslo, Council for Social Science Research.
SPC (1988a), *NAVF — En FoU-Partner Med Utviklingspotensiale,* Oslo, Science Policy Council.
—— (1988b), *Towards a Knowledge-Based Society,* Oslo, Science Policy Council.
SRI (1987), *The Future for Manufacturing Industry,* report by Stanford Research Institute to NTNF, Oslo, Royal Norwegian Council for Scientific and Industrial Research.
STRAPIT Committee (1985), *Strategisk Plan for Norsk Informasjonsteknologi — Industri fram Mot Å 2000,* Oslo, Elektronikkindustriens Bransjeforening.

Tøssebro, J. and Buland, T. (1987), *Informasjonsteknologi og Innovasjonspolitikk*, Trondheim, Institutt for Industriell Miljøforskning.

Westgaard, L. (1985), 'National R&D Priorities and Priority Setting in Basic Research', in European Science Foundation, *Rise and Fall of a Priority Field*, Strasbourg, European Science Foundation, pp.127–34.

11 A framework for successful research foresight

1. INTRODUCTION

Our aim in this final chapter is to synthesize the lessons emerging from the eight country studies, and to set out the principal factors determining success and failure in research foresight.[1] This is no easy task since the criteria for judging 'success' vary widely, as we have seen, across different types of foresight. In particular, it cannot always be assessed in terms of how accurate forecasts subsequently prove since the intention may, for example, have been to highlight a potential problem (like the 'greenhouse effect') which is then avoided by taking appropriate action. The only yardstick against which to measure success must ultimately be whether useful insight and guidance was provided,[2] the precise criteria depending on the nature of the foresight activity and the differing needs of sponsors and other stakeholders interested in the results. Thus, where the objective is to inform priority-setting, success can be evaluated in terms of subsequent influence on establishing research policies oriented towards achieving identified goals. Where the aim is anticipatory intelligence, relevant criteria include foreseeing potentially significant threats and opportunities, while success in foresight conducted for educational or communications purposes depends primarily on whether it stimulates debate and consciousness-raising on important future issues.

The wide range of prospective studies reviewed in the country chapters provides a rich empirical basis for drawing conclusions about the various types of foresight. After first outlining the main trends and general lessons arising from international experience, we identify a number of factors to be taken into account in planning a foresight initiative. More specific conclusions are next considered concerning the organization of activities oriented to priority-setting. Finally, we show that an analogy with the far better understood, but no less complex, process of technological innovation is useful in highlighting the overall determinants of success and failure in research foresight.

2. GENERAL TRENDS AND LESSONS

Given the widespread rejection of research forecasting in the late 1970s, the renewed emphasis on prospective analysis in formulating policies for science and technology over recent years is a significant phenomenon. One reason for this change is the recognition that a longer-term planning perspective is necessary for basic science given the magnitude of the impacts anticipated on industrial technology and socio-economic well-being. However, a second factor has been the adoption of a less deterministic foresight model in which the aim is not to predict what the future *will* be, but rather to explore what it *might* be if certain policy choices are made today (Elzinga, 1983). This places a premium on timely information concerning possible options as they materialize, and on systematic assessment of the underlying scientific, techno-economic and other forces likely to structure their future development. Consequently, governments and companies in all eight countries have been investing heavily in setting up global intelligence systems to monitor — and to detect at as early a stage as possible — scientific advances with the potential for meeting technological, economic and socio-cultural needs. In particular, extensive use is being made of information science techniques in exploiting the burgeoning number of on-line patent and literature data-bases which are now available like the *Science Citation Index*.

As in many aspects of research foresight, the Japanese were first to appreciate the strategic value of a computerized national R&D information system, establishing the Japan Information Center of Science and Technology in the late 1950s. France is likewise taking steps to develop a capability for world-wide scanning of the research literature, recognizing that an effective infrastructure for anticipatory intelligence is often a precondition for corporate and government foresight aiming to inform priority-setting.[3] This does not mean a diminution in importance of direct expert-based monitoring approaches such as the system of overseas technical attachés set up by Sweden to identify emerging threats and opportunities in R&D. Indeed, greater use is also being made of researchers (especially by companies) to track developments in strategic fields on a regular basis, a trend accompanied by increased awareness among science-based firms that a healthy in-house basic research effort is essential for tapping into the informal communications networks through which advances in knowledge are diffused prior to publication. Similarly, large Japanese consultancy organizations like Mitsubishi Research Institute have led the way in deploying teams of technical specialists to evaluate systematically the novel technologies incorporated in new products and processes exhibited at international trade-fairs.[4]

Although the results of our review show that foresight is far more common in strategic and applied research, there is also a growing awareness of the potential value of prospective analysis in shaping policy for fundamental science.[5] This is true not only in countries with established planning traditions like Japan and Sweden. Even in the United States, national priority-setting is finally on the political agenda following the recognition that existing science-policy mechanisms are incapable of handling the difficult choices necessary at a time of budgetary constraint, most notably in relation to whether the Superconducting Super Collider should be given priority over investment in other more strategic research initiatives.

In addition, concern has begun to emerge about the potentially adverse consequences for society of work in certain rapidly evolving areas of basic science like molecular design and genetic engineering. This serves as a reminder that foresight can be essential not only in priority-setting, but also in a technology-assessment role, helping 'mitigate the waste, dislocation and fear left in the wake of scientific progress' — that is, in performing what has been termed a 'garbage-can or clean-up' function in which 'society is forewarned about the economic, social, political and ecological problems of science itself'

(Dunn, 1988, p.3). Such worries have been most prominent in Scandinavia, and this is one reason for the emphasis on socio-cultural factors in Norwegian and Swedish foresight activities and for the fact that disciplinary coverage has been extended to encompass the social sciences and humanities. The experience of Japanese agencies like the Council for Science and Technology highlights the fact that social criteria can be given due recognition in foresight without in any way detracting from techno-economic considerations.[6]

The move towards what Ziman (1987) has characterized as 'steady-state science' has been a major factor underlying the adoption of research foresight by academic funding agencies. In particular, it provides a structured framework for assessing the relative merits of claims by different fields for budgetary support, a task which often proves impossible on grounds of intrinsic intellectual merit alone. By injecting broader socio-economic and other criteria into science-policy decisions of this sort, foresight offers a mechanism for mediating conflict over disciplinary priorities at the same time as satisfying societal demands for greater relevance (demands which have grown as the strategic importance of many basic fields has become more apparent). However, it can also help contain the financial pressures which have arisen from the increased technical sophistication of research. Because scientific progress often depends on radical innovation in measurement and experimental techniques (de Solla Price, 1984), the level of investment in instrumentation required to push forward the research frontier has escalated in most fields with the result that choices on priorities can no longer be avoided (Schmitt, 1985, pp.26-7). Moreover, there is a greater need for foresight to cope with uncertainties introduced by longer-term planning horizons for large central facilities. As we have seen, the problems which first arose in traditional big science specialties such as astronomy and nuclear physics are now common in fields like geology and climatology.

Another reason underlying the growing acceptance of foresight stems from the recognition, even by scientists, that traditional discipline-based mechanisms for decision-making have not always proved effective in channelling resources sufficiently rapidly from declining to emerging areas of research. This is especially an issue in funding policy for fusion fields like chematronics where disciplinary committees have generally found it difficult either to identify promising areas or to judge their potential significance. At the same time, they may become embroiled in institutional conflict over administrative responsibility — witness the infighting in Japan over the Human Frontier Science Programme and in relation to sequencing the human genome in the United States. However, perhaps the most intractable problems result from the fact that fundamental paradigm shifts in science often originate at the periphery of disciplines, with the researchers involved engaged in unconventional work which is seen as threatening established cognitive and social structures. In the ensuing struggle, scientific progress is made only when the minority view succeeds in overthrowing that of the majority.[7] This is usually a lengthy process since those defending the established wisdom will do so from positions of institutional power. Research foresight offers a means of confronting and overcoming rigidities of this sort in the mechanisms for redeploying scientific resources, in particular enabling a broader range of inputs to be made by experts with no specific disciplinary interests to defend. A fruitful model for handling foresight in emerging strategic areas is that employed in 1985–6 by the Swedish National Board for Technical Development (STU) in establishing its interdisciplinary programmes on topics like 'micronics'. In more basic fields, much can be learned from the 1982 exercise undertaken by the French Centre National de la Recherche Scientifique (CNRS) to identify intellectual growth-points at the boundaries between nuclear physics, particle physics and astronomy.

Despite such successes, considerable scepticism continues to exist in most countries, particularly among academics, as to whether research foresight is intrinsically feasible.[8] The argument advanced is that basic science generally evolves in too uncertain a fashion for prediction to be possible. Nor can impacts on technology and the economy be forecast any more easily since they tend to be diffuse and to arise only over the longer term (again through a serendipitous process). Researchers in several countries cited the 1986 breakthrough in high-temperature superconductivity as evidence to support this view, arguing that the discovery was entirely unforeseen and indeed unforeseeable. Yet foresight activities in Japan, especially by industry, had already resulted in the field being judged ripe for advance following intriguing but largely unrecognized results published in 1981 by Danish and French scientists. The increased investment and extensive monitoring of overseas activities which followed ensured that Japanese researchers were well placed to exploit the opportunities opened up by the breakthrough even before it was formally announced by IBM. Achieving such success in foresight is not so implausible as sceptics might argue since scientists themselves often agree on likely breakthrough areas, this being one reason why simultaneous announcements of discoveries are sometimes made by researchers working independently. Hence, the unpredictability of basic science should not be exaggerated.

Another factor impeding the introduction of foresight has been the vigorous defence mounted by academics and other scientists keen to protect their long-established autonomy to determine funding priorities free from political and other 'external' pressures for relevance. This is despite a wealth of evidence that the demand-pull perspectives of research users can be valuable in assessing long-term prospects even in relatively basic fields. While governments have usually countered such resistance by introducing targeted priorities from above for a certain proportion of funds, in other cases the response has been to impose radical organizational restructuring — for example, the dismissal of the Science Council of Japan in the mid-1980s or the enforced reorientation of the Australian Commonwealth Scientific and Industrial Research Organisation (CSIRO) towards more strategic areas of work in 1987–8. Although government officials may justify such action as the only available means of dealing with entrenched interest groups, top-down policies have tended to alienate scientists and to reinforce their worst fears about the external regulation of research. A preferable alternative may therefore be to adopt initially the strategy pursued in France of persuading scientists that their interests would be better served by taking the lead in introducing new approaches to foresight and long-term planning. This was certainly true in the Canadian field survey of earth sciences which generated consensus on the Lithoprobe project, a large-scale interdisciplinary initiative regarded by scientists as intellectually exciting and by industry as technologically and economically promising. More generally, our findings suggest that foresight will become accepted as an integral part of decision-making in basic science only when researchers are convinced that it does not lead to 'distorted' priorities, and when they learn that requests for increased funding are likely to be received with greater sympathy if accompanied by a systematic appraisal of the potential long-term benefits (whether these be economic, social or cultural).

Experience in several countries shows, however, that proponents of foresight would do well to heed the concern of scientists about the associated danger of limiting the creative dissent which so often provides the seeds for fundamental breakthroughs. Consensus-seeking Japanese-style exercises which draw on inputs from large numbers of researchers may well facilitate subsequent acceptance of foresight results, but the advantages need to be weighed against the intrinsic conservative tendencies of majority judgements. Resolving this contradiction is difficult, as we have seen, and calls for considerable care in ensuring that competing viewpoints are taken fully into account — especially in research areas where irreconcilable paradigms coexist. One option here is to focus foresight not only on 'what if' questions but also on those addressing 'what if not' possibilities (Van Dongen, 1988, p.9). Another is to employ scenario-based approaches to map and contrast alternative development paths for competing scientific paradigms. The need for conceptual pluralism means that variety should wherever possible be planned into research. A good example is the Japanese Basic Technology for Future Industries Programme where the first more basic phase in projects emphasizes structured competition among different scientific and technological approaches. However, decision-makers have as yet given insufficient attention to such issues, and especially to the implications of strategic targeting of research[9] which has resulted in a high degree of uniformity in national priorities across countries (the selected areas invariably including biotechnology, new materials and information technology). Among the possible solutions is to adopt the approach employed with some success by the Canadian Natural Sciences and Engineering Research Council (NSERC) in including an 'open' area as one of its strategic priorities.

Given the alien nature of foresight to the scientific community, its introduction involves a long learning process. The examples of Australia, Canada, France and Sweden demonstrate the importance of adopting an incremental strategy of initially exploring a range of possible approaches, and through trial and error gradually establishing procedures which work effectively in the prevailing policy environment and are accepted by scientists. Ultimate success tends to come only through perseverance in tackling the problems which arise when efforts are first made to deploy foresight:

> What may in a particular instance be seen as a failure, when analysed is often a step toward further improvement, a revelation of a defect in assumption, in data, or in analytical technique. Therefore, one of the most healthy aspects of foresight can be a steady build-up of knowledge, experience and saliency in relating to policy. [Coates, 1985, p.42]

In attempting to encourage more widespread use of foresight, greatest difficulties are often encountered in initiating a positive learning cycle so that researchers and others are able to see clear advantages in terms of improved science policies. While this entails careful appraisal of failures to determine why they occurred, it is equally important to avoid a situation where early foresight activities have a potentially adverse effect on the scientific community or on other influential stakeholders. Criticism of the 1981 initiative by the Australian Science and Technology Council (ASTEC) to identify nationally important research areas came principally from scientists in fields perceived to be under threat. The utility of systematic foresight had still to be demonstrated, and technical doubts on the validity of this exploratory exercise made it less easy to launch subsequent studies. A more effective strategy for the initial deployment of foresight is to focus on identifying novel research opportunities which, if sufficiently promising, will be pursued using resources additional to those already committed. Experience with establishing the 'mobilizing' programmes in France, and the prospective exercise undertaken in Norway to identify priorities for the 1985 White Paper on Science Policy, shows that the participation of researchers is far more likely to be secured by avoiding any suggestion of a zero-sum game in relation to future funding.

This problem of gaining acceptance from the scientific community reflects a broader lesson emerging from our review, namely that *authority*, *legitimacy* and *credibility* are fundamental to success in foresight.[10] In an exercise oriented to priority-setting, for example, the authority required to ensure policy impact may come from the involvement of high-level officials in framing the terms of reference, or by enrolling those in a position to influence budgetary and policy-making processes. Legitimacy is no less crucial since the endorsement of scientists and their professional associations is essential if foresight results are to be taken seriously — there are many cases where an exercise has ultimately been derailed by cynicism from the wider research community.

Legitimacy is most commonly guaranteed by involving experts and opinion-formers from the main stakeholders, performers and users (government, funding agencies and research). In addition, foresight exercises should be conducted in accord with accepted administrative principles and procedures, with care being taken, for example, to ensure that self-interest plays a minimal part. However, success is also contingent on foresight results being credible, especially when initially appearing counter-intuitive or challenging conventional wisdom. The techniques and analytical approaches employed consequently need to be rigorous and technically sound, and this requires careful validation of the data on which conclusions are based.

Because foresight is concerned with the future, it is wellnigh impossible to adhere to the same criteria for assessing knowledge-claims that scientists traditionally employ in their own work (Van Vught, 1988). This has resulted in a particularly intractable problem within some countries in overcoming what might be termed the 'credibility gap'. It is therefore important to ensure that foresight adheres, as far as possible, to the established norms of social science which stress systematic approaches to theory construction and data generation. There are several examples where the credibility of prospective analysis was enhanced following the recognition by researchers that, contrary to expectation, technical excellence is in principle no less important in the social and policy sciences than in fields like physics. In the case of Japan, a key factor has been the involvement of professional think-tanks held in high esteem by government, industry and the research community — the Japan

Techno-Economics Society in particular played an influential role in establishing the confidence in forecasting methodology that was necessary for its widespread adoption during the early 1970s. The development of improved analytical techniques has likewise become important in West Germany where the work on technometrics by the Institute for Systems and Innovation Research has begun to lay the foundations for what may eventually come to be accepted as a 'rational' aid to research policy by a scientific community historically more sceptical than most. Valuable contributions to bridging the credibility gap have also been made by the academic science-policy community in Sweden and elsewhere — for example, in developing robust R&D indicator systems and in analysing the growth dynamics of emerging fields and specialties.

The final general point we would make is to emphasize the importance of viewing foresight as a social process which both depends upon and influences a wider organizational and political environment. This is highlighted by the problems which have arisen, especially in West Germany, Australia and Canada, as a result of the lack of a conducive climate for foresight. Indeed, a priority for government agencies and industry in most countries is to develop the type of *foresight culture* which, above all, has characterized the success achieved by Japan. This has been a precondition for the comprehensive long-term 'visions' of future possibilities and needs which have proved so important in providing an appropriate context for the effective integration of foresight with decision-making. It has also helped shape the informal networking and consultative processes among academic and industrial researchers that are crucial in framing timely, relevant and informed advice on future policies for science.

3. KEY FACTORS IN PLANNING FORESIGHT

Given that the costs of a single foresight exercise can run into millions of dollars, it is essential to ensure that the approaches and techniques employed are appropriate. A number of useful lessons emerge from our country studies relating to the factors likely to be important in planning different types of foresight. These can best be considered by concentrating in turn on the various elements in the foresight typology set out in Chapter 2: (i) characteristics of the performing organization; (ii) specificity of focus (e.g. macro-level); (iii) function of the foresight activity (e.g. consensus-generation); (iv) orientation and structural characteristics of the research (e.g. stability of disciplinary structure); (v) balance between various 'intrinsic tensions' (e.g. science-push vs demand-pull); (vi) time-horizon; and (vii) methodological approach — see Table 2.1.

It is perhaps a statement of the obvious to begin by noting that, while *organizational characteristics* have little bearing on propensity to engage in foresight, they do limit the types of *function* and the *specificity* of activities likely to be executed successfully. To take just a few of the permutations possible, research advisory boards to government are usually best placed to undertake national direction-setting or priority-determining exercises at the holistic or macro-level. In Canada, for example, these tasks could not be executed satisfactorily by an independent public-sector advisory body (the Science Council of Canada) nor by an individual government department (the

Ministry of State for Science and Technology), and it was therefore necessary to establish the National Advisory Board on Science and Technology. Conversely, experience in Norway and Sweden demonstrates that high-level bodies of this sort should in general leave priority-setting activities at the meso- and micro-levels to the departments and funding agencies administratively responsible if they are to gain the support of the scientific community. This is not to suggest hard and fast rules concerning the foresight activities that different types of organization should or should not undertake. Specific circumstances may need to be taken into account — we saw in the US chapter, for example, that the highly decentralized structure of the National Institutes of Health has resulted in foresight at a meso-level proving more relevant than the macro-level emphasis which might be expected in an agency with such a wide research remit.

In addition, a relationship often exists between foresight functions and level of specificity. For example, anticipatory intelligence normally needs to be carried out at the holistic or macro-level otherwise it is extremely difficult to monitor emerging developments in cross-disciplinary areas. The widespread recognition of the intellectual and technological promise of many fusion fields has resulted in priority being given to foresight at a more macro-level in almost all countries. What has yet to be fully appreciated is the need to coordinate such efforts with activities spanning not only the meso- and micro-levels, but also different sectors and types of institution in the national research system. In this respect, most can be learned from the integrated approach to foresight employed in Japan, where the holistic thirty-year forecasts produced quinquennially by the Science and Technology Agency (STA) combine an anticipatory intelligence function with the task of shaping consensus on potentially important areas of research. These provide a background for macro-level foresight conducted every five to ten years by government advisory committees (like the Council for Science and Technology and MITI Industrial Technology Council) during the preparation of the direction-setting visions which establish overall guidelines for new or follow-on programme initiatives over the next decade. More focussed meso-level foresight is then carried out within agencies, often on an *ad hoc* basis, to assess systematically the merits of different R&D options and develop consensus on priority lines of research. Finally, foresight at the micro-level (which is generally closely linked with the management of R&D programmes) contributes to steering ongoing research activities (e.g. in the Basic Technology for Future Industries Programme). The existence of such an integrated system for planning and evaluating research means that foresight can be undertaken on a continuing basis, enabling national and agency priorities to be revised regularly in the light of changing circumstances. In contrast, macro-level mechanisms like the National Colloquium employed in France proved too cumbersome to be used frequently, with the consequence that the Ministry of Research became locked into a particular set of research priorities.

Another central issue to be addressed in planning an exercise is the relative emphasis to be placed on different foresight functions. The Japanese case demonstrates the science-policy benefits which follow from combining several functions within a single initiative, as do various exercises in Canada and Sweden which were able successfully to integrate anticipatory intelligence and consensus-generation with foresight for priority-setting — especially noteworthy is the recent study in which the Science Council of Canada cooperated with NSERC in determining possible target areas for strategic academic research. It is also significant that problems have arisen when foresight emphasizes just one function like advocacy, as in the large majority of field surveys conducted by the US National Research Council and NSERC in Canada. This was not, of course, the original intention since most of these studies started off with a priority-setting remit, highlighting how important it is to define the aims of foresight explicitly at the outset. Only then can one minimize the likelihood of the scientific community (or some other interest group) hijacking what starts out primarily as a priority-setting exercise for its own purposes.

Despite such experience with foresight, it would be wrong to see Japan wholly as a panacea since socio-cultural factors have, as we have seen, sometimes resulted in the consensus-generation function being stressed too much. The consequent problems of conservatism in the research system are the reason, for example, why project directors with unconventional views have been given so much influence in steering the STA Exploratory Research for Advanced Technology Programme. Sweden is another country in which consensus mechanisms have traditionally been important in policy-making, yet the Ministry of Industry has now recognized the intrinsic dangers like R&D resources being spread too thinly or the adoption of risk-minimizing strategies when identifying future research opportunities. In Norway, too, the reliance on consensus processes has, in the view of some officials, led to a politicization of longer-term planning and limited the scope for making strategic choices regarding selective support for R&D.

Other lessons concerning the *orientation and structural characteristics of research* relate closely to decisions on both the *methodological approach* and *time-horizon* to be employed in foresight. A central issue here is the extent to which the research area in question forms an integral part of a wider scientific and technological system. This was emphasized by many science-based companies, which have tended to rely less on forecasting and more on monitoring and scanning in fields like life sciences and biotechnology where technological progress is less certain because of the heavier dependence on occasional breakthroughs in fundamental research. The time-horizon employed is therefore less than in fields like microelectronics and information technology where technical development trajectories are relatively stable and formal approaches to forecasting can often be adopted. This illustrates a general principle that foresight must be sufficiently long-term both to take into account the outcome of currently emerging trends and developments, and to allow decision-makers time to exert a positive influence on future events (or at least to try) and to insure against unlikely but possible contingencies. Yet the horizon must not be so far into the future, and be subject to so many uncertainties (as happened in many scenario-based exercises in the 1970s), that the results are of little apparent relevance to the choices which need to be made today.

The trend we have already noted towards the use of early-warning indicators stems not only from their relevance to monitoring developments in interdisciplinary research, but also from their greater acceptability in countries like West Germany where free-market ideologies have hampered the adoption of other forms of foresight perceived to be associated with interventionist policies for R&D. The emphasis placed on exploring the use of such quantitative indicators means that a wide range of methods is now available. These include: (i) on-line scanning of research literature abstracts; (ii) the mapping of citation patterns (e.g. using co-citation techniques); and (iii) computerized analysis of patent applications, patent–patent links, and patent citations to basic research. Especially promising progress has been made in France on theoretical studies concerning the evolution of new research fields, while in Japan attempts are under way to develop approaches for identifying and tracking emerging strategic technologies (e.g. by evaluating joint patenting by firms in different industrial sectors). Continuing efforts by funding agencies and other users of foresight are therefore essential to keep abreast of state-of-the-art techniques. Moreover, the significant contribution to methodological innovation being made by academic groups in several countries demonstrates the importance of investment in basic research on foresight.

While formal foresight approaches should, wherever possible, be combined with expert-based procedures, there is overwhelming evidence that traditional panel reviews are generally inappropriate for assessing prospects for research in interdisciplinary areas. Their use in the US field surveys might have been acceptable had the aim been solely to advocate the merits of supporting particular disciplines like physics, or even to perform an educational and communications function. However, they proved less than adequate in identifying potential opportunities in emerging fusion areas. This was explicitly recognized by STU in planning its exercise to determine options for interdisciplinary research, and for this reason the agency decided to employ a comprehensive survey of all Swedish post-doctoral scientists and engineers. The results then formed an input to national workshops attended by academic and industrial researchers with a broad range of expertise.

Decisions on choice of approach and technique are also partly dependent on the need to achieve an appropriate balance between two of the *intrinsic tensions* in foresight: science-push vs demand-pull and top-down vs bottom-up. Indeed, a key factor in the success of the STU initiative was the way those responsible were able to give sufficient emphasis to science-push factors, reflecting the recognition that previous use of company-dominated 'requirements boards' to identify research needs had resulted in pressures towards shorter-term and conservative policies. As a result, the approach employed enabled top-down inputs from the STU steering committee, and from senior agency and industrial officials attending the workshops, to be counterbalanced by bottom-up inputs from the large number of scientists who participated (including those involved in the initial questionnaire survey).[11] Lessons can be drawn, too, from similar exercises to identify promising opportunities in Japan and from the French National Colloquium and Prospectives 2005 initiatives, as well as from industry where companies like IBM and AT&T are no longer exceptions in understanding the importance of a science-push dimension to foresight in strategic research areas.

Although greater weight must be placed on science-push factors when assessing the prospects for curiosity-oriented fields, funding agencies in many countries have found it difficult to tap fully the collective wisdom of researchers without at the same time allowing their interests to become dominant. Foresight activities controlled by the basic science community tend to yield discipline-based projections and, in the rare cases where future socio-economic benefits are given serious attention, judgements rarely reflect commercial or other realities (often overstating the possible impacts and underestimating the time-horizons involved). While these types of problem are exemplified by several of the field surveys conducted by the US National Research Council, it has also been difficult in other countries, notably West Germany and Japan, to persuade academic scientists that the increased strategic nature of their work requires the inclusion of demand-pull factors as part of the equation in setting research priorities.

Finally, one must take into account the intrinsic tension between allocating responsibility for foresight to interested parties (principally the performers and users of research) as opposed to more neutral third parties (i.e. those with less direct material interests at stake). Although this is often among the most difficult decisions to make since it is generally bound up with conflicting political and bureaucratic considerations, certain broad ground-rules can nevertheless be identified relating to the differing functions of foresight and the level at which it is undertaken. Thus foresight for education and communication is almost always more effective when carried out by interested parties. Likewise, it is difficult even to conceive of anyone other than interested parties initiating foresight oriented to advocacy.

At the other extreme, direction-setting exercises aiming to inform national research policy are normally best entrusted to a central advisory body whose institutional location places it above suspicion in terms of favouring specific departmental or sectoral interests. This is most clearly demonstrated by the successes of the Japanese Council for Science and Technology mentioned earlier. For the same reason, third parties like the Norwegian Science Policy Council and US Committee on Science, Engineering and Public Policy have been central in macro-level foresight oriented to priority-setting. Besides being independent, such organizations have the broad institutional linkages necessary to plan and coordinate the foresight process, as well as the authority required to bring together different funding agencies, scientists and research users. Where government is involved, the third party must normally be an official body with high-level reporting links if it is to be effective. The problems likely to be encountered by organizations located outside government are well illustrated by the limited influence achieved by the Science Council of Canada in shaping national funding priorities. Similarly, the types of difficulty faced by agencies that are not perceived to be independent are exemplified by the US Office of Science and Technology Policy (OSTP). Although formally responsible for national priority-setting, its institutional location has proved too close to the President for the results of foresight activities to be accepted as 'neutral' by the different stakeholders in government agencies and the research community (not least because of OSTP's close identification with political initiatives like Star Wars). Another example is the Japanese Science and Technology Agency where increased responsibilities for supporting research slowly compromised its original main task of coordinating agreement on overall R&D priorities among government departments.

Third-party status is also important in other types of foresight. Where the aim is anticipatory intelligence, care may be needed to ensure that a particular group is not given preferential access to information ahead of others. This was one reason why the Japanese Techno-Economics Society recommended that responsibility for the thirty-year technology forecasts be allocated to a government agency rather than a private-sector think-tank.[12] Yet, in other cases independence from government may actually be a precondition for successful foresight, especially when focussing on emerging problems which call into question status quo interests — a certain degree of autonomy is necessary, for example, to avoid undue influence being exerted by ministries concerned with protecting their policies from external criticism. This was crucial in the studies on future environmental and resource problems carried out by the Science Council of Canada. In comparison, the initial location of the Swedish Futures Studies Secretariat within the Cabinet Office proved untenable after it published findings strongly critical of official policy. This said, foresight conducted outside government may at the same time suffer from the disadvantage that results can be more easily dismissed as unreliable on the grounds that they are susceptible to influence from external interest groups. The US National Research Council, as we saw, is regarded by many in government as conducting 'foresight for advocacy' on behalf of the research community, while the Science Council of Canada ultimately suffered severe budgetary cuts in part because of criticism relating to the orientation of its work.

When a priority-setting exercise focusses on the meso- or micro-level, the active participation of stakeholders like the research community will naturally require more emphasis on involving interested parties if their commitment to implementing the results is to be secured. This is one of several points to be considered in establishing appropriate organizational arrangements for foresight aiming to improve decision-making on research priorities. We now turn to this, the central theme of our book, focussing on the lessons emerging from our international review which are specific to the three main phases of the foresight process distinguished in Chapter 2 (see Figure 2.2).

4. ORGANIZATION OF FORESIGHT FOR PRIORITY-SETTING

(a) Pre-foresight activities

Although the importance of the pre-foresight phase might seem self-evident, many of the priority-setting exercises reviewed in the eight countries failed precisely because insufficient attention was paid to initial *preparatory activities.*[13] These are especially important when studies are conducted in an environment generally hostile to foresight such as that confronting most public-sector organizations in West Germany and the United States. As earlier chapters make clear, it is extremely difficult to achieve success in the absence of prior consensus on the need for foresight. Indeed, in Australia and Canada it was only the recognition that longer-term national survival was at stake if science and technology were not harnessed more effectively which led to acceptance of the need for improved policy-making mechanisms, and in turn for greater use of research foresight.[14] Similarly, the success of the CNRS exercise to establish priorities for French astronomy in the mid-1980s was due in no small part to a realization among scientists that choices had to be made. However, the existence of a 'product champion' for foresight in the shape of the then CNRS Director-General was also important.[15] As in other types of exercise, an influential and respected advocate of foresight can help in overcoming the initial resistance of the research community to prospective analysis, and in building the confidence among policy-makers necessary for the results to be acted upon.[16] A high level of commitment to foresight will ultimately be developed only through successful experience — it took several decades for the US Air Force to reach a situation where officials could be confident from the outset that the findings of Project Forecast II would be implemented.[17] In the same way, acceptance of foresight in Japan stemmed at least as much from early positive experiences as from any intrinsic cultural predisposition to forecasting.

The case-studies of priority-setting reviewed in each country reveal the extent to which success depends on the early establishment of an appropriate organizational framework. Among the decisions to be taken is the assignment of institutional responsibility, and here the normal rule is that foresight should be conducted at 'arms-length' from routine decisions relating to the planning, funding and execution of research. As Coates (1985, p.45) stresses, a degree of integration with policy-making processes is essential for the implementation of findings, yet there must be sufficient institutional autonomy that foresight is not unduly swayed by short-term problems and political pressures, but remains open, flexible and innovative. This means that, for individual organizations, more macro-level foresight should be regarded as a strategic responsibility of the board (to use a business metaphor) rather than an executive task to be carried out by management. Only when this was recognized by CSIRO, for example, was the agency able to overcome earlier problems stemming, first, from foresight being devolved to divisions (which led to conservative projections of existing activities), and later from the move to a centralized model where it was primarily undertaken by a long-term corporate planning group (which tended to produce radical recommendations with little chance of being implemented).[18] The situation now is that the CSIRO Board is responsible for foresight oriented to direction-setting and determining broad longer-term priorities for R&D. The Executive then oversees foresight aiming to inform more detailed decisions on new programme areas and lines of research. Apart from coordination and methodological development tasks undertaken by a central corporate planning group, most activities are devolved to the six constituent CSIRO institutes, with the external advisory committees for each institute playing an important role.

The experience of CSIRO also illustrates a general conclusion concerning the allocation of institutional responsibility for foresight within an individual organization like a government agency, research council or laboratory. This is the need to ensure both internal and external legitimacy for the results of priority-setting activities. In the most common case of an exercise focussing on the meso-level, this might be achieved by establishing an *ad hoc* committee under the direction of the governing board or employing an existing research advisory committee (e.g. for a specific area of R&D). In other words, foresight should normally be undertaken by a body accepted as having a strategic role if the findings are to be regarded as legitimate in the eyes of external funders of research and other stakeholders. Moreover, this also enables the exercise to be seen as legitimate within the organization: micro-level foresight is rarely if ever acceptable when carried out by a wholly independent body because of the potential political, disciplinary and other threats associated with the implied loss of control over policy.[19]

As in other types of foresight, it is essential early on to agree appropriate terms of reference and the resources and staff support likely to be required. Here, the administrative model for foresight used by Japanese agencies (see Figure 2.2) could be more widely adopted in other countries. Decisions on objectives are usually made by the institution concerned in conjunction with the high-level steering committee given overall responsibility for organizing the exercise, and take into account guidelines set in prior direction-setting studies or other long-term strategic planning activities. One of the committee's functions is to ensure that the eventual results have sufficient *authority* to be accepted by those involved in determining priorities for research, and its membership is therefore normally comprised of senior industrialists, scientists and officials from public-sector agencies. An important consideration in selecting members is to establish the links with influential policy-makers and interest groups (in industry, government and the research community) that are essential for securing the participation of key actors in the exercise. In this way, it is possible to build in a *commitment* at the outset *to act* on the findings.

Authority is, however, by no means a sufficient condition for the acceptance of foresight. As several case-studies reveal, the pronouncements of senior officials on the steering committee may be authoritative, but they will not necessarily be regarded as *credible* unless informed by in-depth knowledge of recent developments in science. Such credibility is provided in the Japanese model by delegating the main foresight tasks to a specialist working group, the members of which are chosen, first and foremost, for their scientific or technical expertise, knowledge of market and other demand-side factors, and experience in forecasting or science-policy research. One function of the working group is to initiate the process of consultation required to arrive at well-founded judgements on relevant scientific, technical and socio-economic issues. For this, informal 'light-touch committees' may be used to ensure that any recommendations will later be endorsed by influential stakeholders and interest groups. Other tasks include commissioning think-tanks and external consultants to compile any detailed background information that is necessary and to contribute theoretical and methodological expertise. A much wider variety of knowledge inputs is thus available to the working group than in most other countries where the tendency still is essentially to rely on conventional peer-review procedures.

The task of *designing an appropriate foresight process* spans the pre-foresight and main foresight stages in our conceptual model, and various lessons again emerge from analysis of the factors determining success and failure in priority-setting exercises in the eight countries. Aside from issues already considered, like the need to construct a strategy for involving the research, policy-making and industrial communities (along with other 'social partners' as they are termed in France), it is vital to develop an understanding of the requirements of those commissioning the exercise or likely to make use of the results. Indeed, it is inadvisable to invest further effort and resources unless it is clear that a 'customer' exists who is willing and able to make use of foresight in determining priorities for R&D. In several cases in Sweden, Australia and the United States, priority-setting initiatives failed for the simple reason that there was no real user for the results in a position to think and plan long-term and strategically.[20]

Greatest success with priority-setting exercises has tended to come when the structure of the foresight process is compatible with existing institutional mechanisms for decision-making. Where agencies operate almost entirely on the basis of an annual planning and budget-setting cycle, continuity in strategy is virtually impossible and there is limited scope for research foresight with its typical time-horizon of five to ten years (apart from in relation to decisions on major capital investments like big science facilities). Experience in the United States and elsewhere shows that, where foresight is attempted in such cases, severe problems are usually encountered with implementation. Only when agencies move to multi-year budgeting, as they have done in recent years in Australia, France and Sweden, is it possible to begin developing the longer-term research strategies and priorities which can benefit from prospective analysis. Agencies with a triennial budget system and rolling five-year plans (as in Sweden) are perhaps best placed to take advantage of research foresight with a time-horizon of ten to fifteen years (i.e. around two to three times the 'forward look' normally taken in the planning process).

All this is not to imply, however, that there should be a formal link between foresight and priority-setting. The two tasks are separate: the role of foresight is *indicative* not prescriptive. The primary aim is to explore and assess the merits of alternative research options, not to make definitive judgements on priorities. This always needs to be kept in mind when planning a foresight exercise since too close a relationship with resource allocation will, as was emphasized earlier, generally result in conservatism and a stress on protecting status quo interests, whether disciplinary, sectoral or institutional.

(b) Main foresight stage

A large number of the tasks in the main foresight stage will be carried out by the working group, especially the detailed planning of the priority-setting exercise. Analysis of successes and failures in the case-studies reviewed in previous chapters demonstrates how specialist scientific and technological knowledge needs to be complemented by a theoretical understanding of the foresight process and practical experience gained from previous studies. The STU exercise to identify strategic interdisciplinary fields (like a previous initiative by the Royal Swedish Academy of Engineering Sciences) relied heavily on one official who combined many of the latter skills, while in Japan professional forecasters and science-policy analysts are represented among working group members as a matter of course. In common with the equivalent Agency of Industrial Science and Technology in Japan, STU has recognized the benefits of being able to draw routinely on inputs from a wider infrastructure for foresight when conducting *ad hoc* and more regular exercises, and has made use of quantitative analyses of emerging fusion technologies by the Research Policy Institute at Lund University. Similar techniques were employed by the Japanese in providing background information for the Basic Technology for the Twenty-First Century exercise and the subsequent White Paper on *Trends and Future Tasks in Industrial Technology*.

As experience in country after country has shown, successful planning of the main foresight stage is not just a technical question of determining what information and analytical approaches are required to assess the merits of different research options. The foresight process itself is often at least as important as the end-product. Careful consideration therefore needs to be given as to how best to consult and involve the various funders, producers and users of research (including the emphases to be given to top-down vs bottom-up approaches and to science-push vs demand-pull perspectives discussed earlier). This will depend partly upon the level of the priority-setting exercise: at the macro-level, for example, inputs from a relatively wide constituency of disciplinary, institutional and sectoral interests need to be integrated and balanced, and the requirement for independence and neutral detachment will entail adopting various formal and informal approaches to information-gathering and consultation. In more micro-level foresight, it may be preferable to devolve responsibility to the scientific community. One reason for the success of the CNRS astronomy study and the Canadian earth-sciences review was that virtually all the researchers likely to be affected by the results participated in the priority-setting process.

An appropriate mix of skills and experience among working group members is also essential in considering the four sets of factors which must be taken into account in the task of *strategic analysis*: socio-economic needs and threats; socio-economic resources and other comparative advantages or weaknesses; research opportunities; and scientific and technological capabilities and resources. While the relative importance of these factors varies according to the orientation and structural characteristics of the research area in question, rarely if ever is it advisable to rely solely on expert inputs from the scientific community (or, alternatively, from industrial and other research users). Strategic analysis requires creative and sometimes divergent debate on alternative scientific possibilities — as the French in particular have recognized, promising options may sometimes be identified most readily through confrontational dialogue between experts with differing perspectives on a given field. This can be facilitated by involving not only those from the basic and application-oriented ends of the research spectrum, but scientists from neighbouring disciplines as well. Japanese experience in exercises like the thirty-year STA forecasts shows that significant inputs can even be made by knowledgeable journalists and science-fiction writers (whose long-term visions may be a useful catalyst in raising the important 'what if' questions which are indispensable in stimulating non-dogmatic discussion on future alternatives).

The aim in strategic analysis is not just to identify a wide range of promising research options, but also to assess systematically the likely benefits, costs and uncertainties involved. The country studies show that there is increasingly widespread acceptance of the need for systematic foresight techniques as an adjunct to expert judgement in framing longer-term science policy. The following points have all proved significant to differing degrees in successful priority-setting initiatives:

(1) allocating sufficient resources for a thorough review of the findings from previous foresight studies (not least in other countries) at both higher and lower levels of specificity;

(2) the systematic use of socio-economic indicators to make projections of likely trends, foresee possible counter-trends, and assess the implications for R&D of different scenarios (desirable and undesirable, probable and improbable) concerning the future nature and structure of society;

(3) familiarity with the types of formal R&D investment appraisal techniques widely used in companies (e.g. those developed by Arthur D. Little and McKinsey) to evaluate the likelihood of being able to capitalize upon possible breakthroughs in science and technology (i.e. to assess the 'receptor potential' of research users);

(4) a technical ability to use quantitative indicators of scientific *capacity* (e.g. concerning trained researchers and available facilities), *output* (e.g. publications and citations) and *performance* (i.e. outputs compared to inputs) in assessing the relative merits of future investments by a particular nation or institution in different research areas — all these were employed, for example, in the ASTEC exercise to construct a profile of Australian science in 1987–8;

(5) the availability of broad-based expertise enabling information science techniques to be combined with a detailed theoretical understanding of the dynamics of emerging specialties and the nature of cross-linkages between curiosity-oriented research and technology.

It is in this last area that greatest efforts are currently being made to develop improved approaches to the strategic analysis component of research foresight. Leading science-based companies have for some time understood the need to monitor developments not just in current and next-generation technologies (and in the underlying science base), but also in potentially fruitful radical alternatives. However, the feasibility of undertaking systematic analysis to explore the prospects for alternative lines of development in curiosity-oriented research has also now been recognized (e.g. Papon, 1988). As we saw, this is a major reason behind the decision by the French government to create the Observatoire des Sciences et Techniques. Progress in this area will also depend upon parallel work by organizations like CHI Research in deploying patent-citation analysis to map the structure of emerging technologies and the links with basic science (e.g. CHI, 1987).

As in the use of brainstorming to generate creative ideas, it is a cardinal principle in strategic analysis not to foreclose any potential options before they have been subject to more detailed appraisal. Only after the task of weighing up the advantages and disadvantages of each possible alternative has been completed should the next step of *agreeing the most promising research options* begin. An important lesson here is to ensure that the criteria employed are appropriate to the needs of those commissioning the exercise or using the results, and that due account is taken of specific institutional, political and ideological considerations. This is easiest in priority-setting exercises carried out by mission-oriented organizations such as the Swedish Defence Research Institute, where the wider process of strategic planning means that explicit guidelines already exist for initiatives like the quinquennial Technical Prognosis studies. In the same way, macro-level foresight by government needs to take into account broader national concerns — for example, in smaller or specialized economies such as Canada, emphasis must be given to criteria relating to the exploitability of work in niche sectors of national technological strength like forest products and mineral resources engineering.[21] Foresight focussing on novel science-push opportunities, besides being subject to greater uncertainty (and therefore representing an option mainly for countries able to finance a broad portfolio of basic research), is also likely to result merely in overseas exploitation because of technology transfer by the multinationals which control large sections of Canadian industry. Conversely, the criticism of Japan by Western nations for 'freeloading' on their fundamental science, together with threats of protectionism in relation to research in strategic areas, has contributed to greater stress on supply-side factors in foresight.

Whatever emphasis is given to different criteria at this stage of a foresight exercise, it is essential that a sufficiently wide range of options is set out to offer policy-makers a genuine choice depending on the weighting they attach to particular factors. The relative resource costs of each choice also need to be spelled out, together with the type of research-funding mechanisms likely to be required. In addition, information should be provided on the implications of deciding whether to proceed or not with each option, and on the interrelationships between competing alternatives. This may, for example, involve specifying several portfolios of programme initiatives which might be supported assuming different budgetary scenarios. Other relevant considerations are to ensure that any uncertainties involved are clearly stated, and that stress is placed on building diversity into any strategic decisions on overall research priorities.

Several lessons concerning the *dissemination of results* from foresight can also be drawn. Examples may be cited in most countries where the impact of an otherwise successful exercise was limited by insufficient attention being given to the question of how best to present and diffuse the findings (e.g. in the French Prospectives 2005 study). The guiding principle here is that the report (or any other product) from a foresight exercise should be appropriate to the needs both of the sponsors and of those participating, directly or indirectly, in the priority-setting process.[22] This means in particular that:

(1) Dissemination activities must be planned into the foresight process to ensure adequate time and resources are set aside. The considerable efforts devoted to diffusing the findings of the Pimentel survey of opportunities in chemistry carried out in the United States contrast markedly with the rather less successful Brinkman study on physics where the committee, having produced a lengthy series of reports on individual subfields (totalling nearly 2,000 pages), had little energy left to prepare an overview of the main conclusions, let alone for other dissemination tasks.
(2) Care should be taken to address squarely the issues set out in the terms of reference, and to provide policy-makers with sufficient information to enable them to choose from among a broad range of viable research options. As Martino (1977, p.119) stresses, 'the various futures must not all be minor variations on a single theme, nor should the deck be stacked by the inclusion of only one "preferred future" and several totally unacceptable alternatives'. Moreover, the options proposed must be

politically realistic and not focus solely on alternatives implying impossibly large increases in funding — a feature of many National Research Council and NSERC field surveys in North America.

(3) A variety of mechanisms for the dissemination of results is normally required depending on the nature of the intended audience. For example, in the successful exercise in Japan on Basic Technology for the Twenty-First Century, a popularized version of the report was highly influential not just in persuading the government to fund a follow-on initiative to MITI's Basic Technology for Future Industries Programme, but also in informing industry about prospective developments in forty-six key areas of strategic research at the interface between electronics, life sciences and chemistry. An extensive series of briefings and public seminars, together with a popularized version of the report, similarly helped to diffuse the findings of the Pimentel study on chemistry.

(4) Attention needs to be given to the style and format of presentation since these can be crucial in the implementation of foresight results. Because policy-makers are invariably busy, the information presented to them should be brief without being misleadingly incomplete — the impact made by the Emerging Technologies report prepared for the US Department of Commerce was due in no small part to its brevity and the appealing simplicity of its message. Moreover, the main findings must be outlined with sufficient clarity to provide the basis for decisions on priorities. The STU exercise to identify programme options in interdisciplinary fields demonstrates the advantages of involving science journalists in producing professionally written summary reports which appeal to decision-makers and the general public alike. This lesson also emerges from our analysis of various French initiatives — for example, the CNRS *schéma prospectif* and La Révolution de l'Intelligence.

(5) It is vital that the submission of foresight findings is timed to fit in with institutional decision-making cycles if they are to inform policy choices and budget-setting. No less important, however, is to ensure that policy-makers are kept fully up to date with progress in the foresight exercise from the outset. As Coates (1985, p.45) has emphasized, 'the worst single mistake any policy process can make is to dump a surprise, especially an uncongenial surprise, on the would-be user'.

More generally, it should be recognized that the mechanisms by which foresight impacts on policy can be lengthy and diffuse. Continuing efforts at dissemination may be required, especially in making certain that results feed into subsequent foresight at higher and lower levels of specificity. The benefits of effective follow-up procedures are clearly evident in Japan with its comprehensive and integrated approach to foresight. Implementation of findings may in fact only be possible after further information-gathering and analysis, and the iterative nature of the foresight process means that any *post hoc* assessment of success should consider factors other than the immediate influence of an exercise on policy-making (Steed, 1988).[23] This leads us to several related lessons concerning post-foresight activities.

(c) Post-foresight stage

The emphasis given to the post-foresight stage in Figure 2.1 reflects the importance of the interface between foresight and the broader task of planning and managing research. Among the main conclusions from our review is that prospective analysis can be used as much to 'steer' research programmes once they have been established as to identify priorities for new initiatives. This points to the need to combine monitoring of research in progress with continuing foresight to update knowledge on changing scientific strengths, weaknesses, opportunities and threats. Experience in France, Japan and elsewhere demonstrates the appreciable advantages which stem from the successful *integration of evaluation and foresight* within funding agencies, enabling resources to be redirected in response to emerging research possibilities or problems and to evolving socio-economic demands. We have already noted the central role played by foresight in managing the MITI Basic Technology for Future Industries Programme. However, a variety of other approaches exist which, to differing degress, also provide useful models. In Sweden, for example, the remit of the panels of foreign experts employed to assess the health of research in particular fields now encompasses making recommendations on future initiatives in scientifically promising areas, while in France the GEP mechanism gave advisory groups a dual evaluation and 'prospective' role in managing national R&D programmes.[24]

Experience in nations with a tradition of relatively centralized state planning such as France will clearly be of limited relevance in framing implementation strategies for foresight elsewhere. This is especially true in countries like the United States where policies tend to be determined on an adversarial rather than a consensual basis, with the result that the influence of foresight on decision-making is inevitably less pronounced. For example, although the 1981 exercise by the Energy Research Advisory Board was methodologically sophisticated and succeeded in generating clear priorities, its proposals were largely overridden by political factors. Similarly, the ordered priorities arrived at in the Seitz–Eastman survey of materials research facilities were later modified by the Department of Energy so as to maximize support in Congress. The conclusion to be drawn is that it would be unwise to rely completely on rational decision-making models in establishing formal structures for the implementation of foresight. Often of greater effectiveness is the more pragmatic course of ensuring that the principal stakeholders are integrally involved in the foresight process from an early stage, and that any recommendations take account of areas where the influence of sectional interest groups may prevail over even the strongest arguments about scientific promise or future technological importance to the nation.

We should finally reiterate the widespread concern, especially among scientists, about the dangers of placing too much emphasis on foresight in selective targeting of research. Certainly the strategic importance of some fields, coupled with the fact that no country can now afford to support frontier research across the entire spectrum of scientific and technological disciplines, means that choices, whether explicit or implicit, have to be made. It is therefore unavoidable that funds are concentrated to some degree on those lines of research judged most promising from an intellectual, technological or socio-economic point of view. However, the intrinsically serendipitous nature of scientific and technological progress, together with the growing synergy between disciplines, means that limits need to be set on selective support for research. In more basic academic fields, the consensus now emerging seems to be that no more than around 20 per cent of funds should be distributed through targeted initiatives. Higher levels than this may constrain creativity by 'planning away' the variety, pluralism and competition which are so fundamental to the conduct of science.

5. CONCLUDING COMMENTS: AN ANALOGY WITH TECHNOLOGICAL INNOVATION

Any attempt to set down an overly simple recipe for successful foresight would not only be naïve but also counterproductive in relation to the difficult task of securing acceptance by the research community of more formal approaches to framing science and technology policy. While the results of our international review may be of some value to those attempting to introduce foresight for the first time, problems will undoubtedly occur in the process of acquiring the tacit knowledge and experience which come only through a lengthy process of trial and error. As a result, determination and persistence are essential — success is likely to be achieved only when most, if not all, of the many elements in foresight are executed effectively. This is well illustrated by the Australian Science and Technology Council priority-setting exercise in 1981 which ultimately failed despite explicitly combining science-push and demand-pull factors and employing a fairly sophisticated methodology. Problems arose because: (i) there was no prior consensus on the need for foresight; (ii) the initial suspicions of researchers were not adequately allayed; (iii) the reliance upon panels of elite scientists resulted in a bias towards established disciplines; and (iv) there was no institutional machinery in place to translate the findings into specific policy initiatives within funding agencies.

If we were forced to limit ourselves to drawing just one overall lesson for those with little experience of foresight, it might well be the theoretical and practical benefits of studying the literature on technological innovation. Indeed, many of the factors determining success and failure in foresight for priority-setting show a striking parallel with those identified as important in industrial innovation (e.g. Rothwell *et al.*, 1974).These include the following:

(a) Both innovation and foresight must be considered as a *process* taking place over an extended period. In the case of foresight, that process (including negotiations among different interest groups, generation of consensus, and creation of commitment to new lines of research) is often no less important than the end-product (e.g. forecasts concerning promising fields of science and technology).
(b) As in technological innovation, the results of foresight will stand little chance of being implemented unless they address the needs of the principal customers — in this case, the scientific community, funding agencies and research users in industry and other socio-economic sectors. Such needs should be identified at the earliest possible stage.

(c) Both processes must involve all those likely to be affected by the results if commitment is to be generated to implement the findings successfully. Furthermore, this should be achieved in such a way that the interests of individual stakeholders are kept in balance (which may require a measure of third-party involvement, especially in relation to determining more macro-level priorities). It also entails combining top-down and bottom-up inputs, as well as the integration of science-push and demand-pull perspectives without at the same time allowing either to dominate.

(d) Considerable attention needs to be given to dissemination of the 'product' from the innovation or foresight process. This means ensuring effective marketing and education of users. In each case, mechanisms for transferring the findings to users should be built in from an early stage. Closely related to this is the task of establishing a good communications structure, both internally between those directly involved in different aspects of foresight and research planning, and externally with potential beneficiaries and other stakeholders.

(e) Successful industrial innovators devote significant effort to market research, consulting users on their changing needs and tracking relevant fields of R&D to identify emerging opportunities. They also seek to ensure that their longer-term strategies for technological diversification build upon existing corporate strengths relative to competitors (Pavitt, 1983). In the same way, a key input to successful foresight is systematic monitoring of evolving socio-economic demands and scientific opportunities worldwide, along with an assessment of whether adequate research skills and other resources are likely to be available if an initiative is launched.

(f) Just as the R&D involved in industrial innovation requires close links with other business functions (e.g. production, finance and after-sales service), so foresight cannot be conducted in isolation from existing mechanisms for planning and setting budgets, or activities relating to the management, execution and exploitation of research.

(g) In both the innovation and foresight processes, a central role is frequently played by individuals. Experience in France, the United States, Sweden and Norway reveals that 'the initiation, pursuit and success of futures activities seems to be unusually dependent on the presence, personality and persistence of specific individuals' (Hahn, 1986, p.24). In particular, a senior (and enthusiastic) product champion is often necessary to guarantee high-level commitment to the foresight process and to subsequent implementation of the findings, while ensuring that those involved are afforded sufficient protection from conventional wisdom, short-term pressures and bureaucratic restrictions. As with technological innovation, the role of individuals may become less crucial once foresight is institutionalized or where collective approaches to decision-making are favoured (as in Japan).

(h) Successful management of both processes requires numerous skills. In the case of research foresight, they include: (i) expertise in a broad spectrum of scientific and technical fields; (ii) an understanding of the dynamics of technological change and the nature of innovation; (iii) an appreciation of the needs of potential research users; (iv) organizational and interpersonal management skills; (v) competence in survey design and data-analysis techniques; and (vi) a good background in forecasting and in science and technology policy research.

(i) Finally, just as the administrative structure and organizational culture of a company can influence its approach to technological innovation, so foresight can vary widely according to the scientific, institutional and political environment within which it is conducted.

The last point is crucial since concern about the cultural specificity of foresight may discourage other nations from exploring the possibilities for deploying the types of approach used so effectively by the Japanese. This is a common response to suggestions that the West has much to learn from government agencies and companies in Japan (e.g. Dore, 1983; Irvine and Martin, 1984). Hence, one task in our study was to explore the extent to which country-specific factors may limit the scope for overseas adoption.

Undoubtedly, wider socio-cultural, political and ideological factors can be of great importance in foresight — the traditions of consensual decision-making and looking to the longer-term future in Japan (and to a lesser extent in Sweden) have provided a conducive climate for incorporating prospective analysis in decision-making. Conversely, in an environment where policies emerge from an adversarial process largely in response to short-term political pressures from powerful interest groups (as in a federal nation like Canada), research foresight is often difficult. Under such circumstances, the results are unlikely to play a sufficiently important role in policy-making to secure the resources, participation and continuity required to achieve success. The same is true of countries where the pursuit of free-market industrial policies militates against government planning oriented to ensuring that the public R&D infrastructure is capable of meeting anticipated longer-term national needs.[25] Further problems arise when the scientific community has in the past enjoyed a high degree of autonomy in determining the allocation of resources to research, most notably in the Federal Republic of Germany and the United States.

Yet all this does not mean that research foresight is so culturally specific that lessons from overseas are of no relevance; rather, the conclusion is that individual cultural, political and institutional circumstances need to be taken fully into account in determining the optimum approach to be used in foresight. For example, early-warning indicators have proved acceptable to conservative administrations in West Germany and France. Equally, the 'search conferences' organized by the Australian Department of Industry, Technology and Commerce have been instrumental in changing the previously hostile views of companies towards foresight, demonstrating that such exercises can stimulate the development of Japanese-style consensus on national priorities rather than merely providing government with a means to impose its views on the market.

Finally, we would suggest that the analogy with technological innovation may also give us some cause for optimism about the future prospects for foresight. Having seen how previous attempts have all too often met with failure, or at best with only partial success (at least in countries other than Japan), the reader might be tempted to conclude that foresight is so exceedingly difficult that there is little point in even attempting it. The problems associated with understanding and managing the complexities of technological innovation were previously seen as no less daunting, yet steady progress has nevertheless been made in recent decades. In our view, the theory and practice of research foresight can likewise be improved through a combination of perseverance and a willingness to accept the setbacks which will inevitably occur during the necessary task of 'learning by doing'. With such experience should then come the enhanced ability to create the futures of our choice — the ultimate goal of research foresight.

NOTES

1. We also draw upon the results of a review of Dutch foresight efforts reported in TNO (1989), and comments made by the Ministry of Education and Science 'Sounding Board' for our study.
2. As Dunn (1988, p.3) notes, this criterion is also relevant to policy science more generally where the aim is 'to promote the collective deliberation and critical debate of assumptions, interpretations, and perspectives offered to justify and activate proposals for public action'.
3. It is difficult to see how one can avoid drawing the conclusion that government involvement is essential in foresight activities of this nature. The type of private sector solution attempted in the United States has not been particularly successful because competitive market pressures have prevented the development of an integrated national strategy for research intelligence. The result has rather been to encourage the creation of a diverse set of rival scientific and technical information services (like *Chemical Abstracts*) which actually tend to be exploited to greater effect overseas.
4. Concern is sometimes voiced, especially by Western companies, that such entirely legal approaches to foresight have also been accompanied by activities of a less acceptable nature. This was, however, a question beyond the scope of our study.
5. Even in France where government interest in foresight declined during the Chirac administration, the climate changed markedly after the 1988 election. By 1989, plans were well advanced for the Observatoire des Sciences et Techniques which will, among other things, focus on assessing the prospects for basic research.
6. All too often, organizations like the Australian Commission for the Future have been fighting an uphill battle in persuading government, industry and the scientific community that the introduction of a social dimension to research foresight is needed to balance the existing one-sided emphasis on economic and technological criteria.
7. This point is stressed by Van Dongen (1988) using insights drawn from organization theory. The social processes involved in paradigm change have been a central issue in the sociology of science since Kuhn (1962).
8. For example, Van Vught (1988) attempts to demonstrate that, from a Popperian viewpoint, systematic research forecasting is logically impossible. Although the practical feasibility of foresight is also disputed, insufficient account is taken of the considerable differences in philosophy and approach between conventional predictive forecasting and foresight. The counter-argument that foresight is both possible and desirable is put forward by Papon (1988), the former Director-General of CNRS.
9. A small proportion of funds might, for example, be earmarked for a parallel 'counter-targeting initiative' to ensure support for work outside the main lines of research that will be undertaken.
10. Montigny (1989) arrives at a similar conclusion on the acceptance of results from research evaluation and forecasting. He stresses the importance of three factors — competence, legitimacy and authority.

11. In contrast, two examples from Australia illustrate the problems encountered through failure to achieve a balance between top-down and bottom-up approaches to foresight. On the one hand, predominently top-down planning by CSIRO in the early 1980s had the effect of alienating the scientific community while, on the other, over-reliance on a bottom-up approach in the subsequent exercise on Selecting Technologies for the Future resulted in a vigorous defence of existing interests and yielded disappointingly conservative results.
12. A related consideration was that the official status of the Science and Technology Agency which subsequently took over responsibility for the forecast would guarantee a higher response rate to the large-scale Delphi surveys employed.
13. This conclusion also applies to most other types of foresight, although the opportunities for rectifying problems at a later stage are perhaps rather greater in anticipatory intelligence. For this form of foresight, methodological limitations or inadequate diffusion of results may be a more common reason for failure.
14. Most important has been the turnaround in the views of Australian and Canadian firms on the utility of participating in foresight activities initiated by government. Scepticism continues to be a major problem in West Germany and France, where secrecy on the part of companies concerning their long-term corporate R&D strategy has limited the scope for tapping their views in exercises to identify opportunities for precompetitive strategic research. This is especially true in the chemicals sector. In information technology, by contrast, there are some signs that national and EEC programmes are beginning to have an effect in persuading firms of the benefits to be gained from cooperation. The overall lesson to be drawn here is that the foresight process should, wherever possible, be organized in such a way as to encourage change in the cultural attitudes of companies towards collaborating with competitors on matters of longer-term mutual interest.
15. As we saw in France, Pierre Papon was central in initiating exploratory work on foresight for priority-setting in CNRS. Robert Chabbal similarly acted as a 'product champion' in the Ministry of Research. As soon as these individuals left the two organizations, foresight suffered accordingly. That no such product champion has emerged in West German agencies like Deutsche Forschungsgemeinschaft is one reason why foresight has thus far failed to be given serious attention.
16. As Coates (1985, p.44) observes, 'no system of foresight can work without the willingness of the decision-maker to use the process. While outright enthusiasm may be more than one can reasonably expect, attentive forbearance is a minimum condition for a successful system'.
17. Achieving such success with implementing foresight results is by no means easy, especially when the officials responsible are promoted or leave the organization. In particular, it requires continuing attention to ensuring that the approach adopted is appropriate. The experience of the US Air Force in this respect contrasts markedly with that of the Defense Advanced Research Projects Agency. Previously a keen exponent of research forecasting, DARPA has now lost most of its former enthusiasm and makes little use of the results from foresight exercises like Project Forecast II.
18. Similar difficulties arise when too much reliance is placed on external consultants. A common perception among agency officials in all eight countries is that attempts to use consultants in priority-setting exercises often lead to unsatisfactory results. Their responsibilities should, therefore, be confined to: (i) compiling background material in areas where specialist scientific or technological expertise is required; and (ii) providing methodological and technical skills — for example, of the sort made available to CSIRO by McKinsey in developing a framework to assess the commercial prospects for research in different fields.
19. Problems of this type may also be encountered at a divisional level. For example, foresight activities by the Planning Division of the West German Federal Ministry

of Research and Technology (BMFT) have been resisted by the three main R&D divisions because programme managers and their associated external research constituencies (in government laboratories and industry) regard its activities as an 'external' threat. Consequently, foresight by the Planning Division has had little influence on priority-setting within the Ministry despite the ease of access to senior BMFT officials. This represents another reason why foresight activities have largely been restricted to the development of early-warning indicators.

20. The lack of an 'interested receiver' was cited as the main reason for the limited impact of studies undertaken in the early 1980s by the now defunct Long-Term Planning Unit within the Swedish National Board of Universities and Colleges.
21. This has implications for foresight in smaller nations within the European Community. While there have been some benefits from collective international efforts like the Forecasting and Assessment in Science and Technology (FAST) Programme, countries with specialized economies such as the Netherlands probably still need to place greatest emphasis on domestically controlled activities.
22. Holland (1988) arrives at a similar conclusion in a study of the factors affecting the utilization of evaluation results by policy-makers and officials in government agencies. The perceived relevance and level of specificity of recommendations are among the key determinants of successful policy implementation.
23. Indeed, Steed (1988) argues that the diffuse impacts of foresight are often more important than the more direct ones originally intended. The main effect on priority-setting may come, for example, through the use of findings in later policy initiatives, or through raising the awareness of decision-makers about emerging threats and opportunities.
24. However, French experience with using GEPs for both evaluation and foresight proved somewhat mixed in practice. As we saw in Chapter 3, this was primarily because they were not given adequate resources to tackle these tasks effectively, and ultimately had insufficient influence on the programme teams which managed the execution of research.
25. This rejection of foresight on grounds of ideology is increasingly recognized as misplaced in that the type of approach we have outlined has little in common with the corporatist state planning of science which is now being eschewed even in the Eastern bloc.

REFERENCES

ACARD (1986), *Exploitable Areas of Science*, London, Advisory Council for Applied Research and Development.

CHI (1987), *Identifying Areas of Strength and Excellence in UK Technology*, London, Department of Trade and Industry.

Coates, J.F. (1985), 'Foresight in Federal Government Policy Making', *Futures Research Quarterly*, vol.1, pp.29–53.

Dore, R. (1983), *A Case Study of Technology Forecasting in Japan*, London, Technical Change Centre.

Dunn, N. (1988), 'The Policy Sciences in Public Discourse', *Knowledge in Society*, vol.11, no.3, pp.3–5.

Elzinga, A. (1983), *Foresight as Anticipatory Intelligence*, Ottawa, Science Council of Canada.

Hahn, W.A. (1986), *Futures in Politics and the Politics of Futures,* Washington, DC, School of Government and Business Administration, George Washington University (Reprint 86-9R).

Holland, R. (1988), 'Factors Affecting Utilization of Evaluation Findings by Direct Service Staff', *Knowledge in Society*, vol.1, no.3, pp.45-56.

Irvine, J. and Martin, B.R. (1984), *Foresight in Science: Picking the Winners*, London, Pinter Publishers.

Kuhn, T.S. (1962), *The Structure of Scientific Revolutions*, Chicago, University of Chicago Press.

Martino, J.P. (1977), 'Forecasting and its Impact on Policy Making', pp.111–21 in Boucher, W.I. (ed.), *The Study of the Future: An Agenda for Research*, Washington, DC, National Science Foundation.

Montigny, P. (1989), 'The Cassandra Paradox Relating to the Effectiveness of Evaluation and Forecasting Methods', pp.247–64 in Evered, D. and Harnett, S. (eds), *The Evaluation of Scientific Research*, Chichester, Wiley.

Papon, P. (1988), 'Is a Prospective of Science Feasible?', *Futures*, vol.20, pp.402–19.

Pavitt, K. (1983), *Patterns of Technical Change — Evidence, Theory and Policy Implications*, Papers in Science, Technology and Public Policy No. 3, Brighton, University of Sussex.

Price, D. de Solla (1984), 'The Science/Technology Relationship: The Craft of Experimental Science and Policy for the Improvement of High Technology Innovation', *Research Policy*, vol.13, pp.3–20.

Rothwell, R., Freeman, C., Horsley, A., Jervis, V.T.P., Robertson, A.B. and Townsend, J. (1974), 'SAPPHO — Updated: Project SAPPHO Phase II', *Research Policy*, vol.3, pp.258–91.

Schmitt, R. (1985), 'Overview: The Search for Solutions', pp.23–9 in National Science Board, OSTP and Government-University-Industry Research Roundtable, *Academic Research Facilities: Financing Strategies*, Washington, DC, National Academy Press.

Steed, G. (1988), 'Alerting Canadians: The Scope and Use of Policy Research by the Science Council of Canada', *Technology in Society*, vol.10, pp.165–83.

TNO (1989), *Final Report on Dutch Prospectives Project*, Apeldoorn, TNO Centre for Policy Studies.

Van Dongen, H.J. (1988), 'The Social Construction of Futures' (mimeo), Zoetermeer, Netherlands Ministry of Education and Science.

Van Vught, F.A. (1988), 'Forecasting or Pattern Management?' (mimeo), Zoetermeer, Netherlands Ministry of Education and Science.

Ziman, J. (1987), *Science in a Steady State*, London, Science Policy Support Group.

Appendix: list of organizations consulted

1. AUSTRALIA

- Australian National University, Research School of Social Sciences
- Australian Research Council (ARC)
- Australian Science and Technology Council (ASTEC)
- Centre for Technology and Social Change (TASC), University of Wollongong
- Commission for the Future (CFF)
- Commonwealth Scientific and Industrial Research Organisation (CSIRO)
- Department of Education, Employment and Training (DEET)
- Department of Industry, Technology and Commerce (DITAC)

2. CANADA

- Bell Northern Research (BNR)
- Canadian Advanced Technology Association (CATA)
- Canadian Institute for Advanced Research
- Ministry of Education and Science, Government of Quebec
- Ministry of State for Science and Technology (MOSST)
- National Research Council (NRC)
- Natural Sciences and Engineering Research Council (NSERC)
- Science Council of Canada (SCC)
- Social Sciences and Humanities Research Council (SSHRC)
- Statistics Canada, Science, Technology and Capital Stock Division

3. FRANCE

- Centre de Prospective et d'Évaluation (CPE)
- Centre National de la Recherche Scientifique (CNRS)
- Centre National d'Études Spatiales (CNES)
- CNRS Laboratoire d'Évaluation et de Prospective Internationales (LEPI)
- Comité National d'Évaluation
- Commissariat Général du Plan (CGP)
- Compagnie Générale d'Électricité
- Conseil Supérieure de la Recherche et de la Technologie (CSRT)
- Conservatoire National des Arts et Métiers, Institut de Prospective Industrielle

— École Nationale Supérieure des Mines, Centre de Sociologie et l'Innovation
— Euroconsult
— Institut National de la Recherche Agronomique (INRA)
— Institut National de la Santé et de la Recherche Médicale (INSERM)
— Institut Pasteur
— Ministère de l'Industrie, des P&T et du Tourisme
— Ministère de la Recherche et d'Enseignement Supérieur
— Organization for Economic Cooperation and Development (OECD)
— Pechiney
— Rhône-Poulenc
— St Gobain
— Thomson-CSF

4. JAPAN

— Agency of Industrial Science and Technology (AIST)
— Association for the Progress of New Chemistry (ASPRONC)
— British Embassy, Tokyo
— Council for Science and Technology (CST)
— Hitachi Ltd
— Institute for Future Technology
— Japan Techno-Economics Society (JATES)
— Long-Term Credit Bank of Japan (LTCB)
— Ministry of Education, Science and Culture (Monbusho)
— Ministry of International Trade and Industry (MITI)
— Mitsubishi Research Institute Inc
— National Institute of Science and Technology Policy (NISTEP)
— National Personnel Authority
— NEC Corporation
— Research Development Corporation of Japan (JRDC)
— Royal Netherlands Embassy, Tokyo
— Science and Technology Agency (STA)
— Science Council of Japan (SCJ)
— Sumitomo Chemical Company Ltd
— Suntory Ltd
— Tokyo University
— Toray Industries Ltd
— University of Saitama, Graduate School of Policy Science
— US National Science Foundation, Tokyo Office

5. NORWAY

— Council for Natural Science Research (RNF)
— Council for Social Science Research (RSF)
— Ministry of Cultural and Scientific Affairs
— Ministry of Industries
— NAVF Institute for Studies in Research and Higher Education
— Norsk Hydro

— Norwegian Defence Research Establishment (NDRE)
— Norwegian Research Council for Applied Social Science (NORAS)
— Norwegian Research Council for Science and Humanities (NAVF)
— Norwegian School of Management
— Norwegian Science Policy Council (SPC)
— NTNF Resource Policy Group (GRS)
— Royal Norwegian Council for Scientific and Industrial Research (NTNF)

6. SWEDEN

— Chalmers University of Technology
— Ericsson
— Government Research Advisory Board
— Institute of Future Studies
— Ministry of Industry
— National Defence Research Institute (FOA)
— Pharmacia
— Royal Swedish Academy of Engineering Sciences (IVA)
— Swedish Council for Planning and Coordination of Research (FRN)
— Swedish National Board of Universities and Colleges (UHÄ)
— Swedish Natural Science Research Council (NFR)
— University of Gothenburg, Department of Theory of Science
— University of Lund, Research Policy Institute

7. UNITED STATES OF AMERICA

— American Association for the Advancement of Science (AAAS)
— CHI Research
— Committee on Science, Engineering and Public Policy (COSEPUP) of the National Academy of Sciences, National Academy of Engineering and Institute of Medicine
— Congressional Research Service (CRS)
— Defense Advanced Research Projects Agency (DARPA)
— Department of Commerce (DOC)
— Department of Energy (DOE)
— Forecasting International
— General Accounting Office (GAO)
— J.F. Coates Inc
— National Bureau of Standards (NBS)
— National Institutes of Health (NIH)
— National Research Council (NRC)
— National Science Foundation (NSF)
— Office of Science and Technology Policy (OSTP)
— Office of Technology Assessment (OTA)
— SRI International
— US Air Force Systems Command

8. WEST GERMANY

— AEG AG

- Arthur D. Little, Wiesbaden Office
- Association of National Research Centres (AGF)
- Baden-Württemberg Ministry for Science and Cultural Affairs
- BASF AG
- Bayer AG
- Centre for Science Studies, University of Bielefeld
- Deutsche Forschungsgemeinschaft (DFG)
- Federal Ministry of Research and Technology (BMFT)
- Fraunhofer Gesellschaft (FhG)
- Hoechst AG
- Institute for Economic Research (IFO), Munich
- Institute for Systems and Innovation Research (ISI)
- Max-Planck Gesellschaft
- Siemens AG
- West German Chemical Industry Association (VCI)
- West German Science Council (Wissenschaftsrat)

Index